Mobile Computing

McGraw-Hill Series on Computer Communications (Selected Titles)

Mobile Computing

A Systems Integrator's Handbook

Chander Dhawan

McGraw-Hill

New York   San Francisco   Washington, D.C.   Auckland   Bogotá
Caracas   Lisbon   London   Madrid   Mexico City   Milan
Montreal   New Delhi   San Juan   Singapore
Sydney   Tokyo   Toronto

Library of Congress Cataloging-in-Publication Data

Dhawan, Chander.
     Mobile computing : a systems integrator's handbook / Chander
Dhawan.
          p.   cm.
     Includes index.
     ISBN 0-07-016769-9
     1. Mobile computing.   I. Title.
QA76.59.D45   1997
004.6—dc20
                                                    96-33611
                                                        CIP

McGraw-Hill

A Division of The McGraw·Hill Companies

1 2 3 4 5 6 7 8 9 0   DOC/DOC   9 0 1 0 9 8 7 6

ISBN 0-07-016769-9

*The sponsoring editor for this book was Steve Elliot, the editing
supervisor was Christina Palaia, and the production supervisor was
Donald Schmidt. It was set in Century Schoolbook by North Market
Street Graphics.*

Printed and bound by R. R. Donnelley & Sons Company.

McGraw-Hill books are available at special quantity discounts to use
as premiums and sales promotions, or for use in corporate training
programs. For more information, please write to the Director of Special Sales, McGraw-Hill, 11 West 19th Street, New York, NY 10011. Or
contact your local bookstore.

To my wife, Bina, who has given me love, encouragement, and support while writing this book. She, more than anybody else, gave me the last nudge that I needed to commit myself to start this book.

And to my late parents, Sukh Dayal Dhawan and Kaushalya Dhawan, who sacrificed personally to give their children the best education they could afford. Without their sacrifice, this book would not have been possible.

Contents

Part 1 Mobile Computing Power: Vision, Potential, Applications, and Economics

Part 2   End-to-End Mobile Computing Technology Architecture

Part 3   Mobile Computing Components

Part 4   Understanding the Vendor Offerings

Part 6   Mobile Computing Challenges, Opportunities, and Trends

Preface

The idea of writing a book on mobile computing came to me soon after I finished an assignment as project manager of a sophisticated mobile computing pilot project for a large public-sector organization. I realized during this assignment (during 1993 to 1995) that there was a dearth of information available in a form that I could use as a technology architect and project manager. Although I have an extensive background in implementing IT networking projects, I had then only a superficial understanding of wireless networks and the integration issues of mobile computing systems. It was tough finding independent and comprehensive end-to-end application and systems-integration expertise and knowledge.

When I started doing technology research, I found that only a few large organizations such as Federal Express and UPS had implemented mobile computing solutions on proprietary private/public networks, but their experience base was not available to other practitioners. Several public safety organizations were just trying to implement old applications on new devices and networks. I wanted an open and long-term solution that would address new ways of doing business with new processes. Vendors were willing to provide information but had very narrow, product-oriented, and biased viewpoints. So I set out to acquire this knowledge and experience the hard way through industrywide technology research, requests for information, requests for proposals, and an actual technology pilot.

In 1993, there were several good books on radio engineering available, and, in 1995, a selection of titles on wireless networks was published. Mobile computing systems integration is a more comprehensive and complex topic altogether; however, it seemed to me that the time was ripe for a book on mobile computing that was comprehensive and addressed the issue from an end-to-end perspective—a book that would deal with business process reengineering, business case development, wireless networks, application analysis, application development, product knowledge, network management, and end-to-end design.

Having been through the process myself, I felt that I should gather together the experiences of early implementers of this leading-edge technology and give others the benefit of our combined knowledge.

Writing a book on mobile computing from a systems-integration perspective was at that point, however, only a dream in my mind. I flirted with the idea for several months but I did nothing about it. Then, in April 1995, I attended a seminar by Bob Proctor, one of the finest motivational seminar leaders in North America. Bob convinced me in his seminar that the hardest thing to do is to make up one's mind. As soon as that is done, everything else follows. Making the decision brings out the necessary creative energies required to fulfill the dream. Bob Proctor asked us to each write down on a piece of paper our dream wish. It was there and then that I decided I was going to write this book.

Subsequently, my resolve was further strengthened by an extremely positive reaction when I broached the idea to Jay Ranade, Series Editor for the IBM, DEC, and Communications Series at McGraw-Hill. Jay encouraged me and reinforced my conviction that a book on the subject was needed.

Prior to this, I had spoken at Comdex and other such seminars and had even written a few technical articles for publication in trade magazines. But writing a book—that was something entirely new for me. I found myself approaching the undertaking with both fear and excitement.

For over eight months I collected background information, researching the subject, surfing the Internet and CompuServe, discussing issues with industry experts, putting ideas down on paper and illustrating them with graphics. I limited my consulting work to just one part-time contract and spent a majority of my time on the book. Although it was a long struggle and I am glad it is over, looking back, I realize I have learned a lot and have enjoyed myself in the process—so much so that I might well do it again.

Organization of the Book

The book follows the life cycle of a typical mobile computing project. It is organized into six parts.

The first part deals with the business vision, potential applications, and economics of mobile computing. Here, I discuss business applications, business process reengineering, and the development of a mobile computing business case.

The second part deals with the development of an architecture.

In the third part, I review various component technologies including end-user devices, wireless LANs, wireless WANs, ISDN, and public switched telephone networks. I also review communications switches, application-development tools, and strategies that are unique to mobile computing.

In the fourth part, I look at the current state of technology implementations in terms of vendor product strategies. Brief descriptions of important products are given in the appendixes.

In the fifth part, I emphasize design and integration issues. This is the glue that binds everything together.

Finally, in the sixth part, I analyze the challenges facing mobile computing and look at technology trends that practitioners need to take into account while planning for future projects.

New Concepts and Ideas

The book introduces several new and not-so-new concepts and several well-established technologies. Many design concepts and products in this industry are still going through evolutionary changes and have not fully gelled. I have given significant substance to the idea of mobile-aware application design and the mobile computing server/ switch and have done so to encourage readers to use my ideas to crystallize their own, and thus to formulate them into functional products and applications systems. I also encourage readers to investigate the client-agent-server concept in mobile computing.

Intentional Repetition

The reader will encounter a certain amount of repetition in different chapters. This is intentional for two reasons. First, it keeps the subject matter together and avoids repeated references to different parts of the book. Second, information has been repeated, where appropriate, in order to emphasize its importance.

Feedback on the Book

Obviously, as mobile computing technology evolves, the information in this book will have to be updated. As well, in spite of the efforts of the editorial staff, errors may creep into the final copy, or perhaps certain statements will be challenged by experts and specialists. Whatever the reason, I would like to receive your comments, feedback, and corrections. My e-mail address is cdhawan@mobileinfo.com; my telephone number is (905) 881-9070, and my fax number is (905) 881-3589.

I hope this book meets its objective of being comprehensive and that it meets your information needs.

Acknowledgments

Writing this book was a long and arduous exercise. I could not have accomplished it without the assistance and cooperation of many professional colleagues, vendors, friends, and family members. Therefore, it is my distinct pleasure to acknowledge the following individuals and groups who helped me in this task.

First of all, I must acknowledge Bob Proctor, the internationally known seminar leader who motivated me to make the writing commitment in the first place, and then to make it happen.

I would also like to acknowledge John Rollock, the former Assistant Deputy Minister, and Ailsa Hamilton, then Development Director (now acting Assistant Deputy Minister), at the Ministry of the Solicitor General and Correctional Services in the Ontario government, who had the confidence to let me run a mobile computing project on their behalf. Were it not for them, I would not have developed the expertise required to write this book.

I owe a special thanks to Jay Ranade, Series Editor, McGraw-Hill Series on IBM, DEC, and Communications, who shared my enthusiasm and introduced me to Gerry Papke, the former Senior Editor at McGraw-Hill. Steve Elliot took over as Senior Editor after Gerry left and shepherded my book through the production and marketing cycles. I must thank both Gerry and Steve.

I would also like to thank Dean Malone, a middleware software specialist, and Doug Goddard, Principal of the Client/Server Factory in Toronto, who reviewed my initial work on the book.

I also wish to note the willing cooperation of the many vendors who provided information on their technologies, white papers, and product information brochures. Special credit goes to Apple, ARDIS, Clearnet, Ericsson, GTE MobileNet, IBM, Motorola, RAM Mobile Data, SkyTel, Telxon, Northern Telecom, General Magic, Norand, Racotek, Shiva, and Xircom. Special thanks to Ewa Akerlind of Ericsson's *Mobile Data News* in Sweden. She assembled several application photographics from the customer base especially for this book.

While I have acknowledged sources in different places throughout the text, I would be taking undue credit if I did not also mention here the authors of the numerous books and articles that served as sources of information. While many of the ideas and opinions expressed in the book are my own, I have undoubtedly learned a great deal from books by Bates, Nemzov, David McGuffin, and Black. In certain instances I have used graphics from their books with minor modifications. My special thanks go to Bud Bates and McGraw-Hill for material from his book—an excellent source for wireless networks, especially PCS/PCN. Thank you, all.

Special thanks go to my lovely daughter, Priya, who did most of the graphics work for the book. She became an expert in Microsoft Powerpoint, and it is her work that stands out more than my writing.

Thanks also to Nicholas Stephens, my Toronto editor, who cleaned up the grammar and improved the structure of my sentences. He refused to allow my impatience with the subtleties of English grammar to show through in the manuscript.

Thank you, Priya and Nicholas.

To the editorial and production staff at McGraw-Hill, who converted my manuscript into a finished product, I want to say, "Thank you for your patience, understanding, and efforts in producing this book." Christina Palaia of North Market Street Graphics was meticulous and thorough in editing my manuscript to meet McGraw-Hill standards.

Finally, special thanks go to my ever optimistic and confident wife, Bina, and my very bright elder daughter, Sonia, who is currently doing her Ph.D. thesis and appreciates how tough it is to write a handbook on an emerging technology.

Although my family thought I was crazy to sit in front of my PC late into the night, only to rise with the sun for more creative writing, they nevertheless held on to the belief that the end result was worth the effort.

Chander Dhawan

Information Update Service

Mobile computing technology is changing rapidly. New technology concepts, networks, and products are being introduced on a regular basis. Many existing technologies are becoming obsolete. While we have tried to give you the latest information when the book went to print, some may become obsolete by the time you purchase the book. This cannot be avoided since publishing a book is a long process involving many organizations, and the technology changes during this time lag. We feel that there is a need to update information in this book on an interim basis before a new revision of the book is published.

In the past this was not feasible and the only mechanism was a formal revision—another long process. Now the Internet offers a mechanism to update this information quite easily. This hybrid combination that updates information in the book with a service on the Internet will make our book unique because it marries an old, traditional form of information delivery with the latest and most modern form of information update.

We are offering to put up an interim book update on the Internet during 1997. The details for this update mechanism are being worked out as we go to print. We shall make this information available at www.mobileinfo.com to those who send in a request to the author through fax or Internet e-mail. Please note that this update will be interim and limited to essential changes and corrections, if any. It will not replace the need for a comprehensive revision to the book in the future.

Chander Dhawan
e-mail address: cdhawan@mobileinfo.com
Fax: 905-881-3589

Introduction

What Is Mobile Computing?

Mobile computing is the discipline of creating business solutions using computers and telecommunications to allow users to work away from the fixed facilities where they normally operate. Mobile computing conveys many different visions to people. Besides mobility and remote computing, it often implies dial-up, ISDN, or wireless networks, pen-based applications, notebooks, handheld computers, PDAs, communications servers, and, of course, the users on the go. But more than anything else, it conveys the untethering of the network users.

Mobile computing solutions cover a variety of applications—occasionally connected to continuously connected, small-scale to large-scale, simple to complex, and single-application to multiapplication. The focus here is on the requirements of larger organizations with hundreds or thousands of potential users who require access to several applications on more than one platform under a multitier technology architecture. This is where the need for systems integration knowledge, skills, and know-how is the greatest. This is also where mobility, accessibility to the corporate information, and value of integrated mobile computing solutions are high, as is shown in Fig. I.1.

The Hype and Reality of Mobile Computing

With notebook computers selling in ever increasing numbers and cellular telephone usage growing at a compound rate of 20 percent per year, mobile computing has been hailed as a hot new technology that will significantly change the way in which we conduct both work and non-work-related activities. It stands to reason that providing a salesperson with a notebook equipped with a modem and a cellular telephone makes that salesperson more mobile; or that giving a dial-up connection to a telecommuter with a desktop PC or Macintosh computer at home (or at the cottage) makes that worker more accessible. Equipping service representatives with two-way pagers, or installing

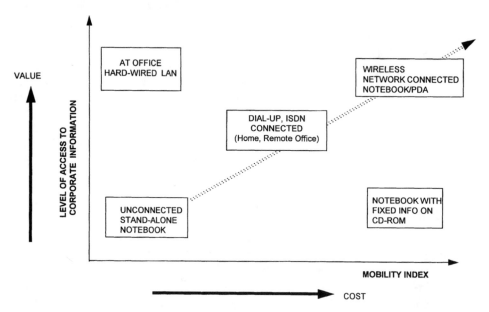

Figure I.1 Relationship of mobility with level of information access.

ruggedized notebooks in service vehicles, are yet other mobile computing implementations. Recognizing this, many computer trade magazines, *PC Week* and *Communications Week* to name just two, have
introduced exclusive feature sections on mobile computing and remote
network access. Specialized wireless network equipment suppliers like
Motorola, and service providers such as RAM and ARDIS, are invest-
ing billions of dollars in their network infrastructure.

Early adopters of mobile computing, UPS and Federal Express
among them, have already demonstrated the potential of the technology. The competitive advantages they have gained as a result of incorporating mobile computing solutions in their strategic business plans
are now leading other businesses to launch pilot schemes based on
wireless network connections. Coca Cola's automated sales route
undertaking is one such example.

It is not surprising, then, that research findings from firms like IDC,
BIS Strategic Decisions, and Infonetics indicate a significant growth
potential for mobile computing, and that vendor marketing executives,
encouraged by these findings, are urging the user community to take
advantage of the growing availability of reliable, high-capacity/wide-
coverage public wireless networks to start implementing solutions.

Yet, despite all this, penetration of mobile computing as a mainstream technology into organizations has fallen behind most industry
forecasts. There are many reasons, which we will explore in detail in

Part 6 of this book. Suffice it to say here that any complex emerging technology must pay its dues in order to become mainstream—a process that includes conducting field trials and working with early technology adopters to iron out bugs and generally assuming a leadership role. Mobile computing is no exception to this rule.

While we are of the opinion that business forecasts quoted in the trade press indicate general trends only, rather than specific rates of growth, it nevertheless does appear to us that the time for planning the implementation of mobile computing solutions by the user community has come.

There are several reasons for our assertion. These reasons can be classified into six categories:

- Business factors
- Economic justification of mobile computing solutions through business process reengineering
- Hardware and software technology affordability
- Wireless and wire-line network infrastructure availability
- Emergence of ready-to-implement applications
- Remote-access industry readiness

Business factors

The following important business factors are having an impact on mobile computing:

- *Societal shifts toward a more mobile work force.* Human beings have always wanted to be mobile. However, we also have always wanted to communicate effectively with each other when necessary. Increasingly, we are spending more time away from our homes and are traveling greater distances from our offices to conduct business. Thirty-three million North Americans (and almost twice as many as that in the rest of the developed world) are estimated to travel regularly as part of their job. Included are salespeople, service representatives, business professionals, and remote workers in the field, at satellite offices, and on project sites.

- *Customer demand for superior service.* In an increasingly competitive world, customers are demanding more concise information and superior service from salespeople and service representatives. Equipping employees with mobile computing workstations that connect them to corporate resources is an effective way to achieve this.

- *Global competition from developing countries.* Because of lower costs from developing countries, industry in developed countries must cut costs at the same time as it provides superior service. Every cost factor, especially sales and administration overhead, is being scrutinized to achieve this goal.

Economic justification through business process reengineering

Innovation-driven companies such as UPS and Federal Express have demonstrated that mobile computing can be an important factor in the elimination or reengineering of many business processes. Other organizations are also benefiting from sales automation via mobile computing. Economic justification through formal business case preparation has started.

Hardware and software technology affordability

These business factors herald profound changes and are creating a large demand for tools and technologies that can make workers more productive and more creative while away from their desks. With advances in end-user devices (notebooks, PDAs, and palmpads), implementation of electronic communication networks (wired and wireless), and availability of shrink-wrapped business applications, industry players are starting to bring forth technologies that can satisfy business needs at affordable price points.

Wireless network infrastructure availability

Billions of dollars have been invested during the past five years and significant investments are still being made to establish a comprehensive coast-to-coast wireless network infrastructure in North America, Europe, and Japan in support of mobile applications. Remote network access through switched network services such as ISDN and frame relay are moving at an ever accelerating pace. On a global scale, satellite technology offers the best potential for a wireless network infrastructure. Several such projects are being implemented.

Gradual appearance of ready-to-implement applications

Many new business applications are being engineered from scratch. Existing applications are being upgraded to support mobile access. Most commercial e-mail applications now have a mobile client component. Paging enhancements based on two-way paging are moving ahead, eliminating the need for office callbacks.

**Remote-access industry creating demands
for true mobility**

The remote access industry is satisfying the basic needs of mobile workers to be able to communicate from home and from other off-site locations through PSTN, ISDN, and frame relay. Once users have tasted this technology, they tend to want to advance to the true mobility of a wireless network. As this demand builds, designers of mobile applications will be forced to develop "mobile-aware" versions of existing and future applications—applications that will use wireless bandwidth efficiently.

Is the Industry Ready to Meet the Challenge?

There still are several obstacles to the implementation of mobile computing, however. In Chap. 19, we shall review the reasons why mobile computing is being held back. Here, we shall explore only one reason.

While individual vendors are gearing up to support their own products, there are not enough third-party systems integration experts trained in multivendor solutions especially for enterprisewide deployment. Education by the seminar industry has begun, but formal and structured education processes have not yet started. Necessary educational resources in the form of books on the subject are not available for systems integrators or implementers. *The key objective of this book is to begin to address this important need.* We hope that this book will assist the adopters of mobile computing by giving them a comprehensive understanding of the business applications, technology components, and planning considerations for mobile computing solutions.

Mobile Computing

Mobile Computing Power: Vision, Potential, Applications, and Economics

Mobile Computing Power: Promises, Potential, and Challenges

Business visionaries conceive the ideas and promise beyond the realm of immediate possibility. Technology missionaries convert these visions into prototype products, and the early adopters bleed through the challenges of these imperfect products so that the followers can benefit from their trials and tribulations.

CHANDER DHAWAN

About This Chapter

In this chapter, we will examine the promises the mobile computing industry has been making to users for the past several years. Bearing in mind that visionaries and enthusiasts tend to reach for the sky, we shall concentrate more on the realistic potential that truly does exist within many of these promises. It is up to business organizations with technology leadership to take up the challenge and bring the potential to fruition. In such a process, use of an emerging technology is continuously refined until it becomes mainstream—provided the fundamentals are right. We think this is particularly true of mobile computing.

1.1 The Promises

Mobile computing is a fascinating concept. Through the use of telecommunications and wireless networks, it connects computers and people on the move. It promises to change the way we think of work and pleasure, home and office, intra- and interorganizational communication. Switching from one mode to another and from one state to another is no longer as discrete as it used to be. The concept of mobile computing gives new meaning to the phrase *instant communication, anytime and anywhere*. It caters to a modern yen to be able to stay in touch, at the touch of a button, so to speak, while at the same time satisfying a basic human need to be mobile and independent. What the cellular telephone did for people on the move to be able to reach out and talk to each other, mobile computing promises to do for remote information exchange.

During the recent trend toward downsizing and cost reduction, businesses started an examination of their internal and external processes—including fundamental ones that had been taken for granted for years. In the process of examining ways to eliminate redundant processes, or to streamline others, analysts are investigating emerging technologies that might assist them in their efforts. Because of its potential for reducing manufacturing costs and improving customer service, mobile computing promises to be one of the most effective technologies available for shortening the time cycles involved in two fundamental areas of business: manufacturing and sales. When vendors saw this potential in the early 1990s, they rushed to the market ahead of the technology with many promises and an assortment of somewhat premature, relatively high priced products. Over the course of the last four years, their promises have grown in leaps and bounds.

Although we can rightly question the optimistic timetable behind many of the promises, it does not follow that we be skeptical about the potential that mobile computing holds for the user community. The fundamentals behind the evolution of mobile computing are strong. In the Introduction to this book, we discussed the reasons why we feel mobile computing is ready to come into its own. Now, in discussing the many promises, we shall take a hard look at mobile computing's great potential.

First promise: the mobile office will make information accessible anytime, anywhere, and everywhere in the world. The availability of mobile computing is spreading rapidly. It is available at home, at the cottage, in the car, in clients' homes, and in customers' offices. The cost of downtown real estate and the diminished productivity that results from travel to and from central offices are just two of the factors that are increasing the number of telecommuters everywhere. Public switched network services are growing rapidly, and more and more wireless network infrastructures are being installed, thereby eliminating barriers to obtaining up-to-the-minute information anywhere and everywhere. Before too long, we will be able to move around, virtually connected on campus, in metropolitan areas, in regions, and in fact throughout entire countries. Eventually, with satellite-based networks such as IRIDIUM, even international wireless interconnectedness will become routine.

Second promise: the mobile computing industry will be worth $15 to $20 billion in the United States by 1999. As we pointed out in our introductory remarks, several market research studies have predicted significant growth in mobile computing industry. The well-known IT market research companies such as IDC, Gartner, Giga, BIS Strategic Decisions, and Link Resources have all been quoted in the press and in industry presentations. There are a lot of differences in the dollar estimates in different studies because they are based on a different set of assumptions. However, they all show upward growth. The following projections in Table 1.1 from Link Resources were published in *Open Systems Today* in 1994.

Third promise: mobile computing hardware components and network services will soon be more affordable.   We shall see the price/performance of hardware and software improving significantly (25 percent or better every year) during the current phase of mobile computing's evolution. Notebooks and handheld devices, the most visible mobile devices, are already affordable. PSTN modems at 28.8 bps or higher are affordable even at home. ISDN and frame-relay digital-interface devices have come down in price and will continue to come down further in the future, as will other components like communication servers, switches, and multifunctional modems—including wireless network radio modems.

Network service charges (ARDIS, Mobitex, RadioMail, for example) have dropped by over 25 percent per year for the past two years. Remote network access through PSTN is affordable today and wireless-based remote networks will become affordable for many applications in the next few years. With increased volume and competition in network services as a result of recent passage of the Telecom Bill in the United States, prices will inevitably come down even further.

TABLE 1.1   Wireless Data Services Market Projections*

No. or $$ (Users in millions, $$ in billions)	1993	1994	1995	1996	1997	1998
Cellular data users	0.6	1.1	1.5	1.8	2.1	2.5
Paging users	17.7	21.2	25.5	30.6	36.7	44.1
Mobile data users	0.35	0.56	0.87	1.1	1.32	1.56
Total no. of data users	18.6	22.9	27.9	33.5	40.1	48.2
$$–Cellular/PCS[†]	13.4	16.8	20.4	22.0	23.5	25.1
$$–Paging	2.9	3.3	3.7	4.9	6.5	8.6
$$–SMR	0.615	0.662	0.714	0.771	0.832	0.900
$$–Mobile data	0.150	0.230	0.315	0.372	0.458	0.544
$$–Total	17.1	21.1	25.1	28.1	31.3	35.2

* Note that these estimates represent only a portion of the mobile computing market because end-user devices like notebooks are not included.
† Cellular/PCS $$ numbers include nondata services as well. Only a small percent (10–15 percent, in our estimates) represents data.
SOURCE: Link Resources as published in *Open Systems Today* (1994).

Fourth promise: the necessary service and support infrastructure will be available.   By the end of 1995, the figure for investment dollars was already in the billions. Several PCS network pilots for voice were started—with the promise of many more to come. Two-way paging applications on narrowband PCS also made their introduction in 1995. ARDIS and RAM networks are being upgraded for more coverage and higher speed. As we go to press, CDPD networks have become operational in metropolitan areas of the United States.

Network providers are promising that trained marketing, technical, and support personnel will be available as networks are installed and become operational.

Fifth promise: mobile and wireless applications will become plentiful. Horizontal mobile applications such as e-mail, sales force automation, paging, and database access are already getting quite popular. The charts in Chap. 2 show growth in market size estimates in the sales force automation and database access submarkets. The e-mail submarket has shown similar growth.

Sixth promise: tools for custom application development will be available soon.   Custom applications are far more difficult to develop than horizontal applications. Mobile applications must be compatible with the pen-based interface now, and soon they will have to accommodate voice recognition as well. The user community needs specific development tools to meet these special needs. Realizing that this is an area where the industry could be held back, companies like ORACLE, SYBASE, IBM, Motorola, RIM, and AT&T Cellular (ex-McCaw) have announced special development engines, SDKs, and APIs. These are discussed in Chap. 13.

1.1.1   Are these promises being fulfilled?

Many of these promises are being fulfilled to a greater or lesser degree, though not necessarily within the time frames that vendors originally quoted. This is due partly to vendors' overly optimistic projections and partly to the fact that changes involving mobile computing tend to be more fundamental, especially when underlying business practices are modified in the process. As the industry grows, however, full realization of most of the promises will be limited only by the speed with which users are able to absorb and implement new technology solutions.

1.2   The Potential

Since users are still in the early phases of a full exploitation of mobile computing, there are only a few actual examples that can be quoted at the time of this writing of promises being fulfilled and the potential of mobile computing being realized. Nevertheless, most experts agree that mobile computing holds tremendous potential for improving business productivity and introducing change. We shall illustrate this potential in terms of the economic benefits to be derived from mobile computing, and the improvements in service that are possible with mobile computing.

- *Improved customer service for sales personnel.*   Since salespeople are effective only when they are closing business, it follows that they must be close to their customers to be effective. If fixed information is available on a salesperson's mobile computer, and dynamic information can be accessed while at a customer's home or office, the sales cycle can be expedited. While it also follows, therefore, that every industry would benefit from such mobile access to information, the financial services and the pharmaceutical industries have been the early adopters.

- *Reduced time cycles through business process reengineering.*   By using mobile computing to rationalize or even eliminate altogether certain business processes, time cycles can be dramatically shortened, and corresponding cost reductions realized. One example is to have orders entered at source and transmitted immediately to shipping and accounting departments. Similarly, manufacturing time cycles can be cut by using handheld computers and RF-based scanning devices on assembly lines. When used with integrated information systems such as SAP, such practices can result in substantial cost savings, which, in turn, can result in lower prices for consumers and increased competitiveness for producers.

- *Increased efficiency in the maintenance service industry.*   Today, many service representatives are dispatched by pager or by professional telephone dispatchers. These dispatch methods are increasingly being replaced by computer-aided dispatch (CAD), which can interface with notebooks, PDAs, or the next generation of two-way pagers.

- *Greater accuracy, fewer complaints, and a reduction in intermediate-tier support staff.*   With information being recorded and accessed at source through easy-to-use pen and/or voice interfaces, there is a reduced need for administrative staff. With source entry, data errors tend to be minimized and the overall quality of information tends to improve—after all, the accuracy (or otherwise) of a piece of information starts with the person who first enters it. This improvement will also have the effect of ultimately reducing customer complaints and the staff needed to deal with them.

- *Improved health services.*   By having monitoring devices transmit critical information directly to portable computers or advanced pagers, doctors and nurses can increasingly be virtually closer to patients, and specialists can be consulted wherever they may be. Eventually, instant bedside access to full medical histories of patients through palmpad computers will lead to improved diagnoses. Information entry and update will become more streamlined.

- *Improved emergency services.* Public safety agencies have been implementing SMR and PMR networks since the late 1970s. Several large agencies among them have already implemented CAD applications on mobile data terminals. With the arrival of new network options like CDPD and public packet-switching networks like ARDIS and Mobitex, CAD will be affordable for smaller communities as well.

- *Automated reading of metered information.* By equipping the next generation of measuring devices with RF transmitters, cost savings can be realized through the unattended reading of meters in homes, test stations, and hazardous places. Truck fleets can be equipped with microprocessors that transmit vital data through wireless networks to central computers, which can then recall vehicles for service as needed.

- *Greater productivity for financial services professionals.* Highly paid sales professionals are already benefiting from the convenience of mobile computing. Notebooks linked by mobile computing technology to stock exchange mainframes help in providing more comprehensive and efficient service to clients.

- *Help for the transportation industry.* Taxis, freight trucks, and courier vans can take full advantage of mobile computing to keep central databases current. Thus, the precise status of a particular item at any given moment can be made readily available to an inquirer, and emergency pickups can be requested from the most appropriately located drivers while in transit.

These are just a few of the more obvious uses of mobile computing. One has only to combine the imagination of business analysts with the creativity of technologists to come up with additional applications. In Chap. 2, we shall discuss both horizontal and vertical industry applications that have either been implemented, or which are under development by third-party software professionals.

1.3 The Challenge of Realizing Mobile Computing's Potential

While there have been setbacks and a gradual reassessment of expectations by the users, some of the potential of niche mobile computing applications is already being realized today in field sales, field service, transportation, manufacturing, route distribution, and, most recently, health care industries. Stand-alone applications that require only occasional connection to central computers through switched network services are relatively easy to implement. More complex and exciting new applications that will take full advantage of the wireless networks face greater challenges during the next few years. From a broader industry perspective, we shall explore these challenges in Chap. 19. Here, we would like to concern ourselves to the challenges of those projects that have a valid business case now. The implementation of these projects requires careful forward planning, phased implementation, and management of user expectations.

Some of the challenges faced by information technology professionals as they harness the power of mobile computing now are described as follows:

1. *Revisiting the technology architecture.* The user needs and application characteristics of mobile computing are quite different from those of office-based computing. Allowance must be made for intermittent users who do not have the time or the inclination to frequently reboot computers or to wait around while computers reconnect to networks. Rather, mobile users are demanding and vocal. They are also very important to the business. The communications component of the technology architecture must be layered and strengthened in order to be impervious to time-outs and momentary losses in physical connections. Additionally, application and data architectures must also be revised to support the demands put upon them by mobile computing.

2. *Reliability, coverage, capacity, and cost of wireless networks.* Public switched network services like dial-up, ISDN, and frame relay have less problem meeting satisfactorily the challenge of mobile computing applications. Wireless networks, on the other hand, tend to be inherently less reliable, have less geographic coverage and reduced bandwidth, are slower, and cost more than the wired-line network services. The challenge, then, is to find ways to use these less-than-perfect network resources more efficiently by designing superior applications. We must devise communications software layers that consume minimal bandwidth and at the same time do not add unnecessary overhead. We must also keep in mind the eventual arrival of video applications on wireless networks sometime in the future.

3. *Integration with legacy mainframe and emerging client/server applications.* Application development paradigms are changing. As a result of the IT industry's original focus on mainframes, we

have accumulated a huge inventory of applications using communications interfaces that are basically incompatible with mobile computing. Now application programmers are building PC LAN file server and client/server applications that take full advantage of the inherent speed of LANs with no thought to the slower speeds of wireless networks. An uncontrolled expansion of LAN applications that do not accommodate mobile workers may prove to be an expensive oversight. Applications must become *mobile-aware**—a phrase coined by the author to denote modifications that conserve the expensive bandwidth of wireless networks and that, in general, give support to hurried (not to mention harried) mobile users.

4. *End-to-end design and performance.*   Since mobile computing inevitably involves multiple networks and multiple application-server platforms, end-to-end technical compatibility, server capacity design, and network response-time estimates are difficult to achieve. No tried-and-true tools exist for the purpose. For progress in this area we must look to the experience of vendors and users who have implemented solutions.

5. *Security.*   Because anybody can access them, wireless networks pose more serious security requirements than wired networks. Law enforcement agencies are particularly sensitive to the security issue. Progress in this area has been and is being made through the use of encryption and security servers with one-time tokens being issued. Readers should refer to Chap. 15 of this book for further information.

6. *Availability of specific know-how and expertise.*   Trained technical resources in wireless networks, multiplatform mobile-computing application programmers, and end-to-end systems integration skills are scarce. Unlike a single vendor who champions a specific emerging technology, mobile computing requires knowledge of multiple disciplines. Systems integration companies or vendor consortiums may emerge as training vehicles for users.

7. *Phased, planned, and managed implementation.*   As with any other emerging technology, mobile computing needs pilot projects that test its capabilities and limitations. Subsequently, rollouts across organizations can be planned with refinements and design changes based on pilot project findings. As the industry matures, the need for pilot projects will diminish.

---

* *Note:* Any similarity of the term *mobile-aware* to current products is unintentional. As far as we know, it does not violate any trademarks.

Summary

In this chapter, we have reviewed some of the optimistic promises made by marketing missionaries in our industry. *Information anywhere and everywhere* is a promise that has not yet materialized. The cost of wireless networks is still high and the infrastructure has not been fully established. The need to design applications that are mobile-aware is still not fully appreciated. The exciting potential of mobile computing still exists, however, as much as when the first promises and claims were being made at industry shows and on convention floors. The fundamentals of the technology are strong, and the move is on. We do not see the obstacles and delays in wireless network installations as causing a prolonged slowdown in the adoption of mobile computing. They are merely adjustments and corrections that inevitably take place in the evolution of a new technology.

Nevertheless, the challenge of bringing the full potential of mobile computing to fruition is a major one. But the challenge is being accepted by vendors and users alike as cheaper network services, greater numbers of easy-to-use, mobile-aware applications, and superior integration expertise in distribution channels become available, and more and more customers adopt the technology successfully and reengineer their business processes. We shall leave it up to BIS, Gartner Group, Giga, IDC, Link, and other well-respected market forecasters to predict the day when it can be said that this challenge has been fully and successfully met.

The author would like to offer just one cautionary note here: I am an information technology practitioner and my real-life experiences in converting my customers' business visions into functioning systems has taught me but one lesson: process changes and implementations of technology tend to take much longer than most emerging technology missionaries would have us believe.

Mobile Computing Applications

Availability of affordable applications is the single most important factor which makes, breaks or delays an emerging technology.
CHANDER DHAWAN

About This Chapter

Providing an exhaustive list of applications for mobile computing is a difficult task, especially with new ones emerging every day. In this chapter we shall discuss some of the more popular applications of mobile computing that have either been implemented by early adopters or have been identified as potential applications by the vendor community. We will start with horizontal industry applications such as e-mail and paging and move on to applications in specific vertical markets. Along the way we will distinguish between packaged applications supplied by vendors and applications custom-developed by in-house programming staff to cater to specific requirements unique to an organization.

We will define the concepts of mobile-worthy and mobile-aware applications. A more detailed description of a mobile-aware application will follow in Chap. 13. Tools for developing custom business applications will also be discussed in Chap. 13.

2.1 Characteristics of Mobile-Worthy Applications

Certain business applications are more suitable for mobile computing implementation than others. We use the term *mobile-worthy* to describe any application that has characteristics that make it particularly suitable for use with remote mobile computers connected to central information resources. These characteristics are listed as follows:

- A significant percentage of users are telecommuters or spend a lot of time away from the home office.

- Remote users are not permanently connected to an organization's information servers.

- The application requires only small, portable, and lightweight carry-on computer devices (often mounted in a user's vehicle, van, truck, or loading unit).

- There is significant economic, public safety, or crime prevention value in the information provided through the application while away from the office: extra travel is eliminated, selling cycles are reduced, patients' lives are saved, information is keyed correctly at source (thereby making for shorter billing cycles), etc.

- Only minimal amounts of data from a central information server need be accessed at the mobile site.

2.1.1 The first-generation nature of mobile computing applications

Most of the applications described in this book are essentially first generation implementations. It is a well-known rule in the application-software-development industry that it generally takes at least two or three major iterations (called software versions) before a piece of software has all the features the users really need and all the software defects removed. Almost always, the software must go through this type of mass customer trial before it can be considered functionally stable and able to meet the performance goals of its users. This evolutionary cycle has yet to be completed with many of the applications we are going to discuss. We are still in the first phase of mobile computing evolution.

2.1.2 Defining a mobile-aware application design

We shall deal with this subject more fully in Chap. 13. However, we would like to introduce the concept here, because application designers who are either building custom applications or buying vendor-developed applications need to keep this design requirement in mind. Applications may simply not fly and may fall to the ground if they are not mobile-aware. Briefly, applications are worthy of a mobile-aware label if they have been designed initially (or modified subsequently) to recognize the following important attributes of mobile computing:

1. The user has the option of sending only essential data over expensive and low-bandwidth wireless networks.

2. The application recognizes that mobile users are often in a hurry and accordingly provides a fast path through the application dialogue.

3. The application provides sufficient amounts of data to enable users to complete tasks without time-consuming and expensive interactions with remote servers.

We will describe mobile computing applications in the following three categories:

- Shrink-wrapped horizontal industry mobile computing applications
- Generic horizontal industry applications requiring extensive customization
- Vertical industry applications

2.2 Horizontal Mobile Computing Applications

In this category we will discuss application suites that can be used in broad segments of various industries. Since these applications are based on common business processes, they are available in shrink-wrapped packages with only minor customization required—or allowed.

2.2.1 Electronic mail

E-mail is the single most popular mobile computing application. In order to provide a high level of customer service, mobile workers and sales professionals must stay in touch with home offices and customers. E-mail is, in many circumstances, the most efficient mode of human communication, where intimate personal interaction is not required and human rapport need not be established. E-mail is precise and leaves an electronic paper trail of messages and responses. It is an excellent complement to voice communication. Even in public safety and law enforcement applications, it can replace routine and nonurgent communication that otherwise would take place on the radio. According to an IDC study quoted in *Communications Week* magazine in 1994, 85 percent of all remote workers use e-mail. Additionally, 83 percent use remote computing to transfer files, 56 percent to access databases, 14 percent to enter data, and 11 percent to use graphics.

E-mail can be received on or sent from a variety of end-user devices, including notebooks, PDAs, and personal communicators. Mobile Products of Modesto, California, has made e-mail available on PDAs such as Motorola's Envoy and Sony's Magic Link running General Magic's Magic Cap platform. Lotus's cc:Mail, Microsoft's MS-Mail, and Radio-

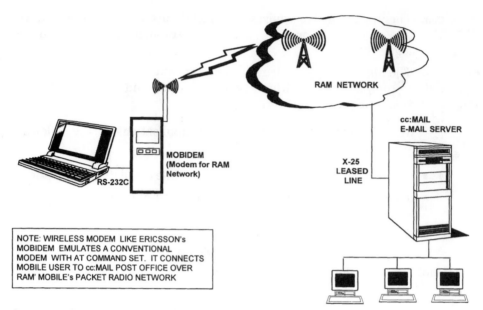

Figure 2.1 Mobile links from mobile user to LAN-based e-mail via wireless modem.

Mail's Eudora (licensed from QUALCOMM and popular on the Internet) are three common e-mail packages available for wireless networks. Figure 2.1 shows an example of notebooks accessing e-mail servers through the RAM Mobile Data network. In this case, Ericsson's Mobidem emulates the AT command set to open a connection between a client and a server modem.

Additionally, since e-mail is now a way of business life, Table 2.1 highlights information about e-mail in a mobile computing context.

Issues for mobile computing e-mail implementations. The following issues should be considered when implementing a mobile computing e-mail application.

Internal versus external e-mail service. If the organization uses an internal e-mail standard that the mobile user community is familiar with, it should be considered as a preferred e-mail platform. The potential should be evaluated for extending the standard to mobile workers. If it is too difficult and expensive to do so, an external e-mail service should be considered.

Mobile versions of e-mail client software. The availability of mobile-aware versions should be investigated. Both Lotus and Microsoft have mobile-aware versions of their products available that offer dis-

TABLE 2.1   Mobile Computing Applications—Electronic Mail

What does an e-mail application do?	■ Allows remote users to send and receive e-mail while away from home offices. Some vendors provide connectivity to in-house desktop e-mail packages; others, especially network providers, offer proprietary e-mail packages.
Desirable features of a mobile e-mail application	■ Extends workgroup e-mail to mobile users through PSTN or wireless networks. ■ Sends pages to remote users when urgent messages are waiting to be received. ■ Connects with Lotus cc:Mail, MS-Mail, and Internet interfaces. ■ Transfers files. ■ Voice mail interface by the telephone with text-to-speech conversion.
Products available	■ Lotus cc:Mobile, Lotus NOTES, MS Mobile, Radio Mail, Xcellnet Mail
Cost—S/W license	■ $75–$100 per user for client software only; server software extra. Some vendors bundle client and server licenses
Usage fees	■ Varies over a wide range; Rule of thumb cost = $0.15 on RAM to $0.60 on cellular for a 600-character message in the United States.
Benefits and payback	■ Among the two most popular mobile computing applications ■ More productive than voice mail—more selection features for high priority messages ■ (5–10% productivity improvement reported in customer studies)
Wireless network support	■ Some vendors have ARDIS, RAM, and cellular network support, e.g., cc:Mobile, Lotus NOTES, MS-Mail.
Typical platforms	■ DOS, Windows 3.1, Windows NT, Unix, NETWARE, LAN Server, Netware.
Compatibility with internal mail systems	■ Many mobile computing e-mail services provided by network providers include interfaces with popular industry packages such as cc:Mail, MS-Mail.
Special considerations	■ Client/server or workgroup versions appearing soon. ■ As much functionality as possible should be supplied to the remote device. ■ Users like to use the same e-mail software in the field (ideally a mobile-aware version with fast pass-through) as they do in the office.

tinct advantages for more efficient use of wireless networks and PSTNs.

E-mail as a starter mobile application.   E-mail is likely only one—albeit, the first—application that professional and sales users may eventually implement. Decisions as to what wireless network to use for e-mail, etc., should be made in the context of other applications that will also be offered to mobile users. (This issue is similar to the one that telecommunications managers have to deal with when building desktop networks for operational and decision-assist business applications.)

2.2.2   Electronic messaging via paging

Paging was one of the earliest wireless network applications. Being relatively fast, portable, reliable, and affordable, it took off when it was first introduced and a paging industry was instantly born, with paging

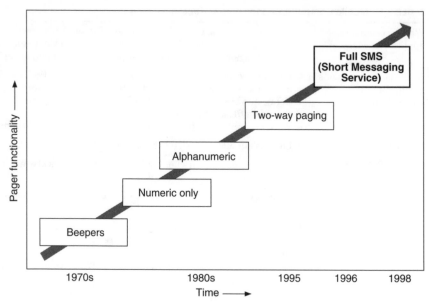

Figure 2.2 Progression of paging—from beepers to full two-way short messaging service.

stores and outlets springing up everywhere (Fig. 2.2). According to Yankee Group estimates, there were over 22 million pagers in use in the United States and Canada alone in 1995 (SkyTel estimates this number to be 27 million, according to its home page on the Internet).

The earliest units, usually called beepers, simply gave a tone alert—a signal to the wearer to call an answering service. Beepers were followed by enhanced pagers which displayed numbers. With these units, paging networks send brief messages consisting of a telephone number and an optional message code (a mutually agreed-upon code number that indicates the type of message). For example, 555-2222-01 might mean to call telephone number 555-2222 at the user's convenience. A 9-1-1 attached to the telephone number, i.e., 555-2222-911, might indicate top priority and the need to call immediately.

More recent pagers, called *alphanumerics,* display complete written messages, such as, "Please call home, you have a letter from the IRS." The original paging protocol, TAP-7, has been modified to include the transmission of text information along with numeric information. There are also a few voice pagers which enable voice transmission through the pager via a telephone. Typically they are limited to local area use (e.g., a factory), or are used by groups such as volunteer fire departments.

Two-way paging application.   Paging has moved from a simple one-way beeper service to a full function two-way messaging service. The most advanced pagers on the market today are those based on two-way paging applications implemented on narrowband PCS paging networks. With this implementation, pagers become transceivers. There are two types of two-way pagers being developed. The first type returns basic acknowledgments such as "yes," "no," or "page received." Brief return messages can be composed and sent back with the second type.

The first-ever two-way paging service based on narrowband PCS was introduced by SkyTel in the United States in October 1995. Eventually, the full implementation of the two-way paging technology will make paging a true mobile computing application for electronic messaging. Please see the description of Skytel's 2-way paging service in Chap. 14.

With two-way paging, delivery is guaranteed and response is immediate. Customer service response times will improve and users will benefit from savings in time and money, as well as increased personal effectiveness and responsiveness.

Users will have four ways of responding to messages:

- *Automatic message confirmation:* Two-way pagers will automatically confirm receipt of a message.

- *Multiple choice.* All messages can include a set of optional responses that recipients can use in reply.

- *Predefined responses.* SkyTel pagers come preprogrammed with a standard set of responses. These can be customized to meet the needs of individual customers. Users can then pick from this response set to respond to messages they receive.

- *Free-form responses.* Users can create free-form responses or initiate wireless messages by using a short cable to connect SkyTel two-way pagers to palmtop or laptop devices such as the HP-LX.

Integration of paging with mobile computing.   The integration of paging applications with mobile computing is relatively recent (1992 to 1993). Now pagers are becoming a peripheral device to the mobile computer carried by remote workers. MobileComm makes a messaging card for Newton and BellSouth's Simon PDA devices. Several PC Card-based pager cards for PCs have recently appeared on the market. It is quite likely that a paging card will become an integral component of future notebooks and PDAs rather than an optional peripheral. The following factors have facilitated this:

- Availability of software-based paging gateways on inexpensive platforms such as the PC
- Availability of PC-based paging software from companies such as Machina and Delrina
- PC Card-based paging cards for notebooks
- Introduction of PDAs such as Newton, Simon, and Marco

A paging application is often implemented in one of the following two ways:

1. As a completely bundled service from a paging service provider

2. As an add-on to a business application such as e-mail or a help desk

In the first case, a specialized paging carrier provides a complete service, including the paging hardware and the network for a fixed or variable monthly charge. The initiator of a paging message dials a pager number that is routed through the public telephone system to a paging computer where the message is stored temporarily. As soon as the paging computer is able to, it broadcasts the message through the paging network. While all the pagers attached to the network listen to the broadcast, only the pager with the appropriate cap code retrieves it. Because of the store-and-forward procedure, only two-way pagers can acknowledge receipt of messages. This process is illustrated in Fig. 2.3.

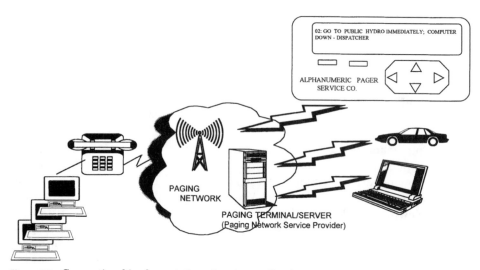

Figure 2.3   Conventional implementation of paging application.

Under the second implementation, messages can be sent to users from an e-mail application. In this scenario, if a user is not logged on to the office LAN and a message is urgent, a mail server can send a page. Even if the recipient does not belong to an in-house workgroup, a page can still be sent. Most modern help desk applications are able to automatically alert a service vendor or a specific support person in case of a problem. With the advent of workgroup applications such as LOTUS NOTES and NOTES paging gateways, the initiation of a paging function has been greatly simplified and can be invoked on the basis of a predefined filter (essentially a codification of paging criteria, such as an urgent message or an emergency) (see Figs. 2.4 and 2.5).

Paging protocols.   The Telocator Alphanumeric Protocol (TAP) communications protocol used for paging is primitive compared to today's communications protocol standards. TAP is the paging industry standard for submitting alphanumeric pages over dial-up lines. The protocol consists of check-summed blocks of data containing the message and the destination PIN number. It allows 7-bit data input. While TAP is capable of sending a certain amount of text information, the hardware used in pagers and pager networks limits the length of messages. Older devices are able to send a maximum of 80 alphanumeric characters; their more recent cousins and PDAs can transmit up to 240 characters.

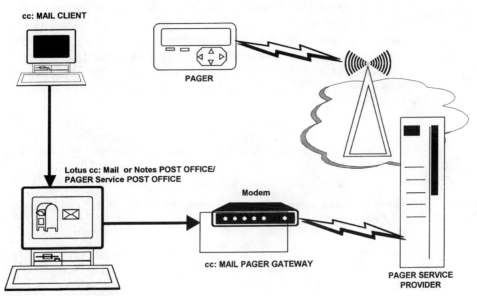

Figure 2.4   cc:Mail to pager application.

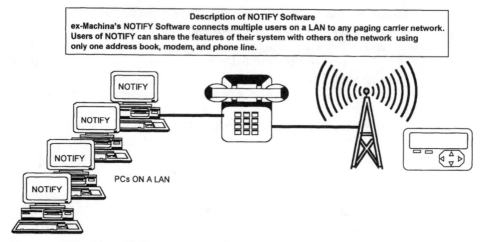

Description of NOTIFY Software
ex-Machina's NOTIFY Software connects multiple users on a LAN to any paging carrier network. Users of NOTIFY can share the features of their system with others on the network using only one address book, modem, and phone line.

NOTIFY

NOTIFY

NOTIFY

NOTIFY

PCs ON A LAN

Figure 2.5   PCs with ex-Machina's NOTIFY paging software for LAN.

2.2.3   Sales force automation

During the first three decades of computer automation (1960 to 1990), information systems departments failed somewhat to address the needs of the basic revenue-generating workers in any organization—the sales force. However, the focus has shifted recently and Sales Force Automation (SFA) is now an important mobile computing application. Salespeople who were once reluctant to involve themselves with computers now enthusiastically embrace SFA and its benefits. In our view, since it impacts an organization's bottom line, SFA is a truly mission-critical application. IDC, as quoted in a special mobile computing issue of *Communications Week* in 1994, expects SFA to grow by 30 percent from $180 million in 1994 to $435 million in 1999 (see Fig. 2.6).

Since many of its requirements are generic across the industry, SFA applications are generally available in shrink-wrap versions with certain customization capabilities. Important aspects of SFA are described in Table 2.2.

Issues with sales force automation applications.   The following issues should be considered:

Interfaces with back-end operational applications on mainframes or other servers.   Note that in a typical scenario, the organization probably has previously implemented business applications that process sales data entered by inside sales staff. SFA essentially replaces this arrangement with source order entry by mobile sales-

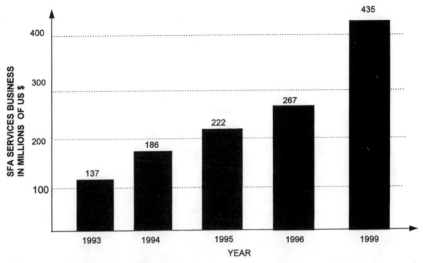

Figure 2.6 Sales force automation application. (*Source: IDC as published in* Communications Week, *April 1995.*)

persons. Software interfaces between the new mobile computing front-end SFA applications and these back-end systems should be carefully evaluated. Otherwise, software bridges with the necessary audit controls will have to be built to accept this data.

Sales force automation through workgroup computing platforms such as Lotus NOTES.   Many organizations are building SFA applications through Lotus NOTES. If this is a strategy adopted by your organization, you should look into the application design issues discussed in Chap. 13.

Open application design.   Sales professionals are well known for their lists of features and functions that they expect to see in future versions of business applications. More often than not, they have good reason for their requests, based on sound business practices. Accordingly, we advise application designers to keep their functional design as open as they can, and to choose those packages that hold greatest promise of further enrichment in future versions.

2.2.4   File transfer application

This is a generic application that enables mobile computers to transfer a variety of files. One common requirement that sales professionals have is to be able to update pricing information on their notebooks on a regular basis.

TABLE 2.2   Mobile Computing Applications—Sales Force Automation

Description	■ Provides computer-based automation for sales professionals. Many of the functions of SFA are implemented as local applications on mobile computers. SFA applications are also often used by sales staff in conjunction with other network-based mobile computing applications
Functions and features	■ Contact- or client-management—names, addresses, telephone nos., product portfolios, etc. ■ E-mail and communications—enables the sending of e-mail from mobile workstations ■ Product catalogs and price lists ■ Order entry, order status inquiries, database updates ■ Sales reporting ■ Sales analysis and forecasting ■ Lead tracking, call tracking ■ Generation of custom letters to clients ■ Personal Information Management—expenses, commission tracking ■ Multimedia and sales presentations
Vendors	■ Aurum Software's Salestrak ■ Brock Control's TakeControl ■ Siebels ■ SalesKit ■ National Management Systems' SalesWorks
Cost range	■ $200 to $700 per seat
Benefits	■ Increases in sales productivity gains ■ Shortened sales cycles leading to increased productivity (10–30% reported improvements) ■ Minimization of errors as a result of order entries at source ■ Improvements in professional image
Typical platforms	■ DOS, Windows 3.1, Windows NT, Windows 95, UNIX, NETWARE, LAN Server, Netware
Wireless network support	■ Some vendors (e.g., Brock's TakeControl, SakesKit's SalesKit, Sales Technology's Virtual Office, Xcellnet's RemoteWare) have ARDIS, RAM, and cellular network support
Databases supported	■ Many single-user versions support Btrieve files or Xbase databases ■ Modern multiuser versions run on SQL Server, Sybase, Oracle Informix
Customization	■ Most packages have capabilities to develop custom forms; some have macro capabilities
Compatibility	■ Usually capable of exchanging information through file transfer protocols
Special considerations	■ Client/Server or workgroup versions appearing soon ■ Mobile computing should move as much functionality to the remote device as possible ■ SFA should have interfaces into operational applications such as order entry, accounting, corporate sales analysis
Typical example	■ RJR Nabisco: 2000 salespeople outfitted with Fujitsu Poqet PC plus Palmtop computers

SOURCE: Vendor information and author's research.

We should remember that in spite of improvements in compression technology, it is still not economical to transmit long files over wireless networks, except in critical situations, such as the following:

- The transmission of news reports/photographs in time to meet printing deadlines
- The transmission of emergency diagnostic information to medical specialists (e.g., electrocardiogram information transmitted from an ambulance en route to a hospital)
- The transmission of identification material to/from police on patrol

In other, noncritical situations, remote users should await the opportunity to access a wired connection before making major transfers. There is an abundance of communications software that provides file transfer capabilities. Most e-mail packages have built-in file transfer features. Important considerations in selecting file transfer software are compression capabilities, checkpoint restart capabilities, and support for PSTN and wireless networks of choice. If you plan to use a specific network such as CDPD for a file transfer application, an appropriate software driver would also be required.

2.2.5 Multimedia

Until now, sophisticated sales presentations meant marching into boardrooms equipped with expensive projection gadgetry. This is changing fast. Modern, high-tech sales professionals now carry their presentations in notebooks equipped with integrated high-performance CD-ROMs, stereo speakers, and MPEG-compatible full motion video. Through this new technology-in-a-briefcase, salespeople are able to get the attention of hard-to-get customers like physicians, senior executives, and computer equipment buyers. If necessary, for bigger audiences, they can project to larger screens. With a little extra work, they can even display up-to-the-minute information direct from the head office, right before their customers' eyes. In our competitive world, it is the salesperson with the sophisticated presentation who is asked to come back and who closes the deals. The following type of multimedia applications are being increasingly implemented in notebooks:

- Sales presentations
- Product demonstrations
- Sales catalogs

In most cases, multimedia applications are stand-alone single-user applications that run on a new breed of mobile notebooks equipped with multimedia features (e.g., ThinkPad 765CD and NEC 4000C; Fig. 2.7).

Figure 2.7   Multimedia application. (*Courtesy of NEC.*)

Mobile multimedia applications are becoming increasingly popular in the pharmaceutical industry. According to *PC Week,* Bristol-Myers Squibb Company of New York has developed a multimedia application for its pharmaceutical sales staff. The application uses Oracle Corporation's Business Objects and Sybase's PowerBuilder front-end development tool. It uses a GUI interface featuring charts and interactive voice prompts that highlights physicians' prescription histories and provides details of customers' professional practices. Upjohn Company in Kalamazoo, Michigan, is developing a similar application for its sales force. Shared Medical Systems Corporation of Malvern, Pennsylvania, is using a multimedia application developed by CompuDoc in Warren, New Jersey, for marketing a portable radiology machine. According to the company, as reported in *PC Week,* sales have increased 30 percent as a result.

The ability to make multimedia sales presentations from a notebook computer also minimizes the need for support staff.

2.3   Generic Horizontal Applications Requiring Extensive Customization

This category includes those applications that are utilized across many different industries. While the core part of the application does not require changes, an organization needs customization at the front end or at the back end. Database access and service representative dispatch are among the most common applications here.

2.3.1 Database access
from an information server

The ability to access information from a DBMS server on a LAN, from a minicomputer, or from a mainframe is commonly requested by users of mobile computing. Users can interact with several front-end inquiry packages (e.g., Forrest & Tree's Info-Access, Oracle's PERSONAL 2000 client software, Sybase PowerBuilder, Xcellnet's RemoteWare) to submit a query to a server database directly—or through a gateway, if the database access protocol differs from the front-end inquiry tool supported. The queries are custom-developed for specific organizations in most cases. The back-end server software retrieves the information from the database, or acts as a gateway for the retrieval of information from a minicomputer or mainframe (Fig. 2.8).

2.3.2 Computer-aided dispatch (CAD)

Computer-aided dispatch is a very common mobile computing application used for the dispatching of service representatives to customers. By eliminating call-ins, CAD improves the productivity of service representatives. Additionally, with CAD, emergency service calls can be directed to the representative nearest to the customer—a capability attested to by the fact that the largest single application on the ARDIS network in the United States continues to be the computer-aided dis-

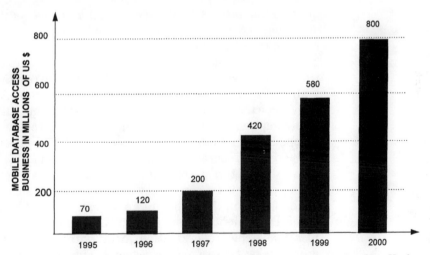

Figure 2.8 Mobile database access market growth. (*Source: Information from Yankee Group as published in* Communications Week.)

patch of IBM customer engineers (see Sec. 8.6.1 on the ARDIS Network in Chap. 8 for more details on IBM's use of mobile computing). There are three common implementations of this application.

Simple service representative model for computer-aided dispatch. Under this model, every service representative carries a mobile computer connected through a wireless network to a CAD System's central console where all service requests arrive. A dispatcher assigns requests to service representatives and sends assignments to the representatives' vehicles. Representatives acknowledge assignments and key-in approximate response times so that the dispatcher can keep the customers informed.

Some of the well-known organizations that have implemented CAD and real-time data collection—or are planning major rollouts—are IBM, Rank Xerox, and Sears. IBM application is briefly described in Chap. 8. Sears, Roebuck and Company started implementing a large ARDIS network–based customer services application in 1995. The company is outfitting 14,000 service technicians with ruggedized Itronix XC600 notebook-style computers and printers that they carry with them on the job. The machines are equipped with built-in wireless modems that let the technicians send and receive information wirelessly over the ARDIS packet radio network. Sears ran a pilot with 600 workers for a year during 1994 to 1995, before it started the rollout. The company expects a 10 percent improvement in productivity, according to a report published in *PC Week* magazine. (See Fig. 2.9 for a Sears service representative dispatch application.)

Complex computer-aided dispatch model for public safety. Public safety agencies like police, ambulance, and firefighters must be able to respond to critical situations fast. Through either a Global Positioning System (GPS) feature or, in some cases, through constant feedback from emergency vehicles, central dispatch knows the location (exact with GPS, approximate in the feedback scenario) of vehicles, graphically depicted on large (21-inch) high-resolution RISC workstations. The computer software recommends which vehicles should be dispatched where. In those cases where the CAD application is integrated with GPS, a Geographical Information System (GIS)-based map database is linked to a street number database. A CAD application can also be linked to a 9-1-1 emergency application. One such integrated application based on this capability and implemented by police services in several jurisdictions results in a 9-1-1 call being routed automatically to a dispatcher's workstation along with an indication as to which cruiser is best positioned to respond. The dispatcher can then confirm the appropriateness of the assignment with one push of a button and send the car on its way, reducing response time to mere seconds. There-

Figure 2.9 Sears service rep application. (*Courtesy of Telxon.*)

after, the dispatcher can monitor the progress of the vehicle as it makes its way to the scene.

There are virtually hundreds of police services and agencies who have implemented mobile technology for dispatch. More information on this subject is available in the discussion at the end of this chapter of an Ontario Government Mobile Workstation pilot project.

Taxi dispatch. Several taxi fleet owners have successfully deployed CAD for dispatching their cars to customers (Fig. 2.10). The payback is a reduction in dispatch staff required, more accurate dispatching, and avoidance of errors in voice communication. With the infrastructure for wireless communications already in place, some taxi services are implementing credit card authorization.

Singapore's Comfort Service, which has 9500 vehicles, is the world's largest taxi operator. The company has introduced a fully automated, satellite-based booking and dispatching system known as Comfort Link. It combines the power of GPS, Mobitex wireless networks, innovative taxi service booking terminals, and CAD software to give Sin-

(a)

Comfort CabLink GPS-based computerized taxi dispatch system

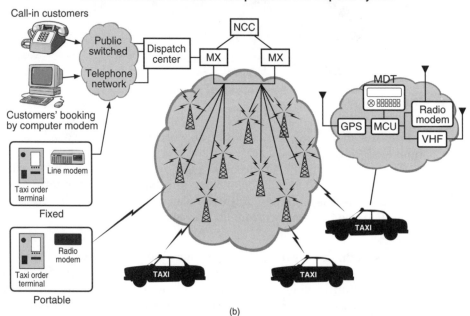

(b)

Figure 2.10   Taxi dispatch application from Ericsson *Mobile Data News:* (*a*) The Comfort CabLink system is based on an innovative booking and dispatching system that reduces waiting time and brings taxis quickly to customers. (*b*) Comfort Transport's fully automated, satellite-based booking and dispatching system enables customers to order taxis from several different types of terminals, including dedicated order terminals. Key features include interactive voice response (IVR) and computer-assisted dispatch (CAD) systems. ST Mobile Data's Mobitex network is used for communications between the CAD system and both taxis and portable order terminals. (*Courtesy of Ericsson Mobile Data News.*)

gapore a state-of-the-art taxi dispatch system, with customer service second to none. By the end of 1995, 5000 vehicles were equipped to operate with Comfort Link.

How does Singapore's taxi dispatch work? Comfort Link customers are given a card with a PIN code. There are special Comfort Service readers located at pick-up and drop-off points in plazas, airports, and other public places (mobile terminals are available for temporary locations such as conferences and exhibitions). When a card is inserted into one of these readers, the system identifies the customer and the location. Using the satellite-based GPS, the dispatching system locates the nearest available car and sends it the request for a taxi via the Mobitex network. Within seconds the terminal prints out a slip listing the customer's location and the identification of the assigned vehicle.

A different taxi-dispatch application has been implemented by RTSI in the United Kingdom. Roadside Pickup and Delivery is a dynamic routing system based on the use of a handheld terminal in vehicles operated by private companies such as Boots (a drug distribution company), British Telecom, Pepsi, and Coca-Cola.

2.3.3 Intrasite and intersite mobility application

There are situations where workers need to move around buildings, warehouses, hospitals, and factory floors, and from one site to another —without losing contact with home-base servers. Examples of applications addressing this need can be found in inventory-control software and in shipping and help desk applications designed for technical support staff who move from building to building. An appropriate mobile device might be a notebook, a palmpad, or a PDA equipped with a wireless LAN adapter or a modem, depending on the situation.

Similarly, there are situations where manufacturing companies have a number of plants together in a campus-type environment, or large organizations have buildings on two sides of a road or highway. In situations such as these, wireless LANs provide more cost-effective, high-performance, and network-transparent solutions than are possible with wide-area links from the telephone company. Wireless LANs are also more reliable than wired-line solutions. The technology for these applications is described in Chap. 7.

2.3.4 Marine ship-to-shore connectivity applications

For boats and ships needing to stay in touch with shore-based information systems, modern wireless data networks based on satellites offer very convenient solutions, especially since satellite coverage extends across sea and lake. These applications are particularly relevant to cruise lines and the needs of their passengers.

2.3.5   GPS- and GLS-based applications

There are two different systems available for tracking moving vehicles or objects. The *Global Positioning System* (GPS) can pinpoint the location of vehicles (or other objects) based on calculations from signals constantly transmitted from at least four of several geosynchronous satellites placed in orbit by the U.S. Defense Services. These calculations are accurate to within 50 meters for commercial purposes and 1 meter for military purposes. To be tracked, a vehicle must have a GPS transceiver mounted in it. Companies such as Socket Technologies supply PC Card versions of GPS transceivers that can be installed in notebooks or other mobile devices. Socket Communications also offers a software development kit that can be used to interface GPS with other applications.

Global Location System (GLS) is an active system based on a GLS transmitter mounted in a vehicle. The transmitter broadcasts a signal that is picked up and relayed by multiple reception towers to a central facility where the position of the vehicle is calculated.

There are many interesting applications of GPS and GLS. Law enforcement agencies are either using or investigating the use of GPS to locate and track patrol cars. GPS-based location tracking saves routine radio calls by officers to the communications center. Other emergency service vehicles such as ambulances and fire trucks are also being equipped with GPS receivers. Car rental agencies are experimenting with GPS to help tourists find their way around. As was mentioned earlier, centralized dispatchers can use a CAD/GPS/GLS combination to direct drivers to emergency pickups or around traffic jams.

Another major public-sector application of GPS is *Geographic Information System* (GIS) mapping and surveying. GPS receivers can provide topological information to appropriately equipped surveyors. Using proprietary GPS/GIS software like MAPINFO, engineers can create maps for many different applications, including road building and laying telephone cables and gas pipes. Several state and provincial departments of natural resources are developing applications for pollution-control sampling purposes. Farmers can use GPS-based mobile technology for recording and analyzing soil conditions and crop yields. Soil analysis information, fertilizer treatment, and crop yield data can be collected in the field and transmitted back to a base computer for further study. Monsanto has developed an application called Infielder on a Newton-based PDA that allows farmers to record and retrieve land and livestock information.

GLS can be used for tracking tagged articles or people within a defined area. One interesting application of this kind is in the paroling of nondangerous offenders. Instead of attending supervised halfway houses, parolees carry GLS transmitters that have been rendered tamper-proof by the use of precise electronic signature frequencies and enabling codes. The GLS tracking of Alzheimer's patients has been talked about as another potential application. Applications such as these will become increasingly more feasible as prices drop.

2.3.6 Field audit and inspection application

There are many situations where head-office staff need to carry out on-site field audits. In such a scenario, a field auditor can carry a notebook connected to a client-information database. The following examples illustrate this type of implementation:

- IRS and tax audits by federal or state agencies (up to 8000 agents in the United States alone are equipped with notebooks connected to IBM mainframes)
- Financial audits in the field by accounting firms such as KPMG, Deloitte & Touche, and Ernst & Young
- Site inspections by building permit inspectors
- Environmental control inspections
- Automobile insurance adjusters

2.3.7 Disaster recovery applications
of wireless networks and mobile computing

Wireless networks provide a mobility, flexibility, and ease of installation that is far superior to that of wired networks. Because of this, mobile computing is being used increasingly for backup and disaster recovery. Assuming that information servers and other important central components of a LAN or WAN are properly protected against disastrous events, mobile computing can provide the minimum network services needed immediately after a disaster. Microwave radio facilities in the T1 range can be installed relatively quickly. Mobile computing plays an important part in the following conventional disaster recovery scenarios which are integral to any IT infrastructure security plan:

- Support personnel can be connected through a wireless network to central servers.
- Wireless LANs can be set up quickly with no delays while rewiring and recabling must wait for police and fire department investigations to be completed.

- Disaster recovery e-mail services can be provided via wireless networks.

In many past disasters, wireless applications have provided temporary telecommunication services. The following natural disasters, emergencies, and military operations are typical examples:

Hurricane Hugo in United States	Wireless e-mail
Riots in Los Angeles	Wireless e-mail, cellular fax/data
Hurricane Andrew in Florida	Same as above
World Trade Center bombing in New York	Same as above
Operation Desert Storm in the Persian Gulf	Wireless e-mail, wireless LAN, wireless PBX

2.4   Vertical Applications: Industry Specific

In this category, we include applications that are specific to industries like insurance, banking, airlines, government, utilities, and transportation. Usually there are business-process characteristics unique to a particular industry that make certain vertical applications inherently mobile-worthy. As a result, some vendors have developed turnkey solutions.

2.4.1   Financial industry: Insurance and financial planning

The insurance and mutual fund industries have begun equipping their sales agents with notebooks that can display and analyze current product and price information right before their customers' eyes (Fig. 2.11). Sales illustrations can be created, printed, and presented to customers on the spot. With a modem connection, additional information stored in central databases can be retrieved from the client's office or home. The overall effect is a measurable increase in sales productivity.

In an extensive business-process reengineering, Merrill Lynch is spending over $500 million to equip 25,000 agents with Pentium-class multimedia notebooks connected to UNIX servers running Sybase. The notebooks are loaded with Windows NT and a financial planning application called Trusted Global Advisor. The software features new analytical and presentation tools that help customers understand different investment scenarios more easily. One of the key features of the new application is the ability to integrate customer data residing in multiple accounts into a single image and to provide reports in an ad hoc fashion as needed. The notebooks will make Merrill Lynch's financial consultants fully mobile, freeing them to visit their customers' offices or homes.

(a)

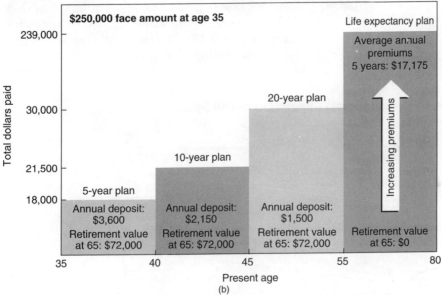

(b)

Figure 2.11 Financial planning and insurance application: (*a*) financial planning agent using a notebook with a customer; (*b*) investment and protection projections. (*Courtesy of Imperial Financial Services, Canada.*)

Wireless Telecom of Colorado, along with Compaq, AT&T, and two other vendors, have introduced a turnkey wireless application called *Mobile Financial Workstation* for the financial industry. Based on the Aero subnotebook, an Ericsson Mobidem adapter, RAM Mobile Data's

Mobitex network, AT&T's EasyLink communications software, and TeleScan financial software, the solution allows users to access stock data from an information provider.

2.4.2   Financial industry: Banks

Many banking industry customers are developing wireless applications to improve bottom-line costs. Even the big banks are realizing that their salespeople must leave their offices to sell directly to customers.

Swiss Bank corporation has implemented a wireless application for integrating e-mail and paging. While this is a straightforward application, it makes Swiss Bank's trading staff more mobile by giving them RadioMail through ARDIS or RAM Mobile Data networks (Fig. 2.12).

The bank intends to put a cc:MAIL server on the RAM Mobitex network.

2.4.3   Financial industry: Stock trading

The New York Stock Exchange has made a significant change to the classical methods used by traders in the past. No longer do stockbrokers clutching pieces of paper race to phones to collect orders; now they tap away at handheld PDAs connected to a wireless network, accessing information from HP servers (Fig. 2.13). The two-year pilot was begun in 1995 with production rollout during 1996 and 1997.

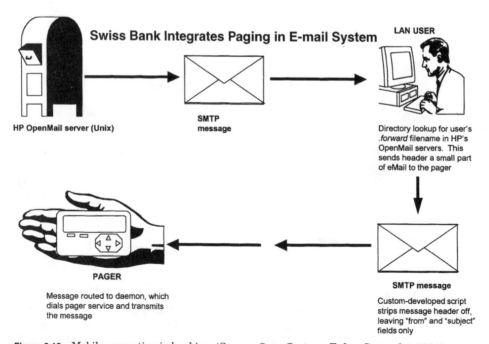

Figure 2.12   Mobile computing in banking. (*Source:* Open Systems Today, *September 1994.*)

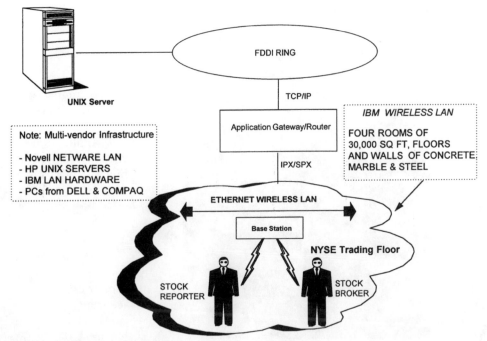

Figure 2.13 Mobile stock trading application. (*Source: IBM home pages on the Internet.*)

It is critical to provide stock traders with as much information as they require to close a sale as quickly as possible. Chicago Mercantile Exchange has already installed a wireless pilot project for mobile computing trading. It is called *Automatic Data Input Terminal* (AUDIT). The rollout of 5000 traders with wireless PDAs started in 1996.

2.4.4 Mobile computing in the retail and distribution industries

Based on a Computer Sciences Corporation study, 33 percent of retailers plan to implement wireless networks in their stores during the 1995 to 1997 time frame (Figs. 2.14 and 2.15). During the same period, a similar percentage of POS vendors will implement wireless POS systems.

The following are examples of mobile computing applications in the retail industry:

- Retailers reconfigure sales departments and move POS terminals without rewiring or recabling as sales trends change.

- Retailers increase check stands using wireless POS terminals connected to store controllers during high-volume sales, seasonal periods, or special sales events.

Figure 2.14   Retail inventory application. (*Courtesy of Telxon.*)

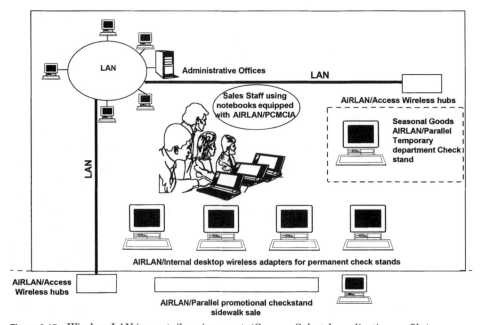

Figure 2.15   Wireless LAN in a retail environment. (*Source: Solectek application profile.*)

- Sales staff use wireless devices to perform on-the-spot markdowns and inventory counts.

- Wireless terminals are used to improve customer service by answering questions anywhere on the floor in large home improvement stores.

The Food Fair supermarket chain in Winston-Salem, North Carolina, was one of the first chains to use a wireless credit card authorization service. The system, based on the RAM Mobile Data Network, offers customers faster service and the option of paying for groceries with credit or debit card. Food Fair started accepting credit cards without the trouble and expense of installing telephone lines. According to market researchers, customers prefer stores that accept credit cards and spend almost 30 percent more per visit than they do if they have to pay by cash or write a check. The faster checkouts that have resulted from the system have also improved customer service. The RAM solution consists of a credit card processing terminal and printer, a protocol converter, and an Ericsson Mobidem M6090 radio modem. Food Fair interfaces with MasterCard's Automated POS Program called *MAPP*. This application shows that a wireless-network-based credit card authorization system can be justified even in a fixed location if it is more convenient to install than a wired solution.

Large retail store chains such as Sears are equipping their service vehicles with ruggedized notebooks with printers for real-time work order and repair receipt printing and the uploading of data.

Automated vending machines become wireless electronic merchandisers. There is a huge market for mobile computing developing as the cost of outfitting automated vending machines with wireless modems, microprocessors, and connections to sensory logic for credit card authorization capability decrease to levels where a business case can be made.

As reported in Ericsson's *Mobile Data News,* DataWave, a progressive company based in Vancouver, British Columbia, has pioneered a wireless electronic merchandiser using the Mobitex network. Based on a Mobidem M2090 radio modem, DataWave's electronic merchandiser built for Kodak dispenses single-use cameras, sunglasses, CDs, and films and accepts credit cards for payment. It registers and verifies each credit card transaction, while supporting real-time supervision and maintenance. Regular sales and inventory information can be retrieved by a central computer for replenishment of the stock. Equipment malfunctions are reported to the central site (Fig. 2.16). Kodak is planning to install 700 units.

Encouraged by the success of the Kodak electronic merchandiser, DataWave has developed the DTM 2001 wireless electronic merchan-

Figure 2.16 Retail application: the Kodak Merchandiser. (*Courtesy of Ericsson and DataWave.*)

diser for dispensing telephone calling cards that are activated at the time of sale. One major U.S. reseller has ordered 5000 units.

Greatly increased implementation of this application will doubtless occur as more and more retailers turn to the automated concept for a host of merchandising items.

Distribution industry applications.   Several route accounting and distribution warehousing applications have been implemented by vendors such as Telxon and Norand. See Chap. 14 for description of vendors' product descriptions.

2.4.5   Airline and railway industries

American Airlines has initiated a pilot scheme to implement a cellular circuit-switched network that will provide any of its flight assistance staff without a physical terminal access to the entire reservation system. The following types of information will be available on IBM 360 ThinkPads used by agents at the gates and on the floor:

- Ticketing
- Scheduling
- Maintenance, fueling, and de-icing information
- Baggage handling

American Airlines is testing these applications on a CDPD network. The objective is to reduce passenger lineups and takeoff delays and to improve customer service.

Figure 2.17 shows Symbol Technology's LRT 3800 handheld scanner/terminal being used by Scandinavian Airlines (SAS) to match individual bags with passengers in real time. The objective was to improve productivity, speed up baggage handling process at transfer points, reduce lost bags, and improve passenger security. Similarly at the Frankfurt airport, baggage handlers are using an Eagle handheld terminal that contains a code reader for scanning baggage tags and a Mobidem M2060 integrated radio modem. The scanner is connected to the host system. This terminal is being used in a new baggage reconciliation system called FRA-BRS, which is designed to improve secu-

Figure 2.17 Airline industry application; Symbol Technology's airline baggage scanner. (*Courtesy of Symbol Technology.*)

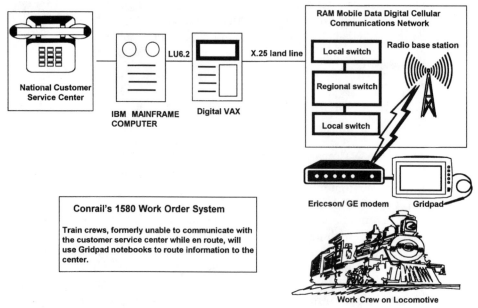

Figure 2.18   Conrail's work order application. (*Source: Figure adopted from* Computerworld, *1994.*)

rity and service at the airport. The system has replaced manual baggage handling procedures and is resulting in greater efficiency, security,
and reliability. Because of baggage reconciliations with passengers,
opportunities for terrorist attacks are greatly diminished.

Delta Airlines is also using mobile scanners to scan bar-coded information from baggage tickets directly into a database. KLM is using
mobile notebooks in its maintenance operations. Technicians on site
can enter status information, order parts, and schedule routine or
emergency repairs right from their notebook computers.

The railway industry is another vertical industry where mobile computing has found a home. Conrail has implemented a sophisticated pen-
based work order application on Grid System's Gridpad pen computers,
using RAM Mobile Data network. As a result of automating the work
order process, work requests take only minutes instead of 12 hours to be
shipped to the work crew. This application is illustrated in Fig. 2.18.

2.4.6   Wireless networks in the
manufacturing and mining industries

The manufacturing industries are finding that broad-spectrum networks can resolve many difficult LAN extension problems, especially
in remote mining areas. In many cases, installation of a cable inside a

Figure 2.19   Manufacturing industry application. (*Courtesy of Telxon.*)

factory or a mine or between two factory sites is too complex and the leasing of WAN facilities too expensive. If there is a clear line of sight available and distances are short (in some cases, up to 10 miles, or 16 kilometers), wireless bridges such as Solectek's AIRLAN bridge can be used as economical extensions to a LAN. An added bonus is the resulting freedom to use portable computers in vehicles and in mines or on shop floors.

Organizations such as Placerdome Canada (a mining company) and John Deere (a foundry) have used the technology in this way. Placerdome connected three sites (7.5 and 0.5 miles, or 12 kilometers and 800 meters, apart) with wireless bridges. John Deere installed AIRLAN equipment and mobile notebooks in indoor mobile vehicles.

Companies such as Telxon and Norand have implemented many vertical applications in manufacturing plants using RF-connected handheld computers (see Fig. 2.19).

2.4.7 Health care applications

Though physicians tend to have a reputation for resisting the move towards computers, they seem to be intrigued by and interested in mobile computing involving untethered (i.e., wireless) connections and specially designed portable devices. They like the idea of calling up patient records, researching a disease, and ordering prescriptions from anywhere in the hospital or home.

Summit Health Services in Greensburg, Pennsylvania, has developed MobileNurse, software that enables home-care nurses to access inventory, call up patient records and Medicare information on wireless notebooks, and compare notes with other caregivers. MobileNurse uses the Mobitex RAM network to access a centralized database. Similarly, Grace Hospital uses IBM system (see Fig. 2.20) for a similar application. Figure 2.21 shows the Telxon handheld computer display of vital information being recorded in the surgery room.

Additionally, doctors are increasingly using handheld devices like Newton and pen-based Palm Pads connected to hospital LANs through wireless PC (PCMCIA) cards to record patient records. As handwriting-recognition software improves, the use of this technology will grow.

Because of the high cost of physician services, business cases for the use of mobile computing are relatively easy to build. Even small increases in physician productivity or diagnostic accuracy translate into

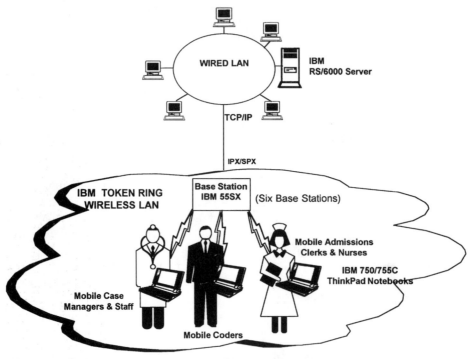

Figure 2.20   Wireless LAN in a hospital setting (Mercy Medical Center). (*Source: IBM application description on Internet.*)

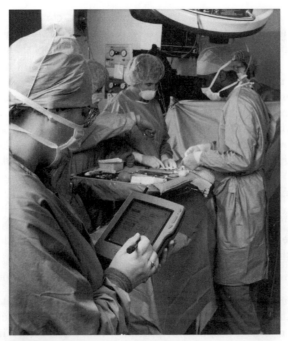

Figure 2.21   Health care industry application: the surgery theater. (*Courtesy of Telxon.*)

savings sufficient to justify the expense of mobile computing solutions. With the current all-pervasive squeeze on health care budgets everywhere, tremendous attention is being paid to the ability of technology to reduce costs.

Telediagnostics.   In certain cases, critical patient data, like EKG and MRI, needs to be transmitted from an ambulance to a hospital or from a small hospital to a medical specialist at a larger hospital. In one such implementation, Systems Guidance, a computer software and hardware supplier in the United Kingdom, has developed a Remote Patient Monitoring System (RPMS). Vital information including EKG waveform, pulse oximetry, heart rate, and blood pressure is captured either at the scene or en route to the hospital. It is then transmitted in real time to a hospital emergency room in a matter of seconds over the Mobitex network. The information is monitored by the consulting doctor, who can make immediate diagnosis and transmit advice back to the ambulance.*

* See Ericsson's *Mobile Data News,* May, 1995 for further details.

Emergency medical services.* Quite often, computer technology needs to be able to follow a medical worker out into the emergency room or a patient's rehabilitation room after surgery. In these situations, pen-based computing can be used to provide a solution that meets the stringent requirements of the medical profession. Westech Information Systems, in Vancouver, British Columbia, has developed Emergency Medical Services (EMS), an application that covers the full gamut of activities—from the time an emergency call is dispatched through to the arrival of the patient in the emergency room. Vancouver General Hospital, British Columbia Service (BCS), and 33 other customers in North America are using or piloting Westech's application.

The Westech application has three objectives:

- To increase the accuracy and completeness of patient data

- To enable data capture at source (that is, at the time and the location of the emergency)

- To improve the sharing of patient information among care givers

On-the-spot medical data capture. Access to accurate and timely information is a critical part of delivering quality emergency care. Westech's EMS makes it possible for paramedic crews to record important trauma information at the scene for later use by emergency room specialists. This information includes:

- Patient demographics, including name, address, birth date, etc.

- Patient condition, including state of consciousness, central nervous system, and cardiovascular system; condition of head, neck, chest; amount of blood lost; etc.

- Medical history, including past history, current medications, known allergies, etc.

- Vital signs, including blood pressure, pulse, respiration, temperature, etc.

- Treatment provided, including IVs, medications, oxygen administered; transportation details

- Trauma and scene details

- Information about the crew and the particular call, including information about expenses

Upon arriving at the hospital, the trauma team in the emergency room also uses a pen-based system to record vital signs and other medical

---

* This application is based on information published in *Pen-Based Computing* online magazine and Westech Information Systems published on the Internet.

information at the patient's bedside. The information from the paramedics is downloaded to an emergency room computer, which then displays both sets of data simultaneously to all members of the team via an overhead monitor.

Benefits of Westech EMS. It replaces several paper forms that previously made finding and entering information difficult. It takes less time to enter all the information, which is entered only once, at the source. Data can be uploaded for billing and other needs, such as the production of legal reports.

Technology used. The graphical user interface was developed with PenRight. It was originally developed on GRiD Convertible and Palm-Pad SL pen computers. Westech also supports IBM 730T and Fujitsu 325 pen computers as platforms of choice.

2.4.8 Public sector: Law enforcement and public safety

Because of the nature of the work, public safety agencies such as police, fire safety, and ambulance services were justified in employing radio networks and implementing mobile computing applications long before they could be cost-justified in the private sector. Vendors such as Motorola worked closely with public safety agencies and terminal manufacturers like MDI (which Motorola eventually acquired) to design customized end-to-end systems that used proprietary radio network technology for the following applications (Fig. 2.22):

- CAD for police, fire, and ambulance vehicles
- Queries to local police databases and state/provincial transport systems for vehicle and driver information
- Records management (in some cases, on site, right in vehicles; in other cases, voice-communicated to transcription centers at home offices for entry into incidence databases)

Early applications were implemented on minicomputers such as DEC VAX and mainframes such as IBM, UNISYS, and Tandem. Hardware used in the vehicles was pre-PC, with limited intelligence and storage for screen formats only. This was done to conserve radio network bandwidth, which was typically 2400 bps in the early 1980s.

Current public safety application trends. Public safety agencies are now piloting and building additional applications on Windows-based PC notebooks and on MAC Powerbooks using open hardware platforms with far more power and flexibility. Stand-alone applications like word processing for records management are being offered on these mobile

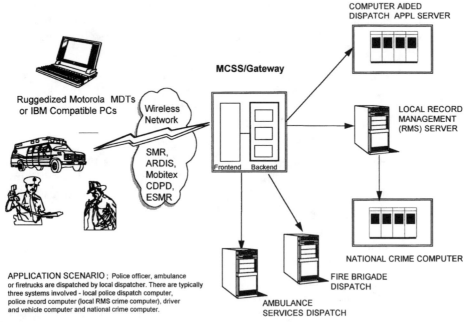

Figure 2.22 Public safety applications in mobile computing.

workstations. Some jurisdictions are experimenting with Electronic Citation (ticketing) applications with on-the-spot payment by credit card and credit authorization links to banks. Several agencies have developed Accident Data Collection systems for recording collision information at the scene, thereby eliminating extensive paper trails.

Because of economic constraints during the 1990s, and the high cost of wireless networks, many agencies are consolidating their network requirements into common infrastructures for public safety applications. Also, they are experimenting with public packet-switched networks such as RAM, ARDIS, and CDPD. While response time is better than on their older 2400 bps radio systems, network coverage is spotty because police and other public safety vehicles travel in rural areas as much as they do in urban areas. For example, a police agency in Groton, Connecticut, is experimenting with a CDPD network and pen-based computers. Groton police have found that CDPD costs for law enforcement applications are less than those of ARDIS or RAM Mobile Data. Another consideration in favor of CDPD is its future potential for higher speeds (56 kbps). The Groton police force has implemented this application in 12 cars and intends to increase the number to 20.*

---

* As reported in special mobile computing issue of *Communications Week,* April, 1995.

We would like to mention two emerging public safety network standards. The first standard, called *APCO 25* (Associated Public Safety Communications Officials), has finally emerged after many years of work and is being promoted by police organizations throughout North America. Vendors like Motorola, E.F. Johnson, and Ericsson have all announced product plans for the standard. The second standard, established in Europe, called *TETRA* (Trans European Trunked Radio) is for the same application.

2.4.9 Public sector: General office

There are many general office mobile computing applications available. We shall mention one or two as examples for illustrative purposes. Elected state/provincial government officials are highly mobile, having constantly to deal with the legislature, constituent offices, and other government officials at municipal or federal levels. While they are mobile, officials also need to keep in touch with their offices. Many public sector staff move around campus-like complexes of offices. Wireless LANs can give them the mobility they need to do their jobs more effectively. For example, the Indiana state legislature has installed a wireless LAN and provided members of the legislature with NEC Versa notebooks equipped with wireless AIRLAN/PCMCIA cards. Along with office automation software and electronic mail, custom software has been developed to provide legislators with timely access to bills and proposed amendments, supporting legal documents, and state and federal constitutions.

2.4.10 Public sector: Natural resources and environmental control

Several environmental control agencies are planning to collect data from the field with the help of mobile workstations. This data collection process will eventually be automated with the help of permanent monitoring devices, especially in hazardous sites. The Ontario Ministry of Natural Resources is using Apple Newton MessagePads 120s. The MessagePads are fitted with GPS cards and forms software. The forms application is used in tracking and fighting fires and recording fire damage. (For more discussion of this issue, refer also to the GPS/GIS application category in Sect. 2.35.)

A state-of-the-art project based on mobile computers installed in the tree harvesters of a Swedish company (Graininge Skog & Tra Forest Products Company) has demonstrated significant cost savings. The key component is an on-board Husky Hunter 16 palmtop computer installed in each harvester. Other components attached to the computers are Mobitex C702 80 MHz modems, printers, and Mobitex wireless network cards. The application provides a communications link between the saw mills and the harvesters. The computer records cross-cut measurements and gathers all the data on the day's timber production. After completion of a felling operation, the Mobitex unit sends the information directly from the harvesters to the head office and on to the sawmills. Knowing in advance how much timber has been harvested and what deliveries are on their way has enabled Graininge to halve its inventories of stored timber. Cost reductions of 10 Swedish krona per cubic meter of raw materials have been achieved. If this application was to be implemented throughout Sweden's forest products and pulp industry, one estimate of the total possible savings is 1 billion Swedish Krona.

2.4.11 Public sector: Miscellaneous applications*

There are many other applications of mobile computing in the public sector. The following sample applications are based on pen computers:

Colorado Department of Energy is using Apple Newton Message-Pads fitted with bar code readers and GPS receivers to locate, test, and classify hazardous (radioactive waste) material. Pen computers can be used even with antiradiation gloves.

The U.S. Department of Defense is using wireless-enhanced Apple Newton MessagePads to access patient information (medical histories and patient records), document lab test results, and schedule consultant visits.

Two New Jersey companies (CDP and Computers at Work) have developed an application for parking control officers. The system is based on Norand PEN*KEY 6100 handheld ruggedized pen computers. Major advantages include improved timeliness and accuracy of ticket writing.

The city of Richmond Hill in Ontario has automated time reporting for its employees. The city's crew supervisors use Fujitsu's pen-based PoquetPads and custom forms created with the PenRight development environment.

Booz-Allen is field testing an interesting solution for U.S. Army Infantry Patrol soldiers. Currently, soldiers must fill out a coded situation report on paper. This has been replaced by an electronic form on Newton MessagePad 120s.

---

* Information based on *Pen Computing,* November, 1995.

2.4.12   Utilities

Utility companies are using mobile computing for collecting and updating inventory information about field assets, such as transmission towers and transformers, which is then uploaded to head-office computers. Previously, this information was brought back from field trips in paper form and only then entered into computers. On-the-spot data entry updates information in databases more quickly, reduces errors, and improves productivity.

Also, field technicians and project staff are now able to stay in touch with head offices through e-mail and messaging applications, while CAD applications are used to dispatch service representatives to repair sites.

A good example of a utility company exploiting mobile computing is the New Jersey utility, PSE&G. It expects to save millions of dollars by equipping 750 field-service workers with pen-based computers that send and receive information via a wireless network. The new network allows field workers to receive real-time information on customers' orders, repair requests, and existing maintenance contracts. Once a job is completed, details are entered into a pen-based computer and fed in real time to the main computer. The dispatch system is based on a client/server paradigm (Fig. 2.23).

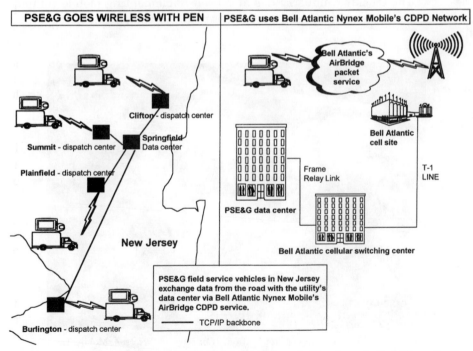

Figure 2.23   Public Service Electric & Gas Company of Newark, New Jersey's dispatch application. (*Source:* Communications Week *and* PC Week.)

Other utilities (e.g., Graininge Energy in Sweden) have implemented monitoring of their distribution systems through wireless networks. From high-powered RISC workstations, operators can view distribution systems at many different levels, from an overall view showing all high-voltage lines to detailed views of individual transformer stations.

Boston Edition's sales force has implemented e-mail from anywhere in the metropolitan area, thereby improving efficiency and timeliness of communication.

2.4.13 Transportation (courier and trucking) industry

Courier companies such as Federal Express, UPS, and DLH were among the leading early adopters of mobile computing technology. In fact, Federal Express implemented a sophisticated parcel tracking application (see Fig. 2.24) based on a wireless network that gave it a business advantage over its competitors. UPS had to respond with a similar application to retain its market share and give its customers superior service. (Please review the UPS application case study in Sec. 2.6 of this chapter.)

The trucking industry has also been implementing mobile computing solutions for several years as a result of the efforts of companies

Figure 2.24 Federal Express application. (*Courtesy of Ericsson* Mobile Data News.)

such as QUALCOMM. Because long-haul delivery and pickup truck costs (capital and operating) are quite high, the industry was easily able to justify the costs of new technology and wireless networks as long as it was able to show that their implementation would result in even a small increase in productivity. Planners needed only to point out that CAD information about emergency drop shipments or en route pickups could achieve significant savings to warrant the implementation of mobile computing solutions. Additionally, the ability to indicate exact delivery times made for excellent customer service that enabled appreciative customers to better plan their day (Fig. 2.25).

Special ruggedized terminals have been developed by companies such as QUALCOMM/CANCOM group (operating in the United States and Canada). These terminals, which utilize satellite-based communication, can send and receive messages to and from long-haul trucks on the highway. According to the company, dozens of freight carriers have installed the equipment. For a fee, CANCOM provides a bundled one-stop service for a fixed number of messages per month.

QUALCOMM also offers VIS (Vehicle Information System), fleet maintenance software that automatically monitors engines through

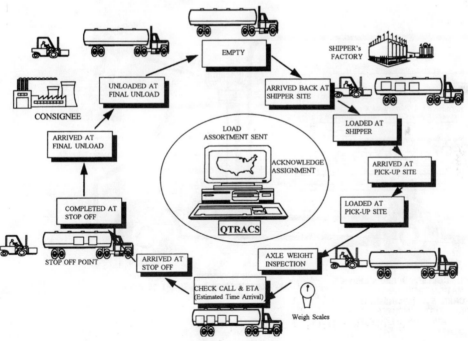

Figure 2.25 QUALCOMM's QTRACS application for fleet communication. (*Source: QUALCOMM software description brochure.*)

on-board sensors and subsystems. Transmitted via satellite, VIS can provide performance statistics reports, vehicle diagnostic data, and trailer status information. With this information, dispatchers can monitor the health of their fleets remotely and arrange preventive maintenance as necessary (Fig. 2.26).

J.B. Hunt Transport in the United States has already installed a customized tablet computer built by IBM with a touch screen interface and satellite communication. The mobile tablets are connected to 13 PS/2 model 77 servers at regional sites, which, in turn, are connected to an IBM 9000 mainframe. QUALCOMM and Rockwell Corporation were also involved in this project. Each computer contains a number of databases that include locations of fuel stops, safety information, and other travel-related information. Engine information regarding idle time, overspeed, overwind, and other features that require routine adjustments are also recorded. The project is already showing major returns. The company has achieved better customer service, more miles per day per truck, more satisfied drivers, and, most importantly, more efficient fleet managers and more efficiently managed fleets. In the past, fleet managers used to spend most of their time calling up truck drivers on the phone.

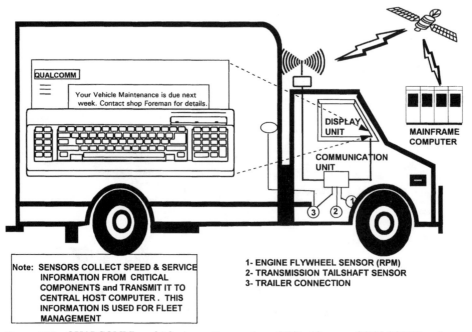

Figure 2.26 QUALCOMM's vehicle inspection system (VIS). (*Source: QUALCOMM software description brochure.*)

In Europe, ASG, a similar transport company, has installed mobile computers in over 600 vehicles operated by 20 different carriers. The ASG system, called *Transport Information Planning System* (TIPS), uses a Mobitex network to provide two-way communication between fleet vehicles and their home bases. The objective of the application is to keep their trucks fully loaded and still provide a high level of customer service. Figure 2.27 is an application diagram of TIPS. With Mobitex-based TIPS, missed pickups are greatly reduced: if the driver has not picked up an order within 30 minutes of an assignment being issued, an alarm signal is sent to the vehicle.

2.4.14 Car rental agency application

Avis was the first car rental firm to introduce GPS-based navigation in rental cars and had installed the equipment in 1000 vehicles in 1995. To use the computer-based navigational system, you press a button on a 4-inch monitor and a message warns you to keep your eyes on the road. Then it asks you where you want to go, allowing you to scroll down a list of preselected points of interest, such as hotels, convention centers, or a specific street address. As you drive, the unit (built by Rockwell International Corporation) displays a color map of the area that you are

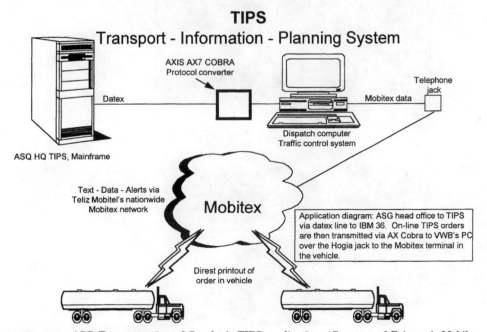

Figure 2.27 ASG Transportation of Sweden's TIPS application. (*Courtesy of Ericsson's* Mobile Data News *and ASG.*)

passing. It also emits an audible alarm prompt, just before you need to turn or take an exit. The feature is available for an extra charge.

2.4.15   Electronic news communication

Most newspersons carry modem-equipped mobile computers these days. Some have cellular modems so that they can send their stories from the field as soon as they get some time to enter the story into the computer. A few are even being outfitted with inexpensive digital cameras. Windows-based packages are available to convert digital photos to DOS/Windows files that can be compressed and transmitted on wireless networks. A picture can be transmitted from any field site without the need for the newsperson to return to a hotel. Similarly, other time-critical news items can be transmitted through wireless networks to newsrooms for editing (Fig. 2.28).

Figure 2.28   News coverage. (*Courtesy of Ericsson* Mobile Data News.)

Journalist 2000 is a joint development effort between PA News in the United Kingdom, RAM Mobile Data, Hewlett-Packard, and AnD Computer Solutions. It comprises an HP Omnibook notebook PC running Copymaster—an easy to use and fully integrated editorial and communications software package from AnD Computer Solutions—and a Mobidem AT modem, which transmits information over the RAM network. The cost of sending digital photographs over wireless networks is still very high. Nevertheless, the cost is considered to be justified when the news event is important enough and timeliness is of the essence.

2.4.16 Hospitality industry

All large hotel chains have teams of salespersons on the road selling meeting and convention space. A few technology-innovative chains have adopted mobile computing solutions to help smaller sales forces sell more space. ITT Sheraton is one such company. About 100 sales staff received 75-MHz, 486-based AT&T Globalyst notebooks with 340 MB or 540 MB hard drives loaded with a Windows 3.1 application suite that includes WordPerfect 6.1, Lotus 1-2-3, and FreeLance Graphics. Communication software is based on pcAnywhere from Symantec Corporation. Also included is Delphi hotel reservation software. Each salesperson's home office also was supplied with an HP OfficeJet—a multifunction device with printing, fax, and copier combined in one unit. Thus equipped, the sales force can now dial into the Delphi system at the main office and make reservations directly from the field. This arrangement incorporates elements of telecommuting, working from remote offices, working from home offices, and mobile computing. Among the benefits, ITT reports cost savings, productivity gains, and better customer service.

2.5 Miscellaneous Industry Applications

There are several interesting applications that fall into the category of miscellaneous vertical industries. We shall describe a few of these here.

2.5.1 Automated toll collection

Mobile Computing technology is now being used for automatic toll collection. The New York Thruway Toll Collection project and the Easy Toll project in the United Kingdom are two such examples. We shall illustrate this application with a description of Easy Toll—a consor-

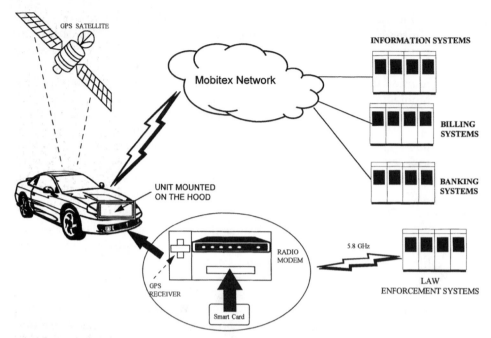

Figure 2.29   ROBIN electronic toll collection in the United Kingdom. (*Courtesy of Ericsson's* Mobile Data News.)

tium of Mobitex, Centre File (UK), and Green Flag. Easy Toll is a research project to test highway toll-collection technology.

The Easy Toll solution is based on the ROBIN (Road Billing Network) system (Fig. 2.29). It uses GPS and a RAM network and its objective is to keep the disruption of traffic to a minimum. Unlike tag-based systems, it does not require the erection of a roadside infrastructure, such as gantries fitted with detectors and interrogation equipment. As a result, it is less expensive to install, according to the consortium.

A dashboard unit for the ROBIN system computes a vehicle's position and determines when a toll is required. The toll is calculated and deducted automatically from a smart card on which prepayment has been made.

ROBIN toll collection system architecture.   The onboard unit (OBU) mounted on the vehicle contains a processing unit and a smart card reader. Connected to the unit are a GPS receiver used for automatic vehicle location and a radio modem for communicating over the Mobitex network.

Instead of the toll bars and bridges used in a conventional toll system, ROBIN uses the satellite-based GPS (Global Positioning System) to determine when a vehicle is on a toll road. A processing unit calculates the fees.

The ROBIN dashboard unit stores data which includes a geographic description of the highway network, tariff structure, toll roads, and toll recipients. This internal data is compared continuously with the vehicle's current position. When a toll is required, the dashboard unit calculates fees according to the distance traveled, time of day, vehicle class, and other tolling information such as entry into a tolled zone of a city.

Two payment options are supported. Prepayment is accomplished using a smart card from which the OBU deducts tolls during the trip. Transport companies and other vehicle operators may also establish a billing account for postpayment with the road operator or a bank. As tolls are incurred, billing information is sent from the vehicle over the Mobitex network.

For occasional users, a simple tag system will be available. The tag system eliminates the GPS receiver and radio modem and instead provides a 5.8-GHz short-range communication system allowing information in the OBU to be queried at special checkpoints. If information transmitted by the OBU is incorrect, the vehicle will be registered as a violator.

2.5.2 Automated information collection: Meter reading

Several utilities are experimenting with the automated reading of electricity meters and gas meters. While the technology is feasible, business cases are still being refined to support implementation. It would seem that the most appropriate time to provide this capability would be when meters come due for upgrade or replacement. The payback, of course, is in the cost savings that would result from reductions in the number of visits by meter readers.

2.5.3 Electronic maps for real estate, insurance appraisers, and others

All Points Software and MapInfo Corporation jointly developed Field-Pack Mobile Professional, which is based on GPS. Using portable computers, workers can input data on site and send information and/or queries to corporate databases. Using *geocoding* (a capability that results from the merging of GPS and map information), database information, forms, and images can be linked to specific geographic coordinates visualized on electronic maps.

This application is useful for real estate agents, insurance appraisers, service technicians, and other workers whose jobs are location dependent.

2.5.4 Special events: Sports, exhibitions, and conferences

Mobile notebooks are being used in sporting competitions, including sailboat races and tennis matches where progress and/or scores are recorded and communicated *wirelessly* to servers, which, in turn, feed information to spectators and the news media. A Mobitex network was used at the International Swedish Yacht Race in 1994 to provide second-by-second progress reports. GPS receivers installed on the yachts relayed information to the spectators on shore through wireless links. It is anticipated that the 1996 Olympics in Atlanta will make extensive use of wireless technology to transmit scores and other information to news rooms and other centers of newspaper or magazine production.

Reporters covering the 50th anniversary of the D Day landings in Normandy in 1995 used mobile data technology for their press coverage. They were able to send copy directly from the scene without a wire-line link by using a Mobitex radio data network.

Leaders at the 1995 G7 Economic Summit in Halifax, Nova Scotia, used Fujitsu Stylistic 500 pen computers linked via a LAN to communicate with their research staff and to verify acts or call up charts to support their discussions. The application was developed by Filbitron, a Toronto-area VAR, using SMART2000 conferencing software from SMART Technologies.

2.6 Application Case Studies

The following case studies provide additional information about customer applications in two different industries: United Parcel Service* and Ontario Public Safety.

2.6.1 UPS parcel delivery status application

When FedEx introduced a wireless network application to keep track of document and parcel shipments, UPS was pressured to respond with a similar or better service. The result was the introduction in February 1993 of the first nationwide cellular-based wireless data service.

---

* Information on UPS case study was provided by Doug Fields, Vice President, Telecommunications of UPS, located in Mahwah, New Jersey. It is based on UPS internal technology backgrounders and articles published in *Cellular Business* and *Computer Reseller News*.

Through cellular technology and a broad alliance of more than 70 cellular carriers, package-delivery information is transmitted from the company's 53,000 vehicles to the UPS mainframe repository in Mahwah, New Jersey, thus enabling UPS to provide same-day package-tracking information for all air and ground packages. Previously, this information was not available until the next day after delivery.

With its *delivery information acquisition device* (or DIAD), a custom-built electronic data collector, UPS is currently the only carrier able to capture both delivery information and customers' signatures. This data is then entered into the cellular network through Motorola-supplied cellular telephone modems. The cellular network provides the connection between UPS vehicles and UPSnet, UPS's private telecommunications network. These systems are set up to be fail-safe, with cellular redundancies, dual access to UPSnet, and multiple connections to the data system.

Competitive business advantages and benefits. UPS believes that its circuit-switched cellular data network service is the most comprehensive radio system available today. It covers a greater area and is more reliable than other mobile radio network alternatives. Key customer benefits quoted by UPS include:

- Immediate access to delivery information on more than five million UPS packages daily.
- The most extensive geographic coverage of any mobile communication alternative. More customer shipments can be given real-time information on delivery status.
- A high degree of reliability as a result of the service's redundancies and backup systems.
- Flexibility to accept future network technologies.

How UPS's delivery status application works. The schematic in Fig. 2.30 shows how delivery-status information is captured and transmitted through the cellular network to the UPS mainframe. The DIAD is inserted into a DIAD vehicle adapter (or DVA, which looks like a notebook computer docking station). The DVA, in turn, is connected to a cellular telephone modem (CTM) that transmits information from the UPS vehicle to the cell sites, where it is routed through the carrier's cellular switch and special primary access equipment. This equipment directly connects to a UPSnet packet switch, which transmits the information to the UPS mainframe in Mahwah, New Jersey. Once the information is incorporated into the delivery-status database, it is available to the company's customer service representatives.

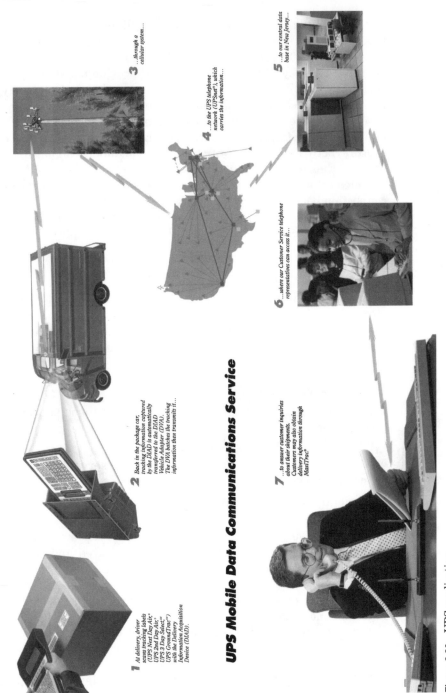

UPS Mobile Data Communications Service

1 At delivery, driver scans tracking labels (UPS Next Day Air,® UPS 2nd Day Air,™ UPS 3 Day Select,™ UPS GroundTrac™) with the Delivery Information Acquisition Device (DIAD).

2 Back in the package car, tracking information captured by the DIAD is automatically transferred to the DIAD Vehicle Adapter (DVA). The DVA batches the tracking information then transmits it...

3 ...through a cellular system...

4 ...to the UPS telephone network (UPSnet®), which carries the information...

5 ...to our central data base in New Jersey...

6 ...where our Customer Service telephone representatives can access it...

7 ...to answer customer inquiries about their shipments. Customers may also obtain delivery information through MaxiTrac.™

Figure 2.30 UPS application.

UPS delivery information acquisition device (DIAD). The DIAD is a hand-held electronic data collector that UPS drivers use to record, store, and transmit information, thereby helping UPS keep track of packages and gather delivery information. A linchpin in UPS's nationwide, mobile cellular network, the DIAD digitally captures customer signatures and package information—an industry first. This capability enables UPS to keep accurate, paperless delivery records. Drivers insert the DIAD into a module in their delivery vehicles to transmit information over UPS's nationwide cellular network for immediate customer use.

The key features of the DIAD are that it:

- Contains 1.5 megabytes of RAM

- Consolidates multiple functions into single keys that save time and space

- Utilizes digital signature-capture technology which is an industry first

- Has a built-in acoustical modem: if a driver cannot access the vehicle, data can be transmitted via telephone

- Has a built-in laser scanner that reads package labels quickly and accurately

- Utilizes smart software that knows the driver's next street

- Interacts with UPS cellular service

Both DIAD and DVA were custom-built for UPS by Motorola. Southwestern Bell, GTE, and PacTel all played key roles in putting together the cellular network consortium of more than 70 carriers.

Project costs and benefits. UPS has estimated the total cost of the project at around $150 million. Senior executives point out that the resulting increase in market share—let alone retention of the company's competitive edge—completely justifies the investment. In addition to the business imperative, UPS cites the following benefits from the application:

- Higher productivity of operational staff resulting from the revamping of processes and reductions in parcel-handling times

- Improved accuracy with the elimination of illegible handwritten records

- Speedier package delivery and tracking

- More information available for customer verification of package delivery and receipt

Unique features of UPS application. The UPS project is characterized by the following features:

- A unique implementation of analog circuit-switched cellular networks that meets UPS's wireless and OLTP file transfer application designs. (The customer service inquiry application is based on wired networks.)

- The business justification for implementation was based on a perceived need to maintain a competitive advantage, rather than on economic considerations.

- The complete reengineering of package-handling business processes.

- A unique ability to capture signatures on handheld, pen-based custom computers.

2.6.2 Ontario government public safety pilot project

An Integrated Safety Project (ISP) was started by the Ontario government in 1993. Several functionally related ministries were involved in analyzing justice system–related business processes in an integrated fashion. These ministries included the Office of the Solicitor General (representing law enforcement agencies), the Office of the Attorney General (representing the courts), Transportation (representing driver and vehicle information), Finance (representing fine-collection agencies), and the Management Board Secretariat (representing central computing and telecommunications infrastructure management). The following five projects were considered critical to the success of an integrated justice system:

- A mobile workstation (MWS) pilot project, which is described later in this section

- A collision data project with aims to automate on-the-spot collection of collision data

- A magnetic stripe driver's license project to introduce new driver's licenses based on digitized photographs

- A central offense database (CODB) project to act as a repository of all offense data, including disposition information available to all police agencies in the province

- A court process reengineering project to streamline court processes and to expedite and handle less formally simple offenses such as parking and minor traffic infractions that are relatively simple but tie up courts

The key business objective of the ISP initiative was to make changes in all the related processes that impact delivery of the justice system to the public. Since the overall implementation promised to be a very complex undertaking, the MWS pilot project was initiated with two key objectives: to validate business case justifications for the roll-out of the entire project, and to validate underlying technology solutions.

The MWS pilot project involved 50 IBM 730T ThinkPad workstations spread across 14 different user organizations, eight different computers, two wireless networks, and multiple transport protocols. The users comprised large provincial organizations, regional police organizations, and local police agencies. Highway truck inspection officers and ambulance drivers also participated in evaluating the technology.

From the perspective of law enforcement agencies, the MWS project formed the heart of the overall undertaking. It had the following business objectives:

- To improve law enforcement processes by providing the most current information from all local, regional, and national databases to police on the road

- To support community policing by reducing the administrative workload on officers and making them more available within the community

- To improve police safety by expediting car-to-car communication and making advance information available on suspects and locations investigated

- To introduce modern business applications such as electronic citations

- To expedite enforcement through on-the-spot payment of traffic offenses

- To improve the accuracy of data through source data entry

- To improve delivery of emergency health services in ambulances

Logical technology architecture of the MWS pilot project. A logical technology architecture for the MWS pilot project (see Fig. 2.31) was developed. It was based on the following considerations:

- The end-user device would be a PC-based ruggedized workstation.

- Initially it would be fixed inside vehicles, but the potential would be left open to switch to a portable handheld device.

- New Windows 3.1 business applications would be built.

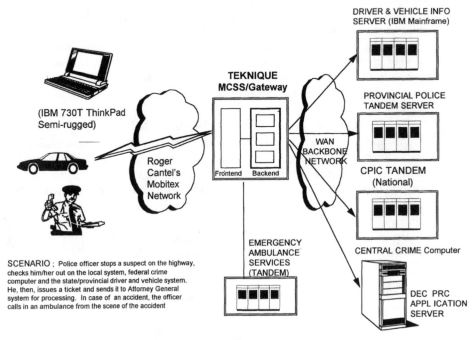

DRIVER & VEHICLE INFO
SERVER (IBM Mainframe)

TEKNIQUE
MCSS/Gateway

PROVINCIAL POLICE
TANDEM SERVER

(IBM 730T ThinkPad
Semi-rugged)

WAN
BACKBONE
NETWORK

Frontend   Backend

Roger
Cantel's
Mobitex
Network

CPIC TANDEM
(National)

CENTRAL CRIME Computer

EMERGENCY
AMBULANCE
SERVICES
(TANDEM)

DEC PRC
APPLICATION
SERVER

SCENARIO ; Police officer stops a suspect on the highway,
checks him/her out on the local system, federal crime
computer and the state/provincial driver and vehicle system.
He, then, issues a ticket and sends it to Attorney General
system for processing. In case of an accident, the officer
calls in an ambulance from the scene of the accident

Figure 2.31   Ontario's public safety application pilot.

- Existing MDT applications based on Motorola terminals would be emulated on the PCs. Major changes to these applications would be postponed.
- A common Windows 3.1 user interface would be used to access all applications.
- All local police agencies would route their information to and from their local systems and would interface with provincial or national systems through a common switch.
- Public shared wireless networks would be used for the pilot. The pilot would investigate both circuit-switched cellular networks and packet-switched networks.
- Common and open transport protocols based on TCP/IP would be implemented, wherever feasible. Protocol conversion from TCP/IP to IBM SNA LU 2 would be implemented to interface with driver and vehicle information systems.
- The overall project would be designed such that the way would be left open for the adoption of new technologies as they emerge.

Applications supported by the pilot project.   The MWS pilot project supported the following applications:

- Queries to the following databases:
  - local police incident report database
  - regional offense database
  - national criminal offense database
- Collision report database
- Emergency health services (ambulance system)
- Ministry of Transportation driver and vehicle information system
- Electronic ticketing

Application integration requirements. Application integration requirements were different for each of the eight organizations involved in the project. Three different interfaces for Tandem, IBM MVS, and DEC from a single Windows client application were designed and developed.

Lessons learned from the pilot project. While the project is still under observation and evaluation for future rollout, the following lessons have already been learned from the pilot project:

- Circuit-switched cellular technology is not suitable for prime-time mission-critical OLTP applications accessed in a moving vehicle. The networks and available communications software interfaces are unable to meet the error recovery and performance requirements of these applications.
- Packet-switching networks such as Mobitex and ARDIS (and CDPD in the future) are more appropriate wireless networks for public safety applications.
- Middleware designed for nonwireless networks could add a significant amount of overhead to application-processing and data-transfer times. The project was forced to drop Tandem RSC middleware used in a wire-line environment.
- Ergonomic considerations in permanently installing ruggedized workstations and printers in modern police vehicles with dual airbag constrains constitute serious challenges to workstation mount designers and equipment-installation vendors. Dedicated attention is needed to take care of safety considerations. Lack of attention to the ergonomic design could ultimately affect the overall success of the project.
- The pros and cons of more expensive, ruggedized, specialty mobile computer devices versus less expensive, off-the-shelf personal notebooks need to be reevaluated.

- Properly implemented solutions can generate tangible economic savings and nontangible benefits sufficient to justify large expenditures on such systems.

Unique features of the project. This pilot project was considered one of the most sophisticated of its type in North America. The following characteristics make it unique:

- It is a large and complex systems integration project involving fourteen organizations, eight different computer systems, and four different computer architectures.
- It integrates multiple existing applications.
- The electronic ticketing application has the potential for reducing a six-week process to a six-minute process through on-the-spot ticketing and payment by credit card.
- Critical response time requirements for OLTP applications impose a great challenge in a wireless environment.

Summary

In this chapter, we reviewed some of the major applications of mobile computing. Both horizontal and vertical industry applications were discussed. Horizontal applications based on electronic messaging have not proliferated, and it is therefore difficult to measure their benefits; e-mail continues to be the most prevalent horizontal mobile computing application and is an essential requirement of a mobile worker's life. On the other hand, vertical applications such as those built by *QUALCOMM* for the transportation industry and Telxon in the manufacturing arena have been extremely successful because vendors have provided complete solutions to users.

The variety of applications described in this chapter demonstrates the reach of mobile computing into almost every facet of personal and business life. As industries and users find appropriate answers and solutions, the adoption of mobile computing will accelerate and we shall see many, many more applications emerging. When that happens, one lone chapter will hardly do justice to the subject of applications. A complete book on the subject, or at least an application software guide will be necessary. After all, if the technology is easy and inexpensive and if the need is there, the scope and availability of applications will be limited only by users' imaginations, programmers' inventiveness, and business executives' desires to use technology as a competitive weapon.

3

Business Process Reengineering: Mobile Computing as Enabling Technology

Do not automate business processes just as they are by using emerging technologies. Obliterate them if possible, redesign them if you cannot eliminate them and leave them alone only if they add significant value to the business. Mobile computing can play a major part in this exercise by eliminating intermediate steps between manufacturing (or distribution or service creation) and sales—two fundamental business activities.   CHANDER DHAWAN

About This Chapter

We discussed the various business applications of mobile computing in Chap. 2. Many of these applications focus on providing information to mobile workers while maintaining back-end processes that are the same as those in the office. Such an approach produces limited benefits only. Organizations who treat mobile computing as an opportunity to reengineer their business processes achieve far greater benefits. The experiences of Federal Express and UPS are among the best examples of such an approach.

In this chapter, we shall discuss the need for business process reengineering and the basic steps involved. We shall then use several examples to demonstrate how mobile computing can be used as an enabling technology for BPR. The objective is not to make you a reengineering expert so much as to make you aware of the need for reengineering when implementing emerging technologies. For additional information, refer to the books on reengineering referenced at the end of this chapter.

3.1   What Is BPR?

BPR is the discipline of first analyzing and then redesigning current business processes and their components in terms of their effectiveness, efficiency, and value-added contribution to the objectives of the business. It draws upon the theory and practice of industrial engineering, information economics, process automation, information technology, and organizational design. It is a multidisciplinary process and should therefore be undertaken by a team composed of individuals from several areas.

According to Davenport and Short of Sloan Management School, "A business process is the set of logically related activities performed to achieve a defined business outcome." Typically, a process is characterized by subprocesses, specific inputs, and specific outputs. Processes can vary from simple tasks such as conducting personnel interviews to entire end-to-end processes for developing, producing, or distributing products or services. Figure 3.1 shows a schematic representing components of a process.

Common attributes of business processes include:

- The involvement of customers and/or their requirements
- The generation of products or services
- Associated time elements
- Associated performance measurements
- The crossing of functional or organizational boundaries

The following are examples of common business processes in most organizations:

- The creation of new products or services
- The construction of products or services
- The launching of products or services

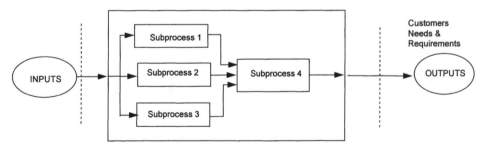

Figure 3.1   Business process components.

- The sale of products or services
- The distribution of products or services
- The servicing of products or services
- Support operations
- Financial and business planning
- Communication with users, customers, and external agencies
- The management of people

3.2   The Need for Reengineering Current Business Processes

International trade, competition, and other factors are causing rapid changes in today's businesses. According to business experts like Hammer, the ability of an organization to respond to these changes is an important indicator of its long-term survival. Many of the world's larger corporations, including AT&T, IBM, Motorola, Xerox, General Electric, Shell, Levi Strauss, and Procter & Gamble have responded to this need by reengineering their business processes to varying degrees. In fact, according to many, the notion of BPR has had an impact on business management such as has not been seen for a considerable time.

Many old, established business processes currently in use were designed with the emphasis on control and efficiency. Given the contemporary industrial environment of the time, only minimal consideration was given to interfunctional relationships. Now customer requirements and the responses to customer requirements are changing at an accelerating rate.

To be fair, it should be pointed out that business managers of just a few years ago were not addressing many of today's modern manufacturing methods and did not have access to the information technology that prevails today. Many of the older processes have the following characteristics:

- Hierarchical organizations
- Linear formats
- Controlled departmental boundaries
- Internal focuses
- Assumptions that individuals understand their tasks

In order to survive in tomorrow's competitive environment, businesses must meet the following new objectives:

- An emphasis on quality and service to retain customer loyalty
- A reduction in costs and cycle (or interval) times for production runs and sales activities
- An improved quality of products/services with corresponding reductions in defects and customer complaints
- An introduction of innovative methods and products

To achieve the customer focus that is inherent in the preceding objectives, we need to empower individual responsiveness and emphasize workgroup organization instead of individual task assignments. Processes can go beyond workgroups and become crossfunctional. Flexibility is a key characteristic of the new business paradigm. A business model should be capable of changing and adapting quickly.

3.2.1 BPR in the mobile computing context

Mobile computing is an emerging technology that has met with varying degrees of success in the business world. The mobile computing experience has been far more powerful for companies who have implemented the new technology in conjunction with BPR than it has been for those companies who have attempted to implement a current suite of business applications in a mobile environment without redesigning back-end business processes. Figure 3.2 shows how returns on investment in mobile computing are higher when combined with reengineered business processes, and lower when the technology is used with standard desktop applications. We shall show in Sec. 3.4 the impact that mobile computing technology can have on a BPR exercise.

3.3 Steps Involved in BPR*

The schematic in Fig. 3.1 shows the basic steps involved in a BPR effort. BPR is in fact based on a methodology that is very similar to the application development life cycle. There is one major difference between the two, however. In an application development project, there is usually a reasonably good understanding beforehand of business requirements and the functionality that the system must deliver. In BPR, fundamental questions about customers, products, and their relationships are asked as part of the reengineering process. In an ideal world, BPR precedes application development.

Figure 3.3 shows the major steps in a flowchart form. We shall describe these steps briefly in the following paragraphs.

---

* *AT&T Reengineering Handbook* (Order No. 500-449, ISBN 0-932764-36-3).

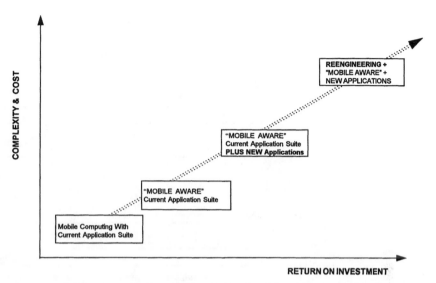

Figure 3.2   ROI and complexity with reengineered mobile computing applications.

Step 1. Create a business model: As a first step, we have to establish a business model that describes the business objectives at a very high level. This should be done by a senior executive who is more concerned with what the business inputs and outputs and what the minimum set of business-process constraints are rather than how they will be achieved. The constraints in question may be based on the current state of production, manufacturing, distribution, or service-delivery technologies.

Step 2. Review current processes: Implicit in this methodology is the second step, in which we review current business processes and applications. It is at this stage that we can identify opportunities for improvement presented by emerging or current technologies such as mobile computing. This review also provides a baseline measurement of productivity levels from which we can determine the future benefits of implementing reengineered business processes.

Process reengineering is distinct from process improvement. While reengineering is a process-improvement technique, it goes beyond the notion of simple or minor changes and emphasizes redesign in a major way. In BPR, one starts with a clean slate, so to speak, and questions each and every process, rethinking all aspects of the process along the way, including organizational and reporting changes, redeployment, job reclassification, etc.

Step 3. Preliminary costs/benefits: After reviewing the current processes, we need to undertake a preliminary costs/benefits analysis. Based on this analysis, we then have to decide whether or not to pro-

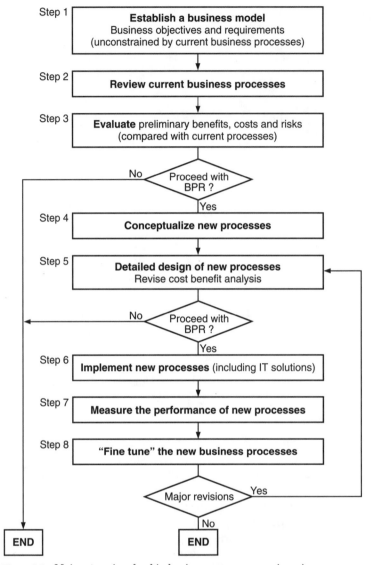

Figure 3.3 Major steps involved in business process reengineering.

ceed with BPR. It is important to be aware that there can be a high margin of error in projecting costs and benefits at this stage, so great care must be taken in the decision to either proceed with or discontinue a project. The risks inherent in introducing major changes should also be considered.

Step 4. Conceptualize the new business: Assuming that we have decided to proceed with reengineering, a BPR team must now be formed. This team, which should represent cross-functional units, should conceptualize the new or modified processes. This step involves the following substeps:

- The obtainment of additional information about customers, products, or services in terms of how existing service levels could be improved compared with world-class benchmark performance levels.
- The generation of innovative ideas for achieving breakthroughs equal to or better than world-class benchmarks. Brainstorming is one of the common techniques used here. The encouragement of lateral and unconstrained thinking by those who are not familiar with existing operations is a good approach. Questions should be asked and assumptions should be identified and challenged. Any tendency to pick the first idea offered or to ignore ideas that appear to be risky should be avoided.
- The conversion of each idea into a process concept consisting of an input, high-level flowchart, estimated requirements of resources (people, facilities), and resultant output.
- The development of concept evaluation criteria and the identification of the most promising concepts based on the resulting criteria.

Step 5. Design new processes or reengineer old ones: Assuming that reengineering is a feasible option, you should proceed to the design stage. In this step, greater details should be developed. Create flowcharts showing all the details to a degree that can be given to a software development team. Consider automation and information technology techniques. Use technology principles and aids for streamlined process design, such as:

- Enter information at the source and only once.
- Ignore geographic dispersal of resources because modern computer and communications technology, especially mobile computing and the Internet, can create virtual offices that are centralized, even if people are dispersed.
- Perform tasks representing the subprocesses in parallel to reduce cycle time and inherent delays in sequential activities.
- Link parallel subprocesses instead of integrating their results after the fact.
- Use technology to automate what you can.
- Shorten intervals by reducing time to perform the process, introducing parallelism, and reducing wait times.
- Eliminate extensive information exchange, data redundancy, and rekeying.

- Remove complexity, exceptions, authorizations, special cases, and audits which can not be justified from economic benefit point of view.

- Look at the entire organization. Introduce integration across departments within the same organization and between organizations (i.e., suppliers, business partners, and customers).

The guiding approach and foremost principle in redesigning new processes is to thoroughly understand the overall business context. We need to identify and prioritize the critical processes that are up for redesign and develop a shared vision for the future. We must then develop migratory steps toward that vision and let the new processes achieve this migration. Since added value is a fundamental criteria in creating processes, we need to be able to define the value of a process. Examples of such values might be "reduction in late pickups by 40 percent in six months and 80 percent in one year" and "responding to customer calls within two rings of the telephone."

Consensus building, collaboration, and ownership of the new processes are all extremely important in a BPR project. Holding a multifunctional workshop is an important way to create an environment for accepting this change.

Step 6. Implement the reengineered processes: Implementation is perhaps the most difficult phase in reengineering. Implementation involves the following subtasks:

- Software development, infrastructure installation, and training (traditional IT project activities)

- Business process changes in terms of personnel redeployment, organizational changes, reporting, monitoring, performance evaluation, etc.

- Migration and cutover considerations

- Management of change

Step 7. Measure the performance of the reengineered processes: BPR goals are typically set in terms of cost, cycle time, reliability, and, for manufacturing organizations, defect rates. For sales and distribution, the goals include higher service levels (measured in terms of fewer customer complaints), higher sales per salesperson, increased market share, and faster turnaround. When the new system has been implemented, performance should be measured in terms of its ability to achieve the stated objectives.

Step 8. Revise the design, if necessary: Based on a pilot implementation, certain processes will require redesigning. This should be done in a controlled setting.

3.4   Mobile Computing as Enabler
of Reengineering

Technology, in general, enables reengineering. This includes manufacturing, distribution, and information technologies. Workflow automation, electronic forms, smart cards, fax gateways, electronic mail, paging, CAD, and electronic messaging are among the different technologies that can assist in this effort. We believe that mobile computing technology can play an important role in process innovation. A very common use of mobile computing is to enable marketing and field personnel to perform office tasks on the road, at customer sites, or in the field. In addition to minimizing administration costs, this can reduce cycle times and improve overall customer service. The following table (Table 3.1) gives additional examples.

3.5   BPR Examples

We shall describe the following two examples of BPR:

- Courier industry application
- Law enforcement—electronic citation/ticket application

3.5.1   Courier industry application

The courier industry, especially large international couriers such as Federal Express, UPS, and Purolator have utilized mobile computing technology extensively to reengineer their businesses and gain a competitive advantage. The example used is generic (Fig. 3.4*a* and *b*).

3.5.2   Law enforcement—electronic
citation/ticket application

This is an interesting example of BPR that has resulted in a new process for a very common act of law enforcement—issuing a citation. The current citation processes for most enforcement agencies are time consuming, bureaucratic, and heavily paper-based. The new process being piloted by a few jurisdictions constitutes an interesting and practical automated approach. The process charts in Fig. 3.5*a* and *b* illustrate this application.

3.6   Principles of Business Innovation

We conclude this chapter with 10 basic principles of process innovation.*

---

* Lotus Accelerated Value Method Course (AVM) material—part of Lotus NOTES training.

TABLE 3.1   Impact of Mobile Computing on Business Processes

Application	How mobile computing can enable business process reengineering
Sales force automation	▪ Eliminate/reduce data-entry processes ▪ Eliminate/reduce call-in requests for messages ▪ Eliminate/reduce calls to customers regarding out-of-stock items ▪ Reduce paper handling as a result of electronic order entry at source
Parcel courier services (e.g., Federal Express, UPS)	▪ Eliminate all stand-alone data-entry activities ▪ Reduce numbers of customer service personnel through automated customer inquiries
CAD Dispatch for service representatives	▪ Eliminate cellular calls to or paging of service representatives ▪ Replace data entry and handwritten documentation subprocesses ▪ Eliminate head office–generated invoicing and payment collection
Health care industry	▪ Replace paper-based patient charts with electronic charts ▪ Eliminate/reduce transcription processes ▪ Expedite back-end support processes with physician-entered patient information
Public safety applications (police, ambulance, fire)	▪ Eliminate/reduce inquiries to communications centers ▪ CAD via notebook computers in vehicles (with voice dispatch used only as backup) ▪ Eliminate/reduce data-entry processes and other intermediate steps through electronic citations ▪ Eliminate manually maintained officers' diaries
Accident/Collision data systems	▪ Eliminate/reduce data-entry processes through real-time electronic entry of accident data ▪ Make information available to police and insurance simultaneously
Taxi dispatch	▪ Replace pickup-request broadcasts to all drivers with specific electronic messages to appropriately located vehicles only
Financial industry (insurance)	▪ Reduce need for visits to the office and instead spend more time with customers ▪ Eliminate/reduce references to rate books
Financial industry (stock trading)	▪ Replace stock exchange floor-based trading processes with computer-based trading
Banking	▪ Replace call-in requests with e-mail and paging ▪ Automate retail branch audit
Airline industry	▪ Increase sales through improved customer service ▪ Reduce/eliminate baggage-tracing costs and missing-baggage complaints ▪ Increase safety as a result of improved recording of maintenance data ▪ Reduce out-of-service time through computer-aided dispatch of parts
Manufacturing	▪ Replace manual entry of inventory information with the use of scanners and hand-held computers ▪ Use scanners to monitor the progress of items on assembly lines
General office applications (public sector)	▪ Reduce the involvement of secretarial staff in routine office tasks and processes by using notebook computers and wireless LANs to provide users with access to such things as on-line regulations and bills
Automatic data collection (meter reading, etc.)	▪ Eliminate/reduce site visits

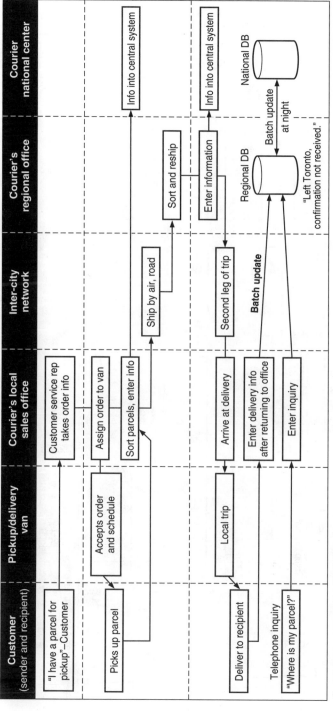

Current **process characteristics**

- Information entered at several points after item picked up or delivered by the driver
- Information entered by data entry staff in batch fashion
- No 1-800-number for customer inquiry into a national information system updated in real time
- Drivers in contact with local office, not regional or national center

Figure 3.4 *(a)* Current business process flowchart for courier industry.

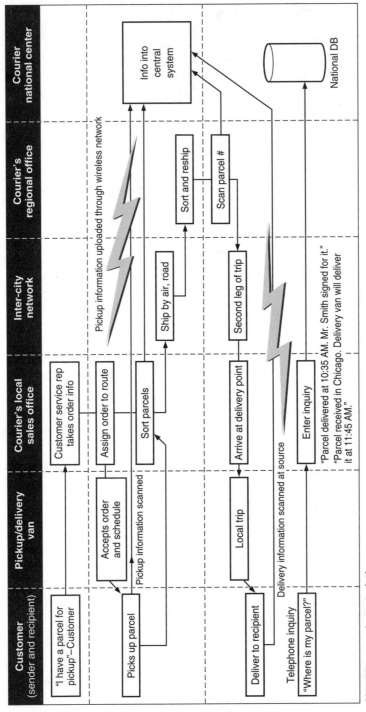

New process characteristics

- Information entered at the source by the driver; no back-end data entry staff
- Information entered by scanning at pickup and delivery points
- Scanned information uploaded via wireless network—cellular
- 1-800-number for customer inquiry into a national information system, updated in real time
- Drivers in logical contact with national information center in real time, always

Figure 3.4 *(Continued)* *(b) Reengineered business process flowchart for courier industry.*

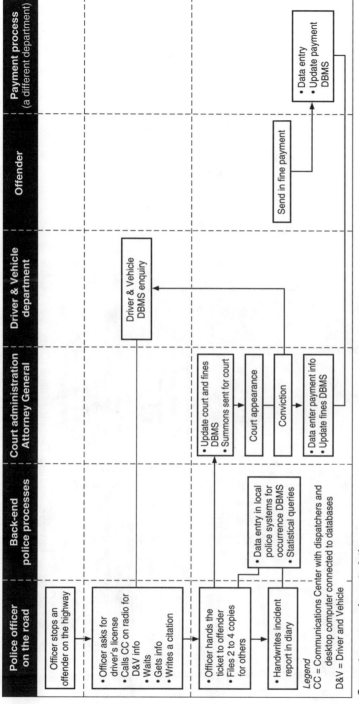

Police officer on the road	Back-end police processes	Court administration Attorney General	Driver & Vehicle department	Offender	Payment process (a different department)

Police officer on the road

Officer stops an offender on the highway

↓

• Officer asks for driver's license
• Calls CC on radio for D&V info
• Waits
• Gets info
• Writes a citation

• Officer hands the ticket to offender
• Files 2 to 4 copies for others

• Handwrites incident report in diary

Back-end police processes

• Data entry in local police systems for occurrence DBMS
• Statistical queries

Court administration Attorney General

• Update court and fines DBMS
• Summons sent for court

↓

Court appearance

↓

Conviction

↓

• Data enter payment info
• Update fines DBMS

Driver & Vehicle department

Driver & Vehicle DBMS enquiry

Offender

Send in fine payment

Payment process (a different department)

• Data entry
• Update payment DBMS

Legend
CC = Communications Center with dispatchers and desktop computer connected to databases

D&V = Driver and Vehicle

Current process characteristics
• Information entered manually at least in four places—police, court system, D&V, and payment departments/ministries
• Significant time spent on administrative duties
• Significant time lags between processes
• Cycle time (start to end of citation process) varies from 3 to 6 weeks
• Lower recovery of fines
• Lot of paper trail
• Different databases out of synchronization for up to 3 to 4 weeks

Figure 3.5 (*a*) *Current business process flowchart for law enforcement.*

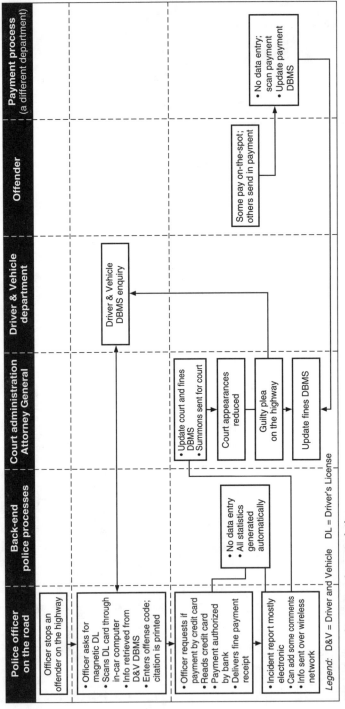

Police officer on the road	Back-end police processes	Court administration Attorney General	Driver & Vehicle department	Offender	Payment process (a different department)

Officer stops an offender on the highway

- Officer asks for magnetic DL
- Scans DL card through in-car computer
- Info retrieved from D&V DBMS
- Enters offense code; citation is printed

- No data entry
- All statistics generated automatically

- Officer requests if payment by credit card
- Reads credit card
- Payment authorized by bank
- Delivers fine payment receipt

- Incident report mostly electronic
- Can add some comments
- Info sent over wireless network

- Update court and fines DBMS
- Summons sent for court

Court appearances reduced

Guilty plea on the highway

Update fines DBMS

Driver & Vehicle DBMS enquiry

Some pay on-the-spot; others send in payment

- No data entry; scan payment
- Update payment DBMS

Legend: D&V = Driver and Vehicle   DL = Driver's License

Reengineered process characteristics

- Police officer's vehicle equipped with a ruggedized in-car notebook-based computer, mag stripe reader, and a printer
- Magnetic stripe driver's license, in-car computer with electronic citation production on in-car printer introduced
- Payment by credit card, if acceptable to offender accepted—incentive to pay by credit card
- Only offense code entered by the officer only once and all incidence info from the computer
- The transaction information transmitted from the car over a wireless network
- All location info, offense code descriptions, by-laws, automatically entered
- Time spent on administrative duties reduced
- For on-the-spot payment, cycle time reduced from weeks to minutes
- Higher recovery of fines; no paper trail
- All databases updated in real time and in synchronization

Figure 3.5 *(Continued)* *(b) Reengineered* business process flowchart of law enforcement.

1. Focus of every process should be on the customer.
2. Processes should be built for flexibility—business does and will change.
3. All activities in a process should add value.
4. New processes should be designed around the outcomes of an existing process rather than around tasks.
5. Complex processes should be broken down into multiple simple processes.
6. A team should be accountable for a process.
7. Decision making should be enabled where the work is performed.
8. Information should be captured only once at its source (or as close to the source as possible).
9. Error correction should be built into processes from the start rather than subsequently by another group or team.
10. Checks, controls, and reconciliations should be justified from an economic viewpoint not from the auditor's perspective.

Finally, we cannot stress too greatly the importance of BPR in mobile computing projects. The success or failure of a mobile computing project may well depend on the extent to which BPR is an integral or related part of it. Expenditures on wireless networks and mobile-aware applications can be justified more easily if the new technology supports the reengineered businesses of tomorrow.

Summary

In this chapter, we have emphasized the role of BPR in the context of mobile computing as an emerging technology. We introduced the concept and described the components of a business process: subprocesses, inputs, and outputs. Next we discussed the basic steps involved in a BPR exercise. We showed the impact that mobile computing can have in many applications. Finally, we illustrated BPR with two examples. We ended our discussion of BPR by enunciating the 10 principles of business process innovation.

The Business Case
for Mobile Computing

*An emerging technology may have a lot of value for the scientist
who invents it but it has very little use for the business person
unless its economic benefits are more than the costs associated
with its implementation.*   CHANDER DHAWAN

About This Chapter

Our discussion in Chap. 2 gave us an idea of the type of business applications that early adopters of mobile computing have implemented. In our case studies we discovered a close relationship between business process reengineering and the success of mobile computing projects. In Chap. 3 we analyzed the need to reengineer the business processes that are involved in mobile computing applications. We discussed how this reengineering exercise can serve as an opportunity to introduce more efficient business processes into an organization—and how that, in turn, can provide additional justification for introducing mobile computing into the organization.

Armed with the information in these chapters, we can now build a preliminary business case for an application of our own. First, however, we need to ascertain the preliminary costs and benefits of the application, remembering always that since they are preliminary in nature, they will have to be revised as more accurate data becomes available.

4.1 Methodology for Developing
a Business Case

Figure 4.1 shows a flowchart of the various steps involved in developing a business case. This flowchart is based on the premise that the development of a business case is an iterative process, with each subsequent iteration producing a more refined scenario. The preliminary case can be based on a rough technical design and on rule-of-thumb costs and benefits obtained from user executives and business process analysts. A pilot project should be created to simulate the new mobile environment so that measurements can be made for improved productivity. The data should then be modified as more precise cost information is made available by vendors and the application's environment is implemented. At each stage, a go/no-go decision can be made that results either in further analysis or moving on to the next step.

4.1.1 Quantifying unquantifiable benefits

Traditional business school ROI analysis can deal with many of the tangible costs and benefits (e.g., productivity improvements). ROI analysis does not, however, deal effectively with intangibles like risk assessment, and benefits such as improved business processes, superior customer service, or increased competitiveness. Increases in market share as a result of such benefits are therefore quite difficult to predict.

We intend to propose a methodology—based on the emerging field of information economics—that attempts to quantify, by a group consen-

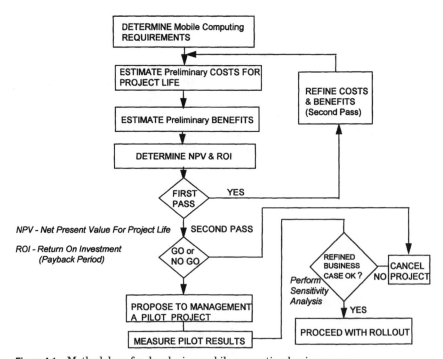

Figure 4.1 Methodology for developing mobile computing business case.

sus process and the assignment of probabilities, many of the more intangible costs and benefits (Fig. 4.1). *We strongly believe that quantification estimates by experienced business executives lead to decisions that are inherently better than those based purely on the intuition of a decision maker.*

The discipline of information economics should be used to identify intangible costs (or risks) and benefits. This is done by assigning weights and then adjusting total cost according to these weights.

4.1.2   Enhanced weight assignment scheme

The analysis can be further enhanced by a weight multiplier that takes into account the degree of support of a specific factor being analyzed. Several management consulting companies have devised different methodologies for this type of analysis. They use the same principles but have implemented them differently. The author has used a scheme based on the conversion of weighted scores into dollar values. This is done by determining the relative weights of tangible and intangible benefits and then, once the dollar values of the tangible benefits are known, using the relative weighting to determine the dollar values of the intangible benefits. Oracle Corporation has developed a similar scheme in their CB-90 financial evaluation model. The reader may obtain the CB-90 model from Oracle Corporation for this enhanced analysis.

4.1.3   Overall process for evaluating tangible and intangible costs and benefits

1. Divide the costs and benefits into the following categories:
   - Tangible costs
   - Intangible costs/risks
   - Tangible benefits
   - Intangible benefits

2. Determine both one-time (i.e., capital) costs as well as on-going (i.e., operational) costs for the entire life cycle of the project. Calculate NPV (net present value) for the entire project.

3. Define a group of major stakeholders in the project (e.g., business-user managers, IT managers, accounting staff, corporate financial managers) that would be affected by an investment in mobile computing technology.

4. Have the group assign weights to each of the categories, with a total for all the categories not exceeding 100. These weights should be

discussed, debated, and agreed upon by the group. Once an agreement has been reached, it will reflect the combined wisdom and judgment of the group.

5. Come to a consensus on intangible benefits and risks. Ask the group members to assign scores between 0 and 10 that indicate their estimation of the degree to which these risks and benefits will be realized.

6. Multiply each score by the previously assigned weight, sum the resulting numbers, and divide the total by 10. This will give a relative cost or benefit number on a scale of 100.

7. Determine the return on investment (ROI) by using conventional means acceptable to the organization, but only after adjusting the tangible costs and benefits by the weights assigned by the group to the intangible factors.

Table 4.1 shows a sample spreadsheet based on this process.

4.2   Costs

We should consider tangible costs as well as intangible costs. Tangible costs can be identified, quantified, and estimated easily. It is the intangible costs that are more difficult to quantify in economic terms. As a result, project sponsors tend to not pay adequate attention to them. Nevertheless, we strongly recommend that intangible costs be evaluated carefully. It is precisely because they require so much more effort to analyze that they, in particular, tend to concern senior management.

4.2.1   Tangible costs

While these costs are relatively easy to quantify, several design options should be considered in order to find an optimal solution. The spreadsheet in Table 4.2 shows component costs and benefits in a typical business case of an information technology project. This can be customized for a mobile computing project. We would like to recommend that the following pointers be considered while undertaking a high-level analysis of this kind.

1. In many mobile computing projects, network costs are among the most important cost items. Appropriate time and attention should be paid to the type of network selected. Also, reliable estimates should be made of the usage patterns and the number of data packets. Remember, if the application is properly designed, network traffic might ultimately be higher than you were led to believe during initial phases. Secondly, not only should user transaction packets be

counted, but also protocol packets and the system overhead. The network provider can assist you in this estimation process.

2. Depending on the area of coverage required, you should consider both public shared and private network options, especially if you are in a relatively small geographic area. If your coverage area is small, your data traffic is on the high side, and you already have a spectrum license, a private network may be cheaper if capitalized over a longer life cycle.

3. Do not ignore the cost of making changes to applications to make them mobile-aware.

4. Do not underestimate user-training requirements, especially if the users are not familiar with the application and the equipment.

5. Allow sufficient resources for comprehensive technical support. Remember that mobile users tend always to be in a hurry and therefore are not willing to tolerate long delays or to hold patiently for the next available representative.

6. Provide adequate estimates for systems-integration costs, both internal and external. The business case is normally built before signing contracts with a vendor. Allow for at least a minimal number of cost surprises.

7. Logistics costs for the permanent installation of ruggedized notebook computers in vehicles are quite significant. They can be as high as $500 or $1000 per vehicle, depending on the type of mount and any special cabinetry required to accommodate dual-air bag constraints and other health and safety considerations, especially in police vehicles. Include these costs after discussion with appropriate installation contractors.

8. Include an appropriate amount for contingency costs. A figure in the range of 20 to 25 percent of the initial budget is not uncommon in order to take unsophisticated cost estimates into account, especially in the first phase of the business case.

4.2.2 Intangible costs

Because mobile computing is an emerging technology, it is subject to the risks inherent in all new technologies. The business case should include the following intangible costs, which are either difficult to quantify or are nonquantifiable.

The multiple network hops and slower supported speeds associated with switched wired networks and wireless networks result in a generally poorer performance than that obtained from private leased-line WANs or LANs. This can give rise to a degree of dissatisfaction on the

TABLE 4.1 Mobile Computing Business Case

	Weight	Score (consensus)	Weighted score	Year 1	Year 2	Year 3	Year 4	Year 5
1.0 Costs								
1.1 Tangible costs	60							
1.1.1 One-time capital								
Hardware, software				$500,000	$250,000			
Development				$350,000	$100,000			
Training				$50,000	$100,000			
Miscellaneous				$250,000	$100,000			
1.2 On-going								
Network				$25,000	$100,000	$100,000	$120,000	$120,000
Maintenance				$25,000	$50,000	$60,000	$60,000	$60,000
Miscellaneous				$25,000	$60,000	$60,000	$60,000	$60,000
Total tangible costs				$1,225,000	$760,000	$160,000	$180,000	$180,000
1.3 Intangible costs/risks								
Resistance to change by users	10	6	6					
Bleeding edge technology	20	5	10					
Poor response time on road	10	4	4					
Total assigned score	40		20					
Costs assigned to intangible factors				$408,333	$253,333	$53,333	$60,000	$60,000
Total adjusted costs				$1,633,333	$1,013,333	$213,333	$240,000	$240,000

2.0 Benefits

	Weight	Score	Weighted					
2.1 Tangible benefits	35							
Headcount savings				$0	$200,000	$400,000	$600,000	$600,000
Other tangible savings				$0	$100,000	$300,000	$300,000	$300,000
Total tangible benefits				$0	$300,000	$700,000	$900,000	$900,000
2.2 Intangible benefits								
Increase in marketshare	20	8	16					
Improved customer service	15	9	13.5					
Future competitiveness	20	6	12					
Misc. benefits	10	8	8					
Weighted value of intangible benefits	65		49.5	$0	$424,286	$990,000	$1,272,857	$1,272,857
Total adjusted benefits				$0	$724,286	$1,690,000	$2,172,857	$2,172,857
Net benefits over costs				($1,633,333)	($289,048)	$1,476,667	$1,932,857	$1,932,857
NPV of costs over 5 years	$2,795,539							
NPV of benefits over 5 years	$4,701,569							
ROI								

EXPLANATION: (1) Weighted score = (Weight*Score/10).
(2) In this example, intangible benefits (wt = 65) are more important than tangible (wt = 35) benefits.

TABLE 4.2 Sample Cost Spreadsheet for a Mobile Computing Project (law enforcement application)

Mobile computing project	Assumptions for costing	1995/1996 Plan $	1996/1997 Plan $	1997/1998 Plan $	1998/1999 Plan $	Total
Staff expenses: salaries and expenses	Provide details in a separate sheet—categories, numbers, and $$	240,000	540,000	430,000	350,000	1,560,000
Staff benefits	Assume 25% of the above Total additional payroll	60,000 300,000	135,000 675,000	107,500 537,500	87,500 437,500	390,000 1,950,000
A. Capital expenditure	1000 in-car mobile computers					
A.1 Hardware	(Gradual phased implementation with a 3-year rollout)					
Mobile workstation hardware acquisitions	# of Mobile Computers = 200, 400, and 400 in 3 years; 20 in pilot 95	100,000	800,000	1,400,000	1,200,000	3,500,000
Radio modems	Decreasing costs for hardware assumed over 3 years		100,000	160,000	140,000	400,000
MCSS switch hardware and software	in-car ruggedized computers at $4000, $3500, $3000, modems at $500, $400, $350		650,000	125,000	125,000	900,000
A.2 Software						
Shrink-wrap software for in-car workstation	Workstation software like Windows, TCP/IP, ARDIS drivers, etc. at $300 per workstation		60,000	120,000	120,000	300,000
Software for switch, network, etc.						
A.3 Software development						
Application development	CAD Application user interface, electronic citation application	300,000	400,000	100,000	100,000	900,000
Communications software changes and APIs	Changes to communications interface for wireless network—TCP/IP interface					
Application software enhancements	Citation back-end system changes plus making applications "mobile-aware"	200,000	400,000	0	0	600,000

A.4 External consulting — Specialized technical expertise, mobile computing architecture development		200,000	200,000	100,000	500,000
A.5 Training — $400 per officer; 2 officers per car plus initial training development		500,000	225,000	225,000	950,000
A.6 Installation — Installation in the car — $700 per car decreasing to $500; includes customization of the car interior		140,000	240,000	200,000	580,000
A.7 Miscellaneous—describe					
A.8 Contingency — For errors and omissions and misc. items, e.g., Tandem HW/SW upgrade		500,000	250,000	250,000	1,000,000
A.9 Initial R&D—pilot installation—20 workstations — $150,000 specialized services + $100,000 for H/W + $100,000 for installation. No private shared wireless network infrastructure costs included. No allowance for enhancing servers in back-end mainframes and minis, etc.		350,000	0	0	350,000
Subtotal—one-time expenditure	600,000	4,100,000	2,820,000	2,460,000	9,980,000
B. Ongoing operations					
Telecommunications network services — $200/car/month going down to $150 in 2d and $100 in 3d year		480,000	1,080,000	1,200,000	2,760,000
Computer processing — Net additional costs only included		200,000	500,000	700,000	1,400,000
Maintenance (hardware/software/equipment) — $300/in car WS/yr on-site, 100K/yr for MCSS—12% of HW (assume 50% only for current year because of gradual installation)		210,000	430,000	550,000	1,190,000
Technical support (outsourced from vendor) — Outsourced from vendor		120,000	360,000	600,000	1,080,000
Omissions, errors, contingency — Assumed at 15%	0	178,235	418,235	538,235	1,134,706
Subtotal—on-going operations		1,188,235	2,788,235	3,588,235	7,564,706
Total capital and ongoing (A + B)	600,000	5,288,235	5,608,235	6,048,235	17,544,706
Total project	900,000	5,963,235	6,145,735	6,485,735	19,494,706

part of users who expect the same level of performance outside the office as they get inside the office.

Mobile workers interface with many different application servers on a multiplicity of platforms. The overall cost of interfacing in such environments is difficult to assess, especially in the early phases of a business case development. We suggest that the business case be increasingly refined as more accurate information is available.

The cost of converting from existing business processes to reengineered ones is also difficult to estimate. The method suggested in Sec. 4.1 should be used to estimate these costs.

4.3   Benefits

The benefits can be classified into two categories: tangible benefits and intangible benefits. Tangible benefits can be quantified more easily than the latter variety. We shall discuss both types briefly.

4.3.1   Tangible benefits

The savings that results from staff reductions is probably the most obvious economic benefit associated with a mobile computing solution. Mobile computing can lead to increased individual productivity, increased sales per salesperson, more service calls per repair person, and less time spent by professionals on administrative work, all of which can ultimately translate into a reduction in total staff required.

However, there are several other tangible benefits associated with mobile computing solutions. A higher order-fill ratio as a result of accessing real-time inventory information at the time an order is submitted can translate into reduced inventory costs. On-the-spot invoice production in service vehicles can lead to shorter payment cycles and better cash flow. The electronic citation/ticketing applications with credit card payment of traffic violations that public safety agencies are experimenting with can lead to a higher ratio of paid fines.

Table 4.3 shows a range of potential benefits that can all be used to justify mobile computing applications.

4.3.2   Intangible benefits

Many of the applications of mobile computing involve automating sales, improving customer service, or gaining a competitive advantage—all benefits that tend to be difficult to quantify. The project team should outline these benefits with as much detail and as specifically as possible. The group should then translate these benefits into percent increases in sales, market shares, and productivity improvements. If

TABLE 4.3 Nature of Benefits from Mobile Computing

Mobile computing application	Type of benefits (tangible and intangible)
Formula for quantification	$ Savings = Number of Users * $ Bundled Wages * % Productivity Improvement + $ Sales Increase * Profit Margin (because of customer service) + Decrease in inventory costs * interest rate
Sales automation	■ Shorter sales cycle; increased sales per salesperson; reduced head count ■ Elimination of order entry staff; more accurate data in database ■ Better customer service; reduced merchandise return ■ Lower inventory costs ■ Increased market share
Electronic mail	■ Less time spent calling the office for mail messages
Computer-aided dispatch for service representatives	■ Increased number of service calls per day ■ On-the-spot invoicing; faster payment cycle ■ Electronic dispatch of parts; reduced administrative and inventory costs ■ Improved customer service
Health care industry	■ Less administration for doctors ■ Better patient care ■ Telediagnostics; faster diagnosis
Public safety applications	■ More efficient deployment of law enforcement staff ■ Reduced time to reach scene of crime, hospital, or fire, safer communities ■ Lower voice-network costs ■ Fewer dispatch or communications control center personnel ■ More efficient records management through officer entry of incidence reports
Accident/Collision data systems	■ Real-time entry of accident data avoids subsequent duplication of effort ■ Information available to police and insurance simultaneously
Taxi dispatch	■ Less cruise time ■ Credit card authorization leads to improved service and more customers ■ Reduced dispatch personnel costs
Financial industry (insurance)	■ Reduced selling cycle ■ Superior customer follow-up ■ Higher dollar sales per sales presentation
Financial industry (stock trading)	■ Faster trades; more trades per hour ■ More accurate trading
Retail industry	■ Faster sales stations during seasonal sales ■ Reduced electrical and wiring costs ■ Improved customer service ■ Automated vending machines with credit authorization lead to higher sales
Airline industry	■ Better customer service leads to higher sales ■ Fewer missing baggage complaints; reduced tracing costs ■ More accurate maintenance data leads to safer plane flights ■ Electronic dispatch of parts results in less time in grounding of planes under service
Manufacturing	■ Lower wiring costs as a result of wireless LANs ■ Accurate inventory control ■ More accurate production tracking
General office applications (Public Sector)	■ Better document control ■ Reduced administrative staff requirements ■ Timely access to bills under debate in the legislature
News communications (sports, conferences)	■ Lower wiring costs as a result of wireless LANs ■ Real-time scores fed to media; score competitive advantage

the benefit can not be quantified with a high degree of reliability (we suggest above 70 percent), it should be handled with weights and scores as described in Sec. 4.1.

4.4 Return on Investment

Once costs and benefits have been identified, return on investment should be calculated using any method that the organization condones or uses. Popular spreadsheet packages like Lotus 1-2-3 and Microsoft Excel have formulae for this purpose.

4.5 Industry Experience of Return on Investment

Management consulting companies such as Gartner have analyzed business cases for mobile computing for several different applications. Their research studies indicate that payback period with mobile computing applications can be as low as 18 months. Customers such as Liebert and Bank of America have indicated that they have been able to achieve up to 20 percent productivity improvement with computer-aided dispatch of service representatives.

Summary

In this chapter, we studied a methodology for developing a business case for a mobile computing project. We discussed the many nonquantifiable benefits of mobile computing that are either difficult or outright impossible to recognize by conventional cost/benefit analysis and proposed a methodology to deal with them based on the assignment of weights for these factors. We looked at cost components and used a spreadsheet example for illustration. Finally, some of the many benefits to be derived from mobile computing applications were enumerated in a table. Information such as this can give you ideas for listing the costs and identifying the benefits of a specific application and can lead to the building of a business case for a project.

End-to-End Mobile Computing Technology Architecture

An End-to-End Technology Architecture for Mobile Computing

Too often, information technology practitioners are in a hurry and start building solutions without developing an end-to-end technology architecture. Let us learn from matured industries such as construction. Developing a high level technology architecture, along with application and data architectures, is an important investment in any IT project. It is no less true in mobile computing.
CHANDER DHAWAN

About This Chapter

We have discussed in Part 1 the business vision, potential, and economics of mobile computing. With this background, you can analyze an organization's business-application needs and build a business case. In Part 2, we move on to provide an overview of an end-to-end technology architecture for a mobile computing solution.

Mobile computing solutions take many different forms. The detailed architecture of any one particular implementation is different from that of other implementations. Nonetheless, we shall discuss the architectural framework of a generic mobile computing configuration that may well share many common elements found in users' organizations. This discussion will comprise the following sections:

- Mobile users' business needs and their influence on technology architecture

- Hardware technology architecture components

- Network architecture components

- Software technology architecture components

- Logical (application level) level architecture

- Interoperability considerations from an architectural perspective
- Technology principles for an integrated architectural framework
- Methodology for developing a custom architecture

While many early implementations of mobile computing projects were often relatively simple and employed only one specific application, one network, and one server on a LAN, we will illustrate a more complex configuration. Our discussion of the configuration's architecture will highlight the technical design issues that arise as such solutions are rolled out across organizations and are integrated with other business applications. First, however, we will discuss basic concepts in a simple configuration before we examine the more complete configuration and its impact on the architecture.

5.1   Mobile Business Users' Interconnectivity Needs and Factors Affecting Technology Architecture

As information becomes an increasingly important weapon in today's competitive business, and as business users become more mobile, there arises an important need to make up-to-the-minute information ever more readily available. The term *mobile user* now includes professional executives making business decisions away from homes and/or offices, salespeople closing deals, service representatives on emergency repair calls, or suppliers' shippers advising shop floor supervisors of exact times of arrival of just-in-time inventory items. This need of sales, service, and telecommuting workers to communicate electronically among themselves or with customers, suppliers, and headquarters is depicted in Fig. 5.1.

Figure 5.2 represents some of the more important factors determining mobile computing technology architecture such as types of remote users, types of business applications, amount of data transferred, currency of data, and geographical coverage of the users.

A typical mobile computing solution and its technology architecture is based on the following assumptions:

1. A large number of mobile users (several hundred to several thousand) need to access information from or send information to one or more application or database servers. Some of these users have access to public switched telephone networks (PSTN), which provide cheaper and more reliable communications. Other users do not have access to regular telephone connections, or they are not conveniently able to use them (e.g., from the factory floor) if they are available. As well, many users who spend a lot of time on the move (road warriors) occasionally require a temporary wireless connection.

MOBILE COMPUTING

(Interconnectivity Between Business Users)

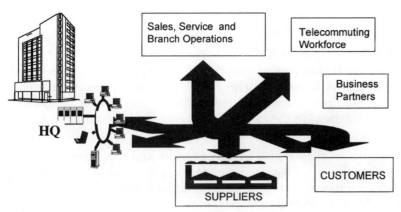

Figure 5.1 Interconnectivity between business users. (*Source: Xcellnet, Atlanta, Georgia.*)

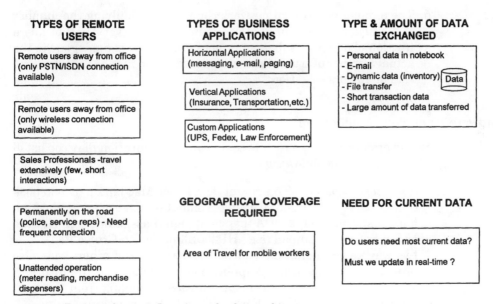

TYPES OF REMOTE USERS

Remote users away from office (only PSTN/ISDN connection available)

Remote users away from office (only wireless connection available)

Sales Professionals -travel extensively (few, short interactions)

Permanently on the road (police, service reps) - Need frequent connection

Unattended operation (meter reading, merchandise dispensers)

TYPES OF BUSINESS APPLICATIONS

Horizontal Applications (messaging, e-mail, paging)

Vertical Applications (Insurance, Transportation, etc.)

Custom Applications (UPS, Fedex, Law Enforcement)

TYPE & AMOUNT OF DATA EXCHANGED

- Personal data in notebook
- E-mail
- Dynamic data (inventory)
- File transfer
- Short transaction data
- Large amount of data transferred

Data

GEOGRAPHICAL COVERAGE REQUIRED

Area of Travel for mobile workers

NEED FOR CURRENT DATA

Do users need most current data?

Must we update in real-time ?

Figure 5.2 Business factors influencing technology architecture.

2. The types of business applications include:

- Messaging applications such as electronic mail and paging (one-way or two-way).
- Updates of local data files (e.g., price list, service schedules for the day) or downloads of data (e.g., sales summaries, presentation materials).

- Mission-critical or operational business applications (e.g., customer order entry and inventory updates, credit card authorizations, and banking transactions). These applications apply their updates in either real time or as deferred updates.

3. The data transferred by remote users may consist of:

   - File transfers of data to synchronize local files on notebooks.

   - Records transfer of transaction data (typically message-based) stored in databases on regional or central servers; elements of data required to complete units of work may belong to different organizations and may reside on disparate databases and platforms in distributed databases.

4. Frequency of connections by remote users and the amount of information transferred between remote users and homes/offices during a working day varies with the type of application used. This may range from occasional use of networks (i.e., once or twice a day) to almost continuous connections (e.g., as in the case of law enforcement applications).

5. The number of physical network connections (implying the number of physical ports required at communications servers) is typically a fraction of actual numbers of remote users. This is analogous to the actual number of trunk lines as opposed to the number of telephone sets in any organization.

5.2   Hardware Technology Architecture

Hardware involved in a mobile computing configuration may consist of one or more of the following components:

- *A mobile computer.*   This might be a PC or Mac laptop or notebook computer, a pen-based portable computer, a special application handheld computer, a PDA (personal digital assistant, e.g., Newton or Simon), or a palmpad (e.g., HP200LX or Zenith Cruise Pad). New two-way pagers being introduced by Motorola and others may also provide limited messaging application capabilities in the future. A variety of hardware platforms are involved, ranging from the most common PC/Mac CPUs to Power PC and RISC microprocessors.

- *A suitably configured modem or an appropriate digital network interface device.*   This might be a switched network modem, a dial-tone generation adapter for cellular connections (or a corresponding device for ISDN), a specialized unit for wireless networks like ARDIS, RAM, CDPD (e.g., ERICSSON Mobidem 901 for RAM, IBM's CDPD modem). A single universal modem that supports all connection scenarios does not currently exist, though PC (PCMCIA) Card based Ethernet/fax/cellular modems are appearing on the market: Where wireless networks are concerned, modems can work only with mobile computers that have radio transmitter equipment installed. In cellular connections, cellular telephones have this functionality built-in.

- *A communications server and or a wireless network switch/gateway* (e.g., Shiva's NetRover for asynchronous remote network access; TEKnique's TX-5000 for specialized wireless network switch/gateway). This type of communications server/switch provides the following functionality:

  - Asynchronous wire-line session connection and disconnection services

  - Session router management across large numbers of remote mobiles

  - Mobile identification

  - Network login, i.e., security management

  - Asynchronous wireless network connection and disconnection services

  - Protocol conversion (i.e., a gateway function)

- *An application and/or database server on a LAN, a minicomputer, or a mainframe where information resides.* Normally there is a LAN (Ethernet or Token-Ring) connection between the communications server/switch and the hardware platform on which information servers reside. This connection can also be accomplished either through high-speed buses or channel connections or wide area network private lines.

- *An electronic mail server on an appropriate platform, typically a LAN.*

A basic hardware configuration for mobile computing is reflected in the schematic in Fig. 5.3.

5.2.1   Characteristics of a basic mobile computing configuration

1. Most mobile computing end-user devices are relatively inexpensive and are portable, lightweight, and easy to use in a variety of non-office environments. Specialized devices are designed for permanent installation in vehicles. Others are simple paging devices carried in the pocket, or handheld computers carried by factory personnel. In short, mobile computing devices should be affordable and should meet the business needs of their users. Thus, ergonomics and portability are essential considerations.

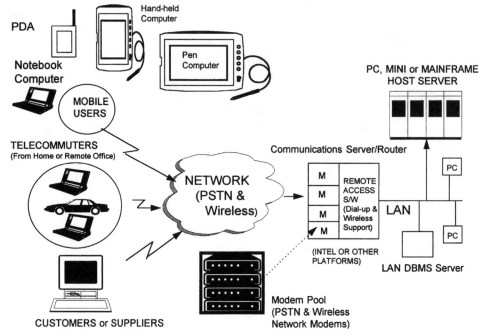

Figure 5.3   Hardware architecture for a basic mobile computing configuration.

2. Mobile computing solutions may consist of wired (through PSTN) or wireless networks, or a hybrid of the two. In fact, users should have access to information resources through either type of network, depending on where they are (e.g., wired access from a hotel and wireless on the road).

3. Communications servers and back-end server platforms may be on identical hardware platforms.

As mobile computing requirements grow in terms of numbers of users, types of networks supported, and application processing platforms accessed, hardware configurations become more complex, as depicted in Fig. 5.4.

5.2.2 Characteristics of a large mobile computing configuration

1. There may be thousands of mobile workstations as an organization rolls out mobile computing across its entire remote population. The architecture should support this user population.

2. While current wireless networks support 4800 bps for ARDIS and 8000 bps for RAM, there is a definite need for higher speeds (19,200

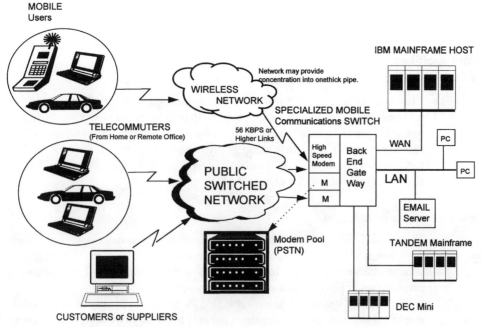

Figure 5.4   More complex mobile computing hardware configuration. (*Note:* Vendor names are for illustration purposes only. This does not imply an exclusive role of named vendors in architecture.)

bps for ARDIS and 16,000 bps for RAM) for applications that are response-time sensitive. Law enforcement, ambulance dispatch, and credit authorization are examples of such applications.

3. Switch hardware has to be scalar in capacity. Scalability achieved through multiple distributed switches may not be the best strategy from a maintenance and support perspective. Switch-hardware performance must go beyond the capabilities of PC platforms that have limited I/O capacities presently before SMP implementations become a norm for communications control applications. Thus, switch hardware should be based on proven high-capacity platforms like VME bus, Tandem cluster, or UNIX implementations.

4. Redundancy should be an integral part of good hardware architectures, since mobile computing applications quickly become mission-critical. Experiences in the financial, banking, and trading industries have shown that current switching architectures should be considered as base models. Such models can, of course, be modified by incorporating restrictions on wireless networks and mobile users.

5. Switch hardware and software have two logical components: front-end and back-end. Front-end components deal with remote mobile workstations and wireless networks. Back-end components inter-

face with information servers on LANs, minicomputer servers, and mainframe superservers.

5.3 Network Architecture for Mobile Computing

This is among the most important components in the technology architecture of mobile computing projects. It is important to understand the network participants involved in a physical connection between mobile computer and the information server. First we shall look at a simple network example and then extend the example to a hierarchical model that wireless networks often employ.

5.3.1 Simple wireless network model

Figure 5.5 shows the various components in a network connection between a mobile computer and a remote server. The mobile workstation is either connected to a public switched telephone (PST) network (dial-up or ISDN) or to a wireless network.

The public switched telephone (PST) network connects the mobile workstation to a communications server, normally resident on customer premises. In some cases, a network service provider may install a communications server at its own location with modem pool hardware.

In the case of a wireless connection, the wireless network service provider's base station is connected through a land line to a central sta-

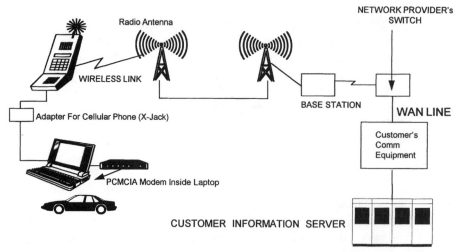

Figure 5.5 Schematic of wireless network connection for mobile computing. *Note:* FOR ARDIS/ RAM Mobitex, cellular telephone + modem combination is replaced by corresponding network-specific modem. Newer cellular modems have been built in connection for cellular phone.

tion, which, in turn, is connected to a customer's communications server. The customer's communications server is connected via the back end to an application and/or database server where business information or mailboxes reside.

The following points are worth noting in understanding the network architecture:

1. A notebook computer must be equipped with a cellular- or network-specific modem (or equivalent device) that has enough power (wattage) to both receive weak signals and to transmit signals of its own of sufficient strength under all circumstances (including, for example, the less-than-optimal circumstances found inside a moving vehicle).

2. Wireless networks are generally slower and more expensive than their wired counterparts. Therefore, their usage must be justifiable from a cost and benefits analysis perspective. Systems designers and application developers need to use wireless networks more efficiently than wired-line networks.

3. Specific network selections should not pose significant constraints on technology components like communications and applications software. Communications interfaces should be modular enough to allow for replacement of older networks with newer, more economical, and better-performing networks as they become available.

4. The extent of a wireless network's coverage depends on a number of factors, including number and power of switches, atmospheric conditions, and physical obstructions like hills and valleys. These factors should be studied carefully and users should be advised of problems.

5. There may be several network-specific telecommunication switches in the path between an end user device and a destination server where information resides. In each of these switches, analog signals are attenuated again on to wired (or a microwave radio) communications links. Similarly, digital information signals may be assembled and processed several times. This gives rise to much higher response times for end users than is the case with wired-line connections, where there may well be fewer hops and faster switches with lower propagation delays.

5.3.2 Hierarchical architecture of a wireless network

Figure 5.6 shows the hierarchical nature of a wireless network connection. Lower-level base stations are connected to higher-level switching centers. This has obvious implications for network design, as multiple hops increase network transit time.

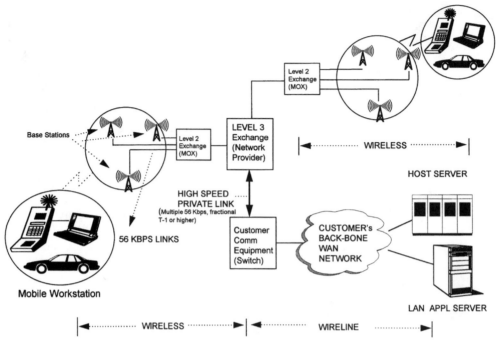

Figure 5.6 Typical network architecture: wireless and wired line.

5.3.3 Wireless network characteristics

Wireless networks are different from conventional wired-line networks in the following respects:

1. Wireless networks provide entry points from mobile workstations to wired lines through a series of hops and switches. A typical end-to-end connection for mobile computing consists of both a wireless (switched) portion and a wired (WAN) portion.

2. Since end-to-end physical connections include, therefore, the complex software used in switches, conventional network design and performance calculation techniques are no longer viable.

3. Wireless networks are more susceptible to transmission errors and temporary losses in connection than terrestrial networks. This is due in large part to the fact that mobile devices "roam" from one cell to another, and to atmospheric conditions. These susceptibilities can have a significant impact on communications software layers like TCP/IP which were designed for more reliable wired WANs.

5.4 Software Technology Architecture

Having discussed the hardware and network architecture, we shall explain software architecture now.

5.4.1   Current state of software architecture
for mobile computing

In the current state of the mobile computing industry, no single, open, standards-based mobile computing architecture for large enterprises exists. There are several divergent efforts by various vendors, user associations, and standards organizations to stake early architectural positions (e.g., the APCO 25 standard, ACM/IEEE initiatives on mobile computing, wireless TCP/IP RFC). Technology strategists and product planners from the vendor community are attempting to put forward a cohesive architecture to which they can give some credibility by introducing products with functions, features, and delivery timetables.

Remote network access industry providing software for mobile computing. Remote network access vendors are addressing the immediate needs of remote workers who must access information resources from remote locations. They have started a bottoms-up approach with Intel-architecture-based communications servers and accompanying software. Typically, they extend the resources of LAN and WAN applications to remote workers who can dial in via a modem. They provide gateways to minicomputers or mainframes where large reservoirs of legacy applications and databases reside. Recently, they have been turning their attention to wireless (in particular, cellular) networks. From a communications perspective, there are essentially two prevalent approaches to remote access.

Remote node. *Remote node* refers to the ability of remote users to dial into LAN-attached servers and to function as full nodes or peers on the network. This permits users the same access to LAN-based resources as they would have if they were actually in the office. When connected to the LAN, standard network packets are transmitted to and from remote users over the phone lines. Shiva's NetModem and Microtest's LANMODEM are examples of this implementation.

Remote control. *Remote control* refers to the ability of a remote user to dial into and control a host computer at another site. The host then acts as a surrogate computer. With this technology, all applications reside on the host computer and only screen images, key strokes, and mouse clicks are transmitted over the phone line between the host and the remote user. Packages like pcANYWHERE, Co-Session, and Carbon Copy are examples of such implementation. This remote control mode of access is particularly appropriate for wireless networks because it uses wireless bandwidths efficiently.

Several hybrid implementations exist that use both remote control and remote node access. Novell's Netware Connect software supports both implementations.

Pros and cons of remote access solutions. Most remote-access solutions are relatively inexpensive to implement, especially for switched wireline connections. They are also becoming more scalable by providing as many as 64 ports through a single communications server.

A major drawback with the remote-access approach to mobile computing on wireless networks is that most implementations are relatively proprietary. Vendors tend to focus on LAN applications only and to not address legacy applications on non-LAN platforms. In addition, users may also find that communications software does not work with a chosen wireless network. For a limited set of vertical and LAN applications, however, existing solutions do meet the needs of mobile workers quite effectively.

Two modern paradigms from an application development perspective

The client/server model. The client/server model provides the flexibility required in mobile applications. This model has in fact become the de facto standard in the application development paradigm. In mobile computing, it is very important to give as much freedom as possible to remote users by making available as much data and functionality as is possible without compromising security. This does mean, however, that common secure data should be stored on the server. Similarly it is necessary to store e-mail on regional or central mail servers.

The agent based client/server model (also called client-agent-server). This is a paradigm in which the basic client/server model is enhanced with the *intelligent agent* software concept. The agent portion of the software works on behalf of mobile users while they tend to other matters, such as dealing with customers. The agent organizes incoming information in the background (inquiry responses, e-mail messages, etc.) and forwards it to the mobile user whenever a connection is made.

The agent based client/server model has the following advantages when it comes to mobile computing:

- It makes possible single, automated communications sessions that cater to all information needs at any given time.

- It reduces costs and the need for resources since sessions do not have to be maintained while mobile users are inactive.

- It boosts user productivity because users are free to pursue other matters with the agent working in the background.

- It takes advantage of both client/server and store-and-forward technologies.

- It shares control responsibility between IS and the user organization because the server portion of the agent can be centralized under the management of the IS organizations.

Framework of a software architecture for mobile computing. We shall articulate in Sec. 5.7 a few architectural principles that can provide a framework for a technology architecture. First, however, we shall discuss software components that must be glued together to construct working solutions.

5.4.2 Software components

Just as Fig. 5.4 illustrates several hardware components, there are many software components involved in mobile computing solutions. These are depicted in Figs. 5.7 and 5.8. The components are:

1. Mobile client software suite

2. Mobile communications server/switch (MCSS) software

3. Application and/or database server software

Figure 5.7 shows the software component from a mobile client's perspective. We shall describe each of these components now.

1. Client software suite in mobile client nodes provides the following functionality:

   - Remote mobile workstation operating software (DOS/Windows 3.1, Windows 95, OS/2, Mac OS, Solaris UNIX) along with network software drivers control hardware devices and peripherals (display, pen, voice input, PC Cards, etc.). A suitable operating system must also be selected to support the remaining components and the programmed business functionality.

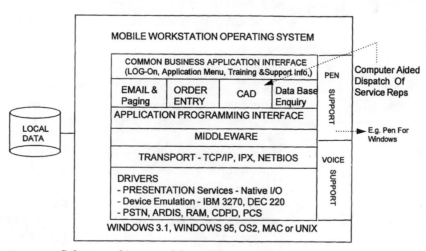

Figure 5.7 Software architecture of the mobile workstation.

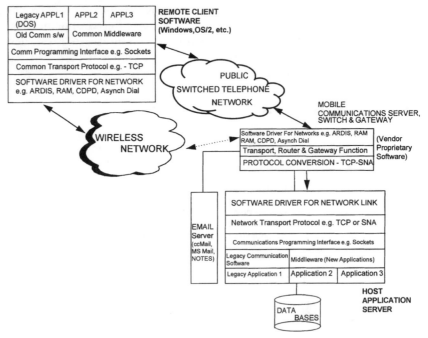

Figure 5.8 End-to-end software architecture for a mobile computing solution.

- A user interface or shell might be included in the operating system itself, as in the case of Windows 3.1 and Windows 95, or it may be distinct from the operating system, as is the case in UNIX with MOTIF.
- The client portion of the communication server software (e.g., Netware Connect, Novell's asynchronous remote access software) provides connectivity through PST as well as cellular circuit-switched network. Almost all remote access software vendors have a proprietary client component that runs on a laptop.
- Communications transport layer software like TCP/IP. TCP/IP has become the most common transport protocol for LAN/WAN internetworking.
- Middleware software that sits between business applications and transport layers such as TCP/IP is optional.
- Software drivers for specific networks (e.g., PPP or SLIP for PSTN, SMR, RAM Mobitex, ARDIS, CDPD, or the emerging PCS network). Specialized mobile radio networks based on proprietary switches from vendors like Motorola may have unique software drivers along with APIs (e.g., INFOTAC on Motorola switches).

- Pen and voice recognition interfaces are becoming *de rigueur* in new mobile applications. Pen not only facilitates the inputting of data, but the interface can also be used as a mouse replacement. Similarly, voice-based inputting and outputting will introduce a whole new set of mobile applications.

2. Communications server/switch software installed on hardware at home offices provides the following functionality:

- Network connection assignments to remote users
- Communication port or modem pool management
- Security verification
- Multithreading of physical connections from different users on multiple ports on a single logical communications channel such as COM1 and COM2,
- Communications protocol handling
- Logical connections to back-end application processors or DBMS systems

3. Application and database server software suites provide business application functionality. This software may reside on one or more than one of the following three platform categories:

- LAN-based database servers like SYBASE, ORACLE, or SQL servers implemented under OS/2, Windows/NT, or UNIX operating systems on Intel, IMPS, or similar hardware platforms
- Minicomputer-based servers like DEC VAX or AS/400
- Mainframe supersavers, including IBM, VMS, UNISYS, and Tandem Cyclone Guardian

Communications support for applications on these platforms may dictate a need for protocol conversions or gateways, or an equivalent function, unless middleware software is implemented with embedded protocol conversion.

A schematic of major software components for client and server implementations is shown in Fig. 5.8.

5.5   Logical Technology Architecture

We shall review logical technology architecture at two separate levels: (1) application data flow level, and (2) the system control level.

5.5.1   Application data flow architecture

Figure 5.9 shows the logical architecture for a mobile computing application. It follows a business inquiry through the various OSI network model layers. Although all the OSI layers are identified, some of the components depicted in boxes handle or incorporate more than one OSI layer. The main point of the diagram is to illustrate the symmetry of the layers of processing between the client and server platforms, which are completely different configurations.

Application layers are implemented in both the client and server platforms. Typical Windows client software may be written in Visual Basic, C++, or PowerBuilder. On the other side, application software on an IBM mainframe may be written in COBOL with a GUI front end.

This application software could access the transport services of an underlying network infrastructure through common middleware, especially if it is necessary to interface with disparate platforms. Please note that the middleware adds overhead, on the one hand, but simplifies the communications programming interface, on the other. The middleware uses the services of transport layers such as TCP/IP. TCP/IP packets are transmitted over the media access layer via wireline or wireless networks, using appropriate software drivers.

5.5.2   System control flow architecture

Figure 5.10 shows system (non-business-application-related) flow control through various functional blocks that may be implemented in switching the server or host/server platforms.

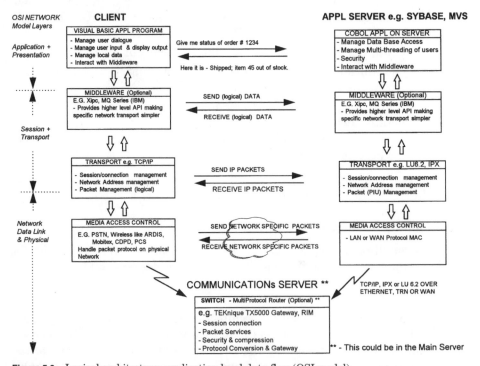

Figure 5.9   Logical architecture: application-level data flow (OSI model).

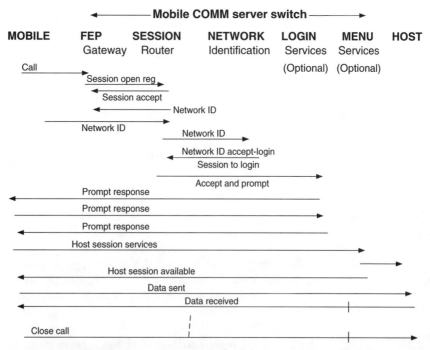

Figure 5.10 System control flow: establish session, send data, and close. (*Source: TEKnique, Scaumberg, Illinois.*)

This diagram is from a switch perspective. Please note that some of these switch software functions have a corresponding peer software module in the mobile client node or in the server node. This *peer module* deals with requests for the performance of specific functions by the switches or handles responses to requests from the server. In some cases, the switch may just provide a pass-through or transport function through the network.

It is worth noting that a significant number of lower-level I/O go through the switch. In fact, if there are intermediate implementations of middleware in the client and server nodes, there are many interactions before an application-level packet is transmitted. For example, in RSC (remote server call implementation on Tandem), we have seen up to 50 such interactions in a TCP/IP environment. This is an inevitable overhead with higher-level application development tools and therefore may not be suitable for wireless networks.

5.6 Interoperability Considerations from an Architectural Perspective

One of the greatest challenges in networking across heterogeneous platforms is to find out what works with what in different components.

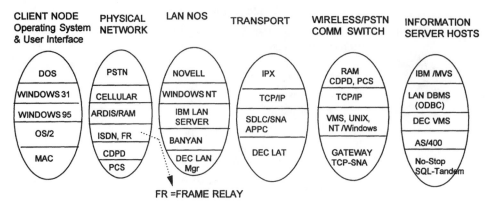

Figure 5.11 Interoperability considerations for mobile computing solutions.

Figure 5.11 shows a variety of client operating systems and user interfaces, physical networks, LAN operating software, transport layers, and server platforms. Any preferred combination must be analyzed as to its feasibility as a workable end-to-end communications solution.

In spite of vendor claims, system designers must watch for all the disclaimers, exceptions, and other constraints that may be imposed by nonstandard combinations of these layers. Finally, they should pilot and test the precise hardware, network, and software combinations they intend to implement.

5.7 Technology Principles for an Integrated Architectural Framework

Presently, it is not possible to define a universal technology architecture for mobile computing. This is because several architectural components are evolving and architectural design concepts are being validated by vendor implementations and user acceptance. Some of these implementations may be discarded as industry rationalization takes place during the next five years. However, we can enunciate certain basic technology principles that should be considered in any architectural investigation by the reader. These principles, combined with technology components described in this chapter, can assist users in developing architectures for specific organizations.

1. Technology architecture for mobile computing must be business-requirements driven. It should be compatible with application and data architectures.

2. Mobile workstations should match the business requirements of their users. Multiple types of workstations should be used only if

they meet specific technical or business needs of the user community. Cost savings achieved by using cheaper and less functional workstations should be weighed against the costs of multiple application development efforts, complex network management, and additional training requirements.

3. Mobile workstations should retain their functionality in offices, homes, hotels, and vehicles. They should support LAN connections, wire-line (switched) connections, or wireless connections. Client software should recognize modes of connection and load appropriate drivers accordingly.

4. Response times observed by mobile clients should be analyzed with an eye for optimizing network performance from an end-to-end perspective. Wireless network usage should be optimized both for communication costs and for user productivity.

5. As much application logic and data should be moved to mobile devices as is consistent with security guidelines and data architecture principles. This will give users maximum mobility and flexibility.

6. Client and switch (à la remote access communications servers) software should support multiple wireless networks as well as wire-line networks such as PSTN, ISDN, and frame relay. Wireless network usage should be optimized both for communication costs and for user productivity.

7. Mobile access security considerations must meet corporate standards. There should be a firewall.

8. If the expense can be justified, applications should be made mobile-aware, if to no other extent than to allow fast path through transaction dialogue and mail retrieval of important messages.

9. Technical support and network management should be primarily centralized, but allowance should also be made for the distribution of certain management functions that may benefit from regional support (e.g., call a service person from a local area for mobile hardware repair).

10. The architecture should be open and based on industry-supported standards (de facto and de jure). Proprietary vendor-specific implementations may be acceptable, provided there is a strong user community with a strong voice. In a similar vein, de jure standards should not be followed if the customer community does not support these standards. One may learn from OSI experience—a perfectly well intentioned de jure standards effort that was not accepted by the user community.

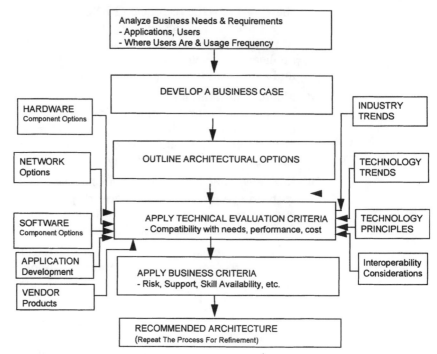

Figure 5.12   Methodology for developing technology architecture.

5.8   Methodology for Developing a Technology Architecture

The flowchart in Fig. 5.12 shows a simple methodology for developing a technology architecture for mobile computing. You should use this methodology as a guideline and adopt it to suit your organizational needs. If you are using a formal application development methodology, you should incorporate this flowchart as one of the submethodologies into the formal methodology.

Summary

In this chapter, we have covered an overview of mobile computing architectures. This overview included an analysis of users' business requirements and their effects on an organization's mobile computing architecture. We discussed each of the architectural components: hardware, network, and software. Application software and systems control software architectures were illustrated graphically. Finally, important technology architectural principles were outlined briefly. The information contained in this chapter should lay the groundwork for users to develop custom technology architectures for their organizations. The methodology illustrated in Sec. 5.8 will help in this exercise.

Mobile Computing Components

6

End-User Devices for Mobile Computing

Mobile computing employs many visible and invisible components which must work together to provide the functionality that end users want. However, nothing is more visible and more important to the end user than the device he/she holds. It is through this device that the end user perceives the benefits of the mobile computing world.   CHANDER DHAWAN

About This Chapter

We discussed the vision, potential, and economics of mobile computing in Part 1 of this book. We also looked at a few of the business applications of this emerging technology. We briefly introduced the discipline of business process reengineering as an important motivation for exploiting this technology. Finally, we described a methodology for developing a financial business case for mobile computing applications.

In Part 2, we introduced an end-to-end architecture for mobile computing. This, in turn, provided a basis for proceeding to Part 3. Here, we shall describe the various building blocks, the components that make up a mobile computing solution. First, we shall start with end-user devices.

Mobile computing end-user devices take various forms and configurations (Fig. 6.1). The packaging, form factors, hardware platforms, operating system support, and functional capabilities vary across these devices. There are, however, many common attributes shared by notebook computers, pen-based computers, handheld computers, and the like, all of which are used in mobile computing. We have divided these devices into the following categories:

- Conventional PC notebooks
- Pen-based computers, also called pen convertibles

Figure 6.1   Assortment of end-user devices. (*Courtesy of ARDIS.*)

- Handheld computers, also called pen tablets
- PDAs as computing devices
- PDAs as personal communication devices
- Pagers
- Specialized and hybrid mobile devices
    - PDA/pager combination
    - PDA/telephone combination
- Mobile printers
- Mobile fax machines
- Mobile scanners

In this chapter, we shall briefly discuss the architecture and features of these devices. We shall also discuss the various attributes that must be considered when evaluating devices for specific applications.

6.1 Notebooks and Powerbooks

In many ways, notebook computers are synonymous with mobile computing. In fact, by the year 1998, more notebook-based personal computers will be sold than desktop computers in North America. However incomplete such a notion may be to the technocrats and systems integrators who design and build end-to-end mobile solutions, the wide-spread and increasing popularity of notebook computers is now seen as an important contributor to the continuing evolution of mobile computing.

"Luggables" become portables and then notebooks. IBM brought to the market the personal computer in 1981, and in so doing introduced what has become the most popular computer product in the world. A few years later, the concept of mobility began to emerge and companies such as COMPAQ introduced unwieldy "luggable" personal computers. Luggables were soon replaced by true portables. However, the mobile computing evolution did not really start until notebooks were introduced by COMPAQ, Toshiba, Texas Instruments, NEC, and IBM. Apple entered the mobile market with Macintosh Powerbooks. Recently, the size and weight of notebooks have been further refined and we have been presented with subnotebooks. All these developments have been driven by market demands for even lighter and more powerful portable computers that can be used in a variety of environments, including the home, the office, in the field, in cars, and in airplanes.

The power of notebooks has kept pace with the times, and Pentium-based notebooks are now commonplace among professional mobile users. Mobility does have a price, however. Miniaturization leads to higher production costs, with the result that notebook computers are traditionally more expensive than their desktop counterparts.

Notebooks overtake desktops as primary PCs. Downsizing—figuratively and literally—has become a major phenomenon of the 1990s. It is a trend that is reflected in the types of personal computers sold. According to PC manufacturers, the sale of notebook PCs will overtake those of desktop computers in 1998. In fact, according to market researcher BIS Strategic Decisions, not long thereafter, notebook PCs will become the desktop PC of choice. The graph in Fig. 6.2 shows the relative percentage of notebooks that will be used as primary PCs. According to this graph, by the year 2000, 80 percent of 21.6 million PC notebooks sold will be used as primary PCs.

6.1.1 Hardware architecture

Intel processors and IBM-compatible PC hardware architectures operating under DOS/Windows 3.1 (see comments on Windows 95 in Chap. 12) are currently the most popular in the marketplace. Macintosh Powerbooks under MacOS 7.5 are used by organizations who have standardized on Macintosh desktops in their offices (Fig. 6.3). Recently, technical workstation manufacturers such as SUN have also intro-

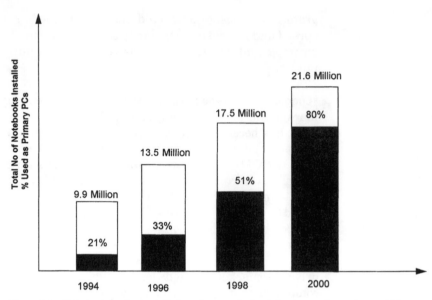

Figure 6.2   Total notebooks installed and percent of notebooks used as primary PC.

duced high-end notebooks based on the RISC architecture (Fig. 6.4). IBM has also introduced Power PC notebooks for specialized applications such as voice recognition.

6.1.2 Considerations in the selection of notebooks for mobile applications

Processing power and capacity. The power of a mobile device must meet the requirements of the mobile application. The processor type and its megahertz rating, the amount of RAM installed, the speed and capacity of the hard disk, the type of graphics engine, the CD-ROM access time, the operating system software, and the application software design all contribute to the overall performance from an end user's perspective. Other factors being equal, the configuration of the

Figure 6.3 Apple Powerbooks. (*Courtesy of Apple Corporation.*)

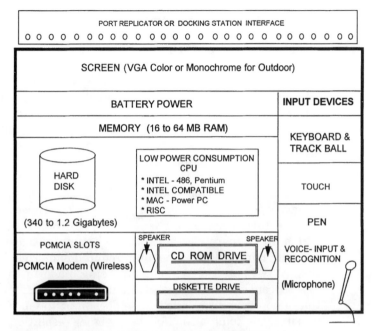

Figure 6.4 Hardware architecture of a notebook for mobile computing.

mobile PC should match the user's desktop configuration, particularly if time is spent both in the office and in the field.

Scalability and modularity. It is important to keep in mind that perhaps the single most important component from a cost point of view is the end-user mobile device. The processing power, main memory, and hard disk installed in the device should be scalable as our requirements increase in the future. Moreover, the device should be modular in design so that you do not have to replace the entire unit in order to accept new peripherals that may become available in the future. Availability of enough PCMCIA (now called PC Card) slots ensures this flexibility.

Size and weight. One of the most important considerations is the size and weight of mobile devices, especially when they are to be carried from place to place during the day. Fully featured notebooks small enough to fit inside a briefcase and weighing less than 4 pounds (2 kilograms), battery pack included, are available today.

Mobile computer type and size of screen. Presently, there are three types of screens commonly available for notebooks: VGA monochrome, VGA dual-scan color, and VGA active-matrix color. Active-matrix color screens are more difficult to manufacture and are more expensive.

From the end user's perspective, especially where sales applications are concerned, active-matrix color screens are more desirable than monochrome screens. Unfortunately, however, color screens do not function well in direct or reflected sunlight. Future developments in monitor technology will hopefully address this shortcoming.

The size of the screen and its resolution are important factors in terms of ease of viewing and crispness of graphics. Standard screen sizes for notebooks currently range from 10.5 to 12.1 inches (26 to 30 centimeters), which is substantially smaller than the average 15 inch (38 centimeters) desktop monitor.

Mobile computer screen resolution.   Most notebooks on the market used to have a VGA resolution of 640 × 480 pixels. However, most notebook manufacturers are now producing SVGA (800 × 600 pixels) screens (standard on many desktops). Since screen sizes are, by their nature, fixed, higher resolutions result in reduced font sizes. To compensate for this, SVGA notebook screens are being made in the 12 and 13 inch (30–33 centimeters) range. Although the fundamental requirement that a notebook fit in a briefcase does limit the maximum screen size possible, the combination of these higher resolutions and accompanying larger screen sizes will help in the future.

Keyboard size.   The keyboard will continue to function as the primary mode of input for another few years, at least until such time as speech recognition becomes mainstream. Space availability, especially in airplanes, cars, or customers' offices, dictates that keyboards be both small and, at the same time, compatible with the desktop variety. The latter requirement is important for end users who spend time in the office as well as on the road. Compatibility can be compromised in favor of small application-specific keyboards if the function keys and numeric pad normally available are not needed.

Peripheral input devices.   The keyboard has long been the predominant method of information input and dialogue manipulation for desktop applications. For GUI applications, the mouse supplements keyboard input. However, both can be awkward to use in the field when no desk or other physical platform is available. To overcome this problem, supplemental input peripherals, including trackball, pen, touch, and voice, are becoming common. While a trackball may be acceptable to sales and professional mobile users, it is not considered favorably by field personnel or public safety agencies. In such cases, a pen-based device is more acceptable for manipulating GUIs and modern mobile applications.

A pen-based device can be either *active* (electronic) or *inactive*. An inactive pen works with a touch screen, where the pen is used instead of a finger. Both the low cost of inactive pens and the frequency with which pens tend to be mislaid should be considered when selecting a method of data input.

Unavoidable keying requirements, on the one hand, and the degree to which pen-based application manipulation is possible, on the other, should determine which device is assigned the greater weight. In some applications, the pen can be the only input device. In such implementations, minimal keyboard input can be provided through a soft keyboard on the screen.

Voice as a method of input has significant potential for future mobile applications. It is expected that as this technology matures, it will find its way into mobile computing end-user devices. Voice recognition is enabled by way of a microphone connected to a sound card, along with a hardware adapter that converts analog voice information into a digital format. In some cases, this conversion can be accomplished by software. Several speech recognition software packages (such as Kurzwell, Holvox, Dragon, and Verbex) are available under Windows and OS/2.

Touch screen as a special input device.   Touch is intuitive and is commonly used in many tasks in daily work. Dialing a phone, writing, using a calculator, and thumbing through an address book are all dependent on touch. Capitalizing on this fact, many field and industrial mobile applications have been designed with touch screens that complement pen-based computing. Often this allows for total elimination of the keyboard, which results in units that are more compact and easier to use in the field.

For simple applications, such as selecting menu items in a restaurant, touch has three major advantages when used on a pen-based computer:

1. It is more natural and intuitive.

2. It is more convenient, even if the pen is tethered to the computer, because the pen does not have to be located or manipulated.

3. It is quicker.

The intuitive nature of simply touching or pointing with a finger is well known and demonstrated by the universal acceptance of touch-screen technology in public kiosks where there is no opportunity to train users to interact with the computer.

In the pen-based computing world, the stylus or the pen serves as a natural method for entering text. But touch input is even more natural and intuitive—not to mention convenient—for many other actions, including pushing keys, selecting menu boxes, or manipulating radio buttons and sliders. This is especially true in the case of many vertical applications, such as a route person making deliveries or a stock clerk checking for stock outs, where input is often intermittent.

Technology behind touch/pen-based digitizers. There are three touch/ pen technologies. The first uses resistive membrane. This touch screen is in fact a transparent plastic switch that is activated by a finger or a pen. It offers good resolution, low power consumption, and a high digitizing rate. It is also low-cost, proven, and easy to integrate into products. On the downside, these sensors suffer from a scratchable surface, diminished light transmission due to multiple layers of plastic, and poor *hand rejection* (the ability to reject signals that result from the hand resting on the screen while writing).

The second technology, employed by IBM, uses a glass sheet with discrete conductive lines. The pen is electrostatically sensitive, while the finger is capacitively sensed. Because both finger and pen signals are separately digitized, this system offers good hand rejection. Other advantages are good resolution, high point speed, and a more durable surface. However, the image quality is reduced due to the visibility of the grid lines.

The third technology is used in MicroTouch products. This technology employs a single continuous conductive coating on a glass sheet and senses both the finger and the pen capacitively. The technique offers excellent optics, low power consumption, high resolution, and high data speed. It also offers automatic hand rejection through a capacitive sensor in the lower barrel of the pen that activates the switch, puts it in pen mode, and invokes hand rejection.

When to use pen over touch. If an application is relatively complex, with a busy screen and many small dialogue boxes, you will need the precision control and pointing of a pen. In the final analysis, the method of input should be based on a combination of user preference after field trials and application design (Fig. 6.5). Development staff must not make these decisions—they can only point out the options to the user community, which should then decide based on the preference of a majority of users. There will always be those who do not like one method or another.

Types of peripheral devices. CD-ROM drives and sound cards for multimedia applications are becoming very popular. In fact, all major notebook manufacturers have introduced multimedia versions of their notebooks. Other newer peripherals, such as video, are also gaining in popularity. We will discuss PC Card (PCMCIA) peripherals separately in Sec. 6.7.

Packaging: The need for ruggedization. Mobile devices generally have to be able to withstand often-rough handling in many different environ-

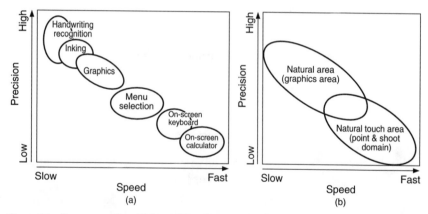

Figure 6.5 Pen computing and touch application characteristics: (*a*) approximate precision and speed attributes of pen-computing input activities; (*b*) natural usage areas of touch and pen. (*Source: Micro Touch White Paper.*)

ments. Because of this, most components are of higher quality and have to meet tighter specifications for shock absorption, temperature and humidity resistance, water tightness, and overall packaging integrity. Many notebooks designed for industrial use are encased in a magnesium shell to withstand shock and movement.

In defense and law enforcement applications, extremely stringent requirements for ruggedization, such as NATO military specifications 810 C&D, may be required. Some of the requirements are as follows:

- Withstand both cold temperatures (to −35°C) and high temperatures (to 45°C)

- Withstand humidity levels in the range of 80 to 90 percent

- Shock resistance of up to 400 G

- Withstand drop test from 3 feet (1 meter) onto a concrete floor

Expense of ruggedized mobile devices.   Until recently, law enforcement agencies have always insisted upon highly ruggedized devices because of the relatively hostile ergonomic environments under which officers have to work. As a result, these proprietary units cost twice as much as other units to purchase and service and are difficult to upgrade. In the current economic climate, senior officers are beginning to question the need for quite such highly ruggedized computers and the price tag that comes with them. Instead they are evaluating the feasibility of buying off-the-shelf semirugged devices and assigning them to individuals rather than to cruisers, thus enabling officers to take units out into the community on foot patrol as well.

Safety and health considerations. With the advent of dual airbag constraints, special safety considerations have to be taken into account when installing notebook computers in vehicles. In police cars, the problem is compounded by the presence of such ancillary equipment as radar devices, weapons, video cameras, and voice communications equipment.

Ergonomics. The following ergonomic considerations should be kept in mind when selecting a notebook:

- Does the unit have an adjustable screen angle to accommodate individual preferences?
- Is there an armrest? This can be an important consideration when extended use is anticipated.
- Is the notebook mount adjustable (swivel sideways and vertically) with respect to the user's seat in the case of a vehicle installation?

Size, weight, and life of the battery pack. While some progress has been made in the power consumption and battery life of most notebooks, the weight and size of the battery pack and the life of the battery charge continue to be sore points with users. Recent initiatives by battery manufacturers hold promise. AER Energy Resources demonstrated at COMDEX in 1995 a 1.8-lb Zinc Air battery that will run up to 12 hours between charges—a significant improvement.

Power management software is a common feature of most notebooks. A recent innovation in power management was introduced by System-Soft Corporation. It is the MobilePRO suite of BIOS and system control software. When used in notebooks powered by "smart batteries," MobilePRO enables users to request the battery life needed for a specific job. The product automatically slows down components as needed in order to meet the requested time. MobilePRO supports hot docking and is available in notebooks from Compaq, HP, and others.

Lighting source for the screen for night use in vehicles. A screen may either be backlit or use reflective lighting. For public safety applications where ruggedized notebooks are used at night, backlit screens are preferred. If this option is not available, special lighting attachments for external lighting may have to be installed.

Logistics of installation in vehicle for in-car applications. While the logistics of permanent installation are obviously not a problem for remote workers and professionals who carry their notebooks with them, they are a very important consideration in many public safety applications where notebook computers are fixed in vehicles. As was touched on previously, with the arrival of dual airbags as standard equipment in post-1994 cars in North America, there is now minimal space inside vehicles for the installation of a computer. Experimental new fixtures and cabinetry have been prototyped by IBM and specialty contractors like Johnson & Gimbel. However, the reaction of users to the prototypes has not always been positive because of their tendency to restrict movement, on the one hand, while putting the computer out of easy reach, on the other. Contractors are looking at the possibility of completely redesigning dashboards in order to work around existing restrictions.

6.2 The Pen-Based Computer: A Slightly Different Notebook*

Most large notebook manufacturers have introduced pen-based versions of their notebooks. The main difference between a regular notebook and its pen-based counterpart is the replacement of the physical keyboard with a pen and a soft keyboard that can be activated by pen or by touch. The pen is used as a replacement for a mouse/trackball and, to a limited degree, for handwriting recognition.

Although it is true that frustrations in trying to use a keyboard under certain conditions gave rise to pen-based technology, handwriting-recognition technology has still not proven itself as a viable and error free method of input. While limited handwriting use in specific applications (such as signature verification) is being pioneered by some users, the main use of pen-based computers continues to be in mobile applications where data must be sent to a server away from a client station (Fig. 6.6). To facilitate this, pen-based computers are often equipped with a wireless LAN connection or a wireless WAN connection (Mobitex, ARDIS, cellular, or CDPD) to a remote server.

Many applications have already been developed for pen-based computers, especially for sales and field-force automation. More applications can be expected as the ingenuity of business process innovators and the skillfulness of software developers come together to solve business problems. The following examples illustrate the diversity of these applications:

- Sales call/client management, e.g., TopSales from IRM for account coverage, order capture
- Field assessment/emergency management, e.g., software from Geofirma for digital photographs of flood-ravaged sites stored with descriptions of damage

---

* *Pen Computing Magazine, Mobile Office,* and *Pen-Based Computing Journal* (Volksware) provide good coverage of current mobile devices.

Figure 6.6   Pen-based computer in use. (*Courtesy of Telxon.*)

- Order entry, e.g., Pen On The Road/RadLink from Ruggedware, which allows a choice of scanning, writing, or typing the order information into a Windows application, with later transmission to a Sybase Server.

- Field inspection, e.g., Inspectwrite from PenFact for maintenance and instrument reading information, and POS/Inventory control application from Telxon

- Health care industry, e.g., Clinsys for patient management, and Digital Nurse Assistant, which automates records for admission

6.2.1 Palmpad version of pen-based notebooks

Palmpads such as CruisePAD from Zenith Systems, GRID AST, and other manufacturers comprise a unique category of end-user devices that provide only the user interface portion of a PC, including display, input, and, optionally, sound. CruisePAD uses high-speed wireless communication (based on digitally scrambled, security-encrypted wireless

LAN technology) to connect to a host PC and its resident processing power, memory, storage, operating system, application software, and external connections to create a complete, lightweight, mobile system. CruisePAD can be 500 to 1000 feet away from the host PC (Fig. 6.7).

Since a palmpad is so much smaller than a notebook, it can be held in one hand and operated by the other. Input choices include a configurable on-screen keyboard and stylus input for entering data while standing or moving about. Either a full or a compact keyboard is optionally available for touch typing.

Several interesting applications in the public safety, health care, manufacturing, and distribution industries have been and are being developed for palmpads.

6.3 Handheld Computers

These are special versions of either pen-based computers or lightweight industrial computers with a bar code scanner (or other attachments for specific vertical applications) for material management, warehouse inventory control, or assembly line monitoring. Because of environmental requirements for these applications, their packaging design quite often must be rugged. In most cases, there is a radio fre-

Figure 6.7 The CruisePad. (*Courtesy of Zenith.*)

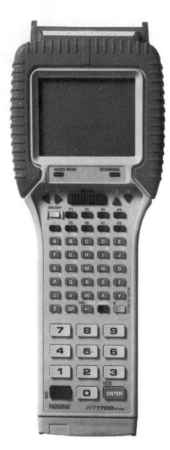

Figure 6.8 A handheld terminal from Norand. (*Courtesy of Norand.*)

quency link with a host PC that can receive data from the handheld computer (Fig. 6.8).

We shall illustrate the common features of handheld computers in Table 6.1, which lists three handheld computers currently available on the market. Refer to vendors and trade magazines such as *Pen Computing* and *Mobile Office* for additional information.

6.4   Personal Digital Assistants (PDAs)

While several research studies predicted the sale of huge numbers of PDAs during 1995 and in future years, actual sales results, so far, have fallen significantly short of these estimates. Although we do not support the optimistic forecasts of industry researchers, we do agree with the general trend these studies indicate in terms of an increased demand for PDAs. One of the reasons for the relatively poor sales performance of Motorola's Marco and Envoy or Sony's Magic Cap is that

TABLE 6.1   Handheld Computers

	Norand	Telxon	Symbol
Model	Pen*Key 6600	PTC 1180-1184	PPT4600
Operating system	MSDOS, WinPen, PenRight	MSDOS, PENDOS, PenRight	MSDOS, WinPen
Processor	486 DX2/50 MHz	F8680/12 MHz or 486SLC/25	486 SLC/20 MHz
Standard RAM/Max RAM	4 to 16 MB, 2–8 MB Flash	3 MB/3 MB, 1 MB ROM	2 to 16 MB
Storage—hard disk	PCMCIA type acceptable	40 MB optional	2–4 MB Flash
PCMCIA	Two Type II and one Type III	One Type I, one Type II/III	One Type II and one Type III
Screen size and type	6.25 in, 640 × 480 VGA	9.5 in diagonal, 640 × 480 VGA	5.5 in, 320 × 480
Input—digitizer pen/touch	Inductive pen	Electromagnetic/ 1000 dpi/150 pps	Passive pen supplied
Comm connection	RF WAN or LAN (optional)	Spread spectrum/ WAN RF	PCMCIA wireless cards
Size/weight	10.1 × 8.5 × 2.1 in, 4 lb	12.25 × 9.5 × 1.5 in, 4.0 lb	5.4 × 9.6 × 3.5 in, 1.9 lb
Comments	Rugged, full screen, pen	Rugged unit, hardened display	Ergonomic pen system for public safety; vehicle mount

SOURCE: Vendor information and *Pen Computing Magazine,* November 1995.

PDAs are still trying to define their product role. Over the next few years, the price, functionality, and application of PDAs will converge to a point where users will begin to identify with them. At that point, software developers will start building business and consumer applications, and corporate planners will be able to create business cases that show an adequate return on PDA investment.

In simple terms, PDAs are lighter, smaller, cheaper pen-based computers with miniaturized screens, fewer keys, less RAM, and less storage, if any. There are already several different models of PDAs on the market and more can be expected as this product develops a role in mobile computing. One way to categorize PDAs is by their operating system:

- Apple's Newton Intelligence
- General Magic's Magic Cap
- Geoworks's Geos
- Simon
- Proprietary OS

We shall describe Newton and Simon in some detail here. Newton, more than any other device, pioneered the PDA concept—and, in the process, took a lot of criticism from the user community for having the sort of problems that are inherent with any new product. Simon represents a competing PDA format, coming, as it does, from the cellular phone arena.

6.4.1 Newton MessagePad

Newton MessagePad 120 is an upgraded version of the original PDA from Apple (Fig. 6.9). It allows you to organize your work and personal life; it allows you to integrate information with your personal computer; it allows you to communicate using faxes, pages, and e-mail. It has a built-in note pad word processor, to-do list, date book, telephone log, and address book for organizing personal and business contacts, as well as Pocket Quicken to help organize personal and business expenses.

You can enter information in many ways: as digital ink (handwriting), printed text, hand-drawn graphics, with an on-screen keyboard,

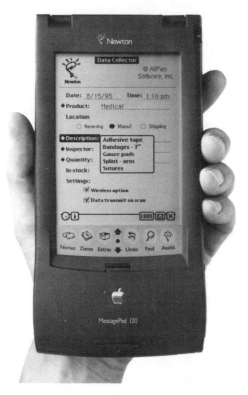

Figure 6.9 The Newton Message-Pad. (*Courtesy of Apple.*)

or with an optional external keyboard. Pen-based navigation and data entry make finding information easy and fast.

Its improved recognition software transforms your handwriting into typed text. You can print information just as it appears on the screen or as fully formatted pages. The MessagePad 120 can receive wireless messages and alphanumeric pages and can send and receive faxes and electronic mail.

Features of the Newton MessagePad 120.   Newton MessagePad 120 has the following features:

- Built-in personal organizer
- Built-in communications: sends and receives e-mail messages and faxes; assists in making phone calls
- Built-in personal computer integration: backs up all data onto a personal computer
- Expandability: works with a large selection of PC Cards and serial peripherals such as modems, pagers, wireless communications, and storage products

Technical architecture of MessagePad

- Newton Recognition Architecture: improved recognition of hand script (printed or cursive, or a mixture of the two) with the assistance of a built-in list of 93,000 words
- Object-oriented database: stores, finds, and links information
- Newton Toolkit: enables the user to create custom applications by way of a graphical toolbox
- Newton Hardware Architecture: ARM 610 RISC processor operating at 20 MHz
- Low-power, reflective LCD display (320 by 240 pixels) 3.8 by 2.8 inches (9.65×7.11 centimeters)
- 8 MB of ROM, 687 K of system RAM, 1361 K of nonvolatile user RAM
- One PC Card Type II slot with 325 mA capacity
- LocalTalk and RS-232-compatible serial port
- Low-power, half-duplex, infrared transceiver works at up to 38.4 kbits/second within 1 yard (1 meter)
- Printer support for Apple Laser Writer, HP LaserJet II, DeskJet 10e, etc.
- Fax support for group 3 fax machines

6.4.2 Simon

Simon is a personal communicator that includes a cellular telephone and additional personal mobile computing functions (Fig. 6.10). It has advanced phone features such as 0.6-watt transmission, a DTMF indicator, a roaming indicator, a call-in-absence indicator, an automatic system retry, up to four NAMs, and a signal strength indicator.

Simon has built-in fax features (9600 bps, group 3). The internal fax/modem sends and receives faxes wherever you are. You can handwrite comments and send the responses back. There are two ways to create a fax: you can write on the screen, or type a message with a built-in keyboard. To get a printed copy of the fax, you attach the unit to a fax machine.

Simon can be used to create, send, and receive e-mail. It operates via LOTUS cc:Mail without any additional software. It can store up to 20 full-page messages. With the appropriate PC Card inserted, Simon can function as a pager, receiving alphanumeric messages in the United States over MobileComm's nationwide paging network. The PCMCIA paging card can receive messages even when separated from Simon.

Simon also has personal information manager (PIM) functions including an address book, a paperless note pad (handwriting or typing), and a personal organizer.

Table 6.2 compares the features of some of the early PDAs.

Figure 6.10 IBM/BellSouth's Simon personal communications assistant (PCA).

TABLE 6.2   Early PDAs

	Apple	Motorola	HP	Sony
Model	MessagePad 120	Envoy	OmniGo 100	Sony Magic Link PIC 1000
OS	Newton Intelligence	Magic Cap	Geos 2.1	Magic Cap
CPU	ARM 210/20 MHz	Motorola 68349	86 compatible	Motorola 68349
RAM/ROM	544 KGB, 512 KGB	1 MB/1 MB max, 4 MB ROM	1 MB/1 MB, 3 MB	512 KB/512 KB, 4 MB
Disk storage	PCMCIA Card	PCMCIA Card	PCMCIA Card	PCMCIA Card
Screen size	3.4 × 5.4 in, 480 × 320 reflective	3.0 × 4.5 in, 480 × 320 pixels	2.38 × 2.38 in, 240 × 240 pixels	3.0 × 4.5 in, 480 × 320 pixels
Input (pen/touch)	Pressure-sensitive, 3 supplied	Pressure-sensitive, 2 supplied	Pressure-sensitive, 1 supplied	Pressure-sensitive, 1 supplied
Comm connection	To desktop PC through infrared	To PC desktop (optional)	Wynd 2-way messaging (opt)	To PC desktop (optional)
Size/weight	7.5 × 4.4 × 1.0 in, 14 ounces	7.5 × 5.75 × 1.2 in, 1.68 lbs	3.75 × 6 × 0.9 in, 11.6 ounces	7.5 × 5.2 × 1.0 in, 1.2 lbs
Handwriting recognition	Yes, letter	No	Yes, Integrated Graphitti	Third party
Options	Forms software	Messaging card, printer connect	Messaging card, printer, memory	Magic Link pager card, keyboard, memory
General use	First PDA with integrated pager	Two-way wireless communicator with Magic Cap	Inexpensive, personal pen/keyboard organizer	Attractive price, for messaging and electronic commerce

SOURCE: Vendor information and *Pen Computing Magazine.*

6.4.3   Evolution of PDAs

PDAs are still evolving, trying to find their personality physically (shape, size, weight) and logically (what functions to provide). There are competing products (notebooks and subnotebooks) on the upper end with full functionality—but they cannot be carried in your coat pocket. The question is, should we change our attire to accommodate PDAs, or should we insist that PDAs continue to fit into existing coat pockets? The answer may turn out to be a little of both.

On the logical side, technology vendors and market research companies are wondering whether a PDA should be a miniature computer or a sophisticated communicating device. Interfaces and software vary widely with the choice. Most industry researchers agree there is a growing need for a small, handheld communicator. Now, with the increasing availability of narrowband PCS, GSM, SMS, and two-way paging, that need may soon be filled. Once the appropriate software application and interfaces with e-mail systems are developed, it is expected that Nokia, Ericsson, and Motorola will introduce palm-sized handsets that combine a cellular phone, a pager, a web browser, and an answering machine in a single device.

6.5 Pagers

Pagers are by far the most popular and cheapest devices for contacting mobile workers. Because paging network providers derive their revenue from the monthly fees they charge for their services, pagers themselves have always been sold at relatively low prices. As a result, these small, low-cost devices enjoyed immediate success as soon as they were introduced.

Paging devices started out as beepers which essentially indicated that there was a message awaiting retrieval. Soon after, they were replaced by numeric pagers and then by alphanumeric pagers. In 1995, SkyTel introduced a two-way paging service for which Motorola and other vendors have since introduced paging devices (Fig. 6.11). The first two-way pager from Motorola is called Tango. We shall see, during 1996 and later, additional incarnations of two-way pagers from several different vendors.

6.6 Mobile Printers

Many mobile computing applications, such as sales illustrations, proposals by financial services salespersons, field workers' service invoices,

(a)   (b)

Figure 6.11 (*a*) SkyTel pagers. (*Courtesy of SkyTel.*) (*b*) Two-way narrowband PCS device. (*Courtesy of Research in Motion, Waterloo, Ontario.*)

Figure 6.12   The Pentax mobile printer. (*Source: Pentax.*)

and law enforcement agents' parking and traffic infraction citations, require that documentation be printed in the field (Fig. 6.12).

Mobile applications used inside customers' homes or offices do not present a problem—there are many portable bubble jet, ink jet, and even laser printers available from companies such as Hewlett-Packard, Canon, Tally Mannesman, and others. Just like notebooks, they need to be ruggedized for mobile use, however. Designing rugged printers for mobile applications out in the elements continues to offer technical challenges to engineers due to the fact that ink freezes in low temperatures. Because of the alcohol base of its ink, however, the HP 340 mobile printer is said to be suitable for temperatures as cold as –35°C.

6.7 PC Card (PCMCIA) Peripherals*

The Personal Computer Memory Card Interface Association (PCMCIA) is an industry group that comprises some 300 members now. It was founded in 1989 for the purpose of setting and promoting industry standards for bus interface cards. The association has introduced specifications for various types of peripheral cards that can be installed and exchanged between portable devices such as PC notebooks, PDAs, and even pagers. PC Cards (PCMCIA) are the current generation of what was originally introduced as a *memory charge card*. These cards are having a major impact on mobile computing because PCMCIA vendors adhere to the following standards:

---

* Refer to *PCMCIA Primer* by Larry Levine (M&T Books, 1995) for additional information.

- PC Cards must be hardware and operating system independent.
- PC Cards must be simple to configure and must support multiple peripheral devices.
- PC Cards must be simple to install.
- PC Cards must be inexpensive.
- PC Cards must be interchangeable.

While there is usually something of a time lag between the introduction of a full-size adapter card and its PC Card (PCMCIA) counterpart, it is a time lag that is steadily being reduced. At the same time, interchangeability and compatibility, although still not perfect, are also improving. (Regarding this issue, we caution systems integrators to actually test a card for PCMCIA compatibility rather than take it for granted simply because it is said to be PCMCIA-compatible.)

A PC Card (PCMCIA) slot can be one of four types, depending on the thickness of the card it accepts:

Type I: 3.3 mm thick; primarily used for RAM and flash memory

Type II: 5 mm thick; includes complete input/output capabilities; used for modems, LAN adapters, and memory cards

Type III: 10.5 mm thick; capable of housing miniaturized hard drives

Type IV: 18 mm thick; not available until 1997; used for high-capacity hard drives

Many notebooks currently available in the market (such as the IBM Thinkpad 750P) have two Type II slots or one Type III slot. Popular PC Cards available on the market include:

- Memory cards
- Removable hard disk cards
- LAN cards—wired and wireless
- Fax/modem cards—wired and wireless
- ISDN card
- Radio modem cards—ARDIS, RAM, or narrowband PCS
- Paging cards
- Global Positioning System (GPS) cards
- Specialty Cards—multimedia and video cards

We shall discuss the uses of each of these cards.

6.7.1   PCMCIA memory cards

The first implementation of PCMCIA cards was for memory—conventional as well as flash memory. Memory cards range in capacity from 1 MB to as much as 64 MB. They are seamlessly integrated into the operating system by file system software and each is assigned a drive letter. There are really two types of memory cards. First, there are *SRAM PC Cards* that can be read or written to without any special memory technology drivers. They are volatile and must be maintained with a built-in battery. Second, there are *ROM PC Cards* that do not require a battery to maintain their data. One type of ROM card that can be written to is an EEPROM or Flash PC Card.

6.7.2   Removable hard disk PC Cards

These cards provide the same function as a regular hard disk but with one obvious advantage: they can be inserted and removed easily. This flexibility provides many uses for these cards. For example, in law enforcement agencies with notebook computers installed in patrol cars, officers can each have individual hard disk cards that have been loaded with the appropriate software.

6.7.3   PC Cards (PCMCIAs) for LAN:
Wired and wireless

These cards are the PCMCIA versions of desktop Ethernet or Token-Ring adapters. With a wired LAN PCMCIA adapter in a notebook and a connection to a LAN transceiver in the office, users are able to utilize LAN resources as though the notebook was a desktop computer. When it is time to take the notebook home or to a customer site, the user simply unplugs it from the transceiver. Now, wireless LAN adapters are available in PCMCIA format as well.

6.7.4   PC Card (PCMCIA) for modem:
Wired and wireless

These are the most popular cards used in notebook computers today. They come in two varieties. The first is a PCMCIA version of a dial-up fax modem card, where it is assumed that the user is able to connect to an RJ-11 telephone jack in a remote office. If an RJ-11 jack is not available, users can be equipped with the second variety, a more expensive cellular modem available from vendors such as Ositech, U.S. Robotics, or Megahertz that attaches to a cellular phone through an X-jack connection.

The first variety is available as a V.34-compatible, 28,800-bps model that works with relatively few errors over land lines. The second variety is not as reliable, and error-free transmissions at speeds higher than 9600 bps are rare.

6.7.5 PC Card (PCMCIA)
wireless radio modems

These cards provide the functionality of old radio modems, which used to be quite bulky and required a lot of space. Radio modem cards are specific to particular networks and therefore ARDIS, RAM, Mobile/Mobitex, and CDPD modems are all different. The PCMCIA modems available for ARDIS and RAM are cheaper than those available for the newer CDPD technology. IBM markets an individual PCMCIA card for each of the three networks (Fig. 6.13).

6.7.6 PC Card (PCMCIA) ISDN cards

For remote network access solutions, ISDN offers much higher bandwidth than PSTN at 28,880 bps, or wireless networks at 4800 to 19,200 bps. Several vendors provide PCMCIA ISDN cards for mobile configurations. IBM's WaveRunner PCMCIA card (56 or 64 Kbps) and SCii Telecom's ISDN DataVoice cards are examples of these cards.

6.7.7 PC Card (PCMCIA) paging cards

A very common PCMCIA card that is becoming increasingly popular is the pager card that can be installed in notebooks and PDAs. With two-way paging becoming available on PCS narrowband, usage of these cards will increase.

Figure 6.13   PCS PCMCIA card by Research in Motion. (*Courtesy of Research in Motion.*)

6.7.8   PC Card for Global Positioning System

Global Positioning System PCMCIA cards are being installed in notebooks for many of the applications described in Chap. 2. We have described a popular consumer application of GPS where upscale car rental agencies are providing cars with built-in GPS features.

6.7.9   Multimedia PC Cards

Sound and video PCMCIA cards enable users to add multimedia capabilities to mobile notebooks. Video PC Cards allow users to attach custom cameras or standard camcorders to PCs to view and record crime and accident scenes.

6.7.10   Encryption PC (PCMCIA) Cards

Security is extremely important in mobile computing. As a result, several manufacturers have created encryption and digital signature PC Cards. Some can even support the Department of Defense's Clipper program.

National Semiconductor has introduced a PCMCIA 2.1 compliant PC Card that embeds RSA digital signatures, message digests, and encryption functions. This product has applications in data security, secured electronic messaging, workflow, electronic commerce, and information metering. Hardware tokens protect the private keys of the RSA encryption algorithm using secure packaging technologies.

The PC Card from National Semiconductor—named *Tessera* after the token ancient Romans used as a ticket or means of identification—provides among the highest level of commercial security (FIPS 140—1 level 3) and can now be implemented at a relatively low cost (less than $100 per user for large orders) compared with previous solutions. Encrypted data can be used to provide positive identification of users, store private medical records, include authorization codes, or even perform secure transaction processing.

When it comes to secure financial transactions, however, there are still concerns that the degree of privacy afforded is not absolute enough!

6.7.11   Multifunction PCMCIA cards

Even though current PCMCIA standards specification 2.1 does not support multifunction cards, several manufacturers have introduced their own proprietary implementations. Most notably, they have integrated fax/modems with LAN adapters. Motorola, Ositech, and Megahertz are among the vendors who have introduced multifunction cards.

It should be mentioned that the PCMCIA 5.0 specification does support multifunction cards.

6.8   Mobile Fax Machines

Just as we have mobile notebook computers, so we have mobile fax machines with wireless connectivity hardware now. Mobile fax machines can either be connected to a PC notebook or they can work in a stand-alone environment on a wireless network.

WalkieFax 200 illustrated in Fig. 6.14 is a mobile fax machine. It can plug into a wall outlet or be run off of a rechargeable battery or the cigarette lighter in a car.

Almost all networks provide support for fax transmissions. PSTN, ISDN, or circuit-switched cellular networks all provide a relatively economical means of transmitting faxes to mobile workers. In contrast, packet-based networks can be expensive.

6.9   Specialized End-User Devices

There are several specialized hardware devices in the market for vertical applications. Point of sale devices (see Fig. 6.15), scanners for automatic data collection applications in manufacturing, distribution,

Figure 6.14   WalkieFax 200. (*Source: Products of Imagination, California.*)

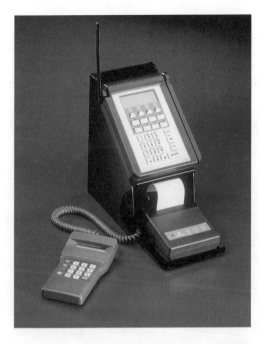

Figure 6.15 Point of sales devices. (*Source: Research in Motion, Waterloo, Ontario.*)

and warehousing industries (see description under handheld devices in Sec. 6.3), and baggage scanners for the airline industry are a few examples in this category (see Fig. 2.17). Also, see the description of the custom-designed data input and acquisition device (DIAD) for the UPS application in Sec. 2.6.1.

6.10 Infrared Links for Peripheral Devices

The whole idea behind wireless networks is to untether end-user devices from other larger devices holding information. Mobility requirements in remote places stipulate that we do without cumbersome cable connections between peripheral devices. The mobile computing industry has now come up with an infrared standard that affords peripherals such as mobile printers the convenience of wireless connections—within limits—to servers or host PCs, thus making remote configurations easier to set up.

Summary

In this chapter, we discussed mobile computing end-user devices that come in different form factors. From notebooks that support all the features of desktop computers to PDAs and personal pagers, these devices have varying amounts of mobile functionality. Careful attention should be paid to the selection of the most appropriate mobile device for the particular application under consideration. Ergonomics must also be considered with care. If intended users do not like a selected mobile device, the chances of a project's success will be greatly diminished.

Wireless LANs: A Platform for Mobile Computing in Metropolitan Areas*

The introduction of the IBM PC started the first wave of a revolution in personal computing. In the second wave, LANs became the de facto standard for connecting these devices. LANs gave birth to a minor industry that specialized in wiring, repeaters, hubs and switches. These devices were tolerated until professionals began to want to extend their LAN resources outside the office and to have the mobility to move freely around a building, campus, or a factory. This user requirement, along with the advent of infra-red and related technologies, led to the development of wireless LANs.   CHANDER DHAWAN

About This Chapter

In the previous chapter, we discussed the end-user devices that are the eyes and ears of mobile computing users. The next building block in constructing a complete mobile computing solution is the local area network (LAN), which connects the end-user devices to shared resources like files, printers, DBMS, and application servers. In this chapter, we shall explore wireless LAN technology as a method of connecting end-user devices and extending wired LANs to mobile computing. We shall also cover LAN-to-LAN bridging in a campus or metropolitan area environment using wireless technology. These extensions to the physical connectivity of users gives greater mobility without the loss of logical access to information resources on LANs and enterprise-computing servers.

---

* For more information on wireless LANs, please refer to *Wireless Local Area Networks* by Davis and McGuffin (McGraw-Hill, 1995).

7.1   The Need for Wireless LANs

LANs have become the de facto standard of physical connectivity among desktop PCs, workstations, and servers. In most cases, the use of copper-based (and now fiber-based) cable is an unavoidable necessity in achieving this physical connectivity. Cable does not add any value or provide any application functionality; it simply is the medium for information transportation.

The silicon industry has made considerable progress in improving the price performances of microprocessors and peripheral hardware through the 1980s and 1990s. A great deal of software has been developed to address various business needs. However, little attention has been paid to solving cabling problems. Coaxial cables, shielded cables, and unshielded cables continue to be strung and re-strung in office building conduits and cable trays. Cabling is a labor-intensive, time-consuming, and expensive activity. And yet, users continue to demand ever greater flexibility in their freedom to move equipment around at will. With cable conduits in many cases already heavily congested, changes, additions, and moves of workstations are less and less feasible. Meanwhile, notebooks and handheld computers are becoming an integral part of an executive's briefcase. The dependence of professionals on the availability of up-to-the-minute information wherever they go has increased their need to stay connected to the LAN.

It is in this context that wireless LANs meet an important business need for specific niche applications. Apart from their ability to connect notebooks or oft-moved desktops to LAN resources, they can also be used to connect LAN segments or LANs themselves.

7.2   The Differences between Wired and Wireless LANs

This book is about mobile computing. Both wired and wireless LANs can be used to provide mobility. Each has a unique role to play in the IT infrastructure. We shall consider the following aspects in a comparison of wired and wireless LANs:

1. *Fixed versus mobile nature of end users' business activities.* Since cables cannot be stretched and are generally a hindrance when it comes to moving equipment around, wired LANs do not provide the mobility that nomadic end users seek. We could, perhaps, take our cue from the electrical and telephone wiring trades and attack the problem by installing spare LAN jacks in every conceivable place we might use a notebook. Unfortunately, such an approach would be prohibitively expensive, on the one hand, and likely would not give us the flexibility and ease of use that we want, on the other. For complete mobility,

a wireless LAN is about the only really feasible solution—and even then, only so long as it is a LAN on which the information we need is available.

2. *Conduits and cable trays.* If the conduits and cable trays in a building are already full, the cost of installing additional trays and cables can be prohibitive. Under such circumstances, making a business case for wireless LANs is relatively easy. Please note that the expense of installing additional cable, in and of itself, is not a justification for installing wireless LANs in an office. Relative costs can be cited as but one of several reasons to implement a wireless LAN.

3. *Building code, fire regulations, factory floor, and historical building restrictions.*   There are building codes and fire regulations in certain buildings which stipulate that all cables must be in plenum, which again is an expensive proposition. Many factory floors, on the one hand, are made out of thick concrete, while overhead cranes rule out ceiling wiring, on the other. Additionally, high-voltage arc welders and electrical motors cause electrical noise that can introduce errors in a conventional wired LAN. For these and other reasons, wireless LANs with infrared transmission technology are becoming extremely popular for use with handheld computers from vendors like Telxon, Psion, and Norand.

Many older historical buildings have no conduits at all. Installation of new conduits may be permitted only in the ceiling—or no alteration to the building may be allowed at all. Wireless LANs make a lot of sense in such cases.

4. *Aesthetics of wiring.*   With their minimal wiring requirements, wireless LANs do not disfigure office landscapes the way wired LANs often do. This is not to deny, though, that a shabbily installed wireless LAN awash with antennas and control modules can be a lot more unsightly than a neatly installed wired LAN.

5. *Error rates.*   Wireless LANs are prone to more errors than wired LANs because of atmospheric conditions and other disturbances.

6. *Speed and capacity.*   Generally speaking, wireless LANs are slower than wired LANs and do not have as much bandwidth. Most popular wireless LANs operate in the 1 to 5.7 Mbps range. There are some infrared-based implementations that match the speed of Ethernet (10 Mbps) or Token-Ring speed (up to 16 Mbps). Vendors have also announced 20-Mbps wireless Ethernet cards, but they are expensive and, in the meantime, the wired LAN industry has set 100 Mbps as its next speed plateau. It will take many years for wireless LAN products to match that speed.

There is one major area where wireless LANs outperform conventional wired LAN connections. This is in site-to-site bridging of LANs. In the wired world, site-to-site bridging is normally achieved through a WAN link leased from a common carrier for a recurring fee. An infrared-based or laser-based wireless connection can be much faster— a 2 Mbps infrared wireless link versus a 128 Kbps ISDN connection. A T-1 leased line would come closer in matching the speed of a low-end wireless LAN bridge, but would be far more expensive.

7. *Cost.*   The total cost of wiring, NIC, and other components like access points or control modules for a wireless LAN must be compared with the costs attached to a wired solution. While rule-of-thumb average costs of a node are given in App. F, actual costs can vary substantially. It is important, therefore, to determine specific costs for each particular environment. Hardware-maintenance costs for a wireless LAN should also be included, since many wired-line LAN adapters carry lifetime warranties.

Table 7.1 lists the key differences between wired and wireless LANs.

7.3   Three Major Scenarios of Wireless LAN Implementations

There are three connectivity scenarios where wireless LANs are used commonly.

7.3.1   True wireless LAN

Site roaming connection of desktops is often used in very dynamic environments, e.g., in a lawyer's office or in a retail plaza where the store owners may have special sale events outside their stores. The ability to provide quick connections for additions, moves, and changes is a significant improvement over a wired system (Fig. 7.1).

7.3.2   Flexible mobile LANs

Adding mobility to wired LANs is often done in situations where end users need to take their notebooks from office to office to demonstrate sales presentations to other colleagues, and, similarly, in situations where roving factory or warehouse personnel capture information by scanning or entering data on the spot. In both scenarios, the objective can be achieved economically through the use of wireless LAN connections to host PCs or servers (Fig. 7.2).

7.3.3   LAN-to-LAN bridging in a campus environment

LAN-to-LAN connections may involve:

- Building-to-building connections (e.g., in universities, manufacturing plants, warehouses, and adjacent offices where the laying of cable would be possible but expensive)

TABLE 7.1   **Comparison of Wired and Wireless LANs**

	Wired LAN	Wireless LAN
Wiring media	▪ Various copper-based cables with or without shielding ▪ Thin coaxial cable and now also twisted-wire-pair are common for Ethernet ▪ Token-Ring LANs use Type 1 cable	▪ No cabling needed—spread spectrum ▪ Radio- or light-based transmission
LAN adapters	▪ Ethernet and Token-Ring	▪ Desktop wireless adapters or PCMCIA cards for notebooks, handheld computers, and PDAs
Costs	▪ Initial wiring costs higher, varies with cable type and distance ▪ Adapter cost lower ▪ Moves, additions, and changes are more expensive ▪ Total cost is lower in many situations	▪ Wiring cost is almost nil ▪ Adapter costs generally higher because of lower volumes and need for miniaturization ▪ Cheaper to connect two LANs by wireless in many cases
Speed and performance	▪ 4–100 Mbps	▪ 1–20 Mbps ▪ Faster than T.1 WAN connection between LANs
Error rate	▪ Very low, depending on the cable used and the environment	▪ Relatively high
Distance limitation	▪ Less than wireless, but can be extended with repeaters	▪ More than wired (300 to 800 meters per access point but can be increased with multiple access points; LAN segments can be bridged up to 25 miles with wireless bridges)
Constraints	▪ Each user needs a LAN termination	▪ Supports fewer users
Most appropriate application	▪ Where users work from fixed locations such as offices ▪ Sufficient conduit capacity assumed ▪ Higher speed is an important requirement	▪ Portable notebooks used as primary workstations ▪ Repeated temporary setups and relocations (retail industry) ▪ Where running cable would be difficult or expensive (e.g., in a factory or an old building)
Major advantages	▪ Cost is generally lower ▪ Speed is higher, especially with 100 Mbps adapters ▪ Commonplace technology	▪ Provides flexibility and mobility to users, making information available anywhere ▪ Easier to install and set up, no tethers ▪ Can be used in buildings with asbestos ▪ Wiring costs are lower
Major disadvantages	▪ Lack of mobility ▪ Changes and reconfiguration are time-consuming and expensive	▪ Adapter costs are higher ▪ Speed is generally slower than wired LANs ▪ Errors are higher generally than with wired LANs ▪ Potential RF health hazards ▪ Interference with other RF devices such as security systems ▪ Limited user density in a building because of lower speeds supported on wireless LANs (30–50 users per access point)

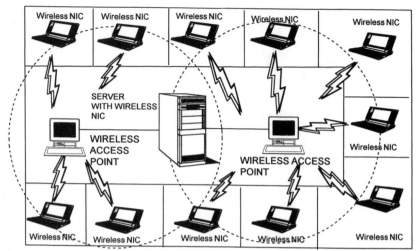

NIC = Network Interface Card

Figure 7.1   Lawyer's office with notebooks connected through true wireless LAN.

- Building-to-building connection in a downtown core (as shown in Fig. 7.3*a*) where buildings may be a few miles apart and where the laying of cable is not a feasible solution

- Building-to-building connection in open rural areas (as shown in Fig. 7.3*b*) where buildings may be up to 25 miles apart

  Wireless LANs are economical in the following circumstances:

- WAN connections through routers and bridges are expensive. Most WAN links involve a monthly fee for the leased line or frame-relay

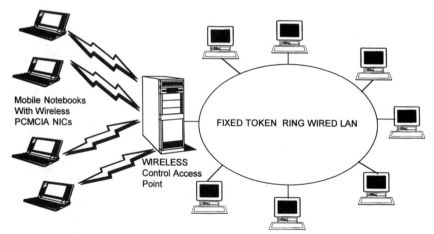

Figure 7.2   Wireless LAN grafted on to a wired LAN.

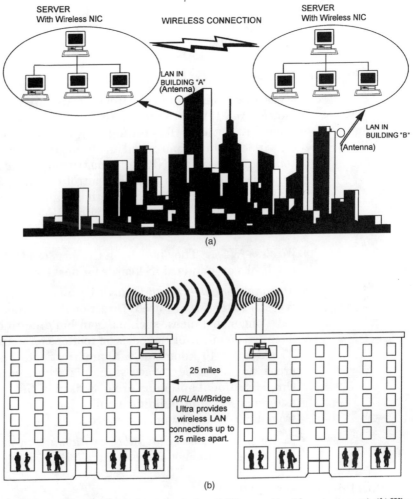

Figure 7.3 (*a*) Site bridging through wireless LAN connection (downtown core). (*b*) Wireless LAN bridge under ideal conditions.

service. There is no equivalent monthly fee for the wireless LAN-to-LAN bridging. (The only charge for a wireless LAN connection, once it is installed, might be a modest maintenance charge for the installed equipment.)

- A right-of-way for cable is either not available or is expensive (e.g., between two adjacent buildings across an intervening parking lot or roadway).

- The required lead time for a wired installation between two sites is too great.

- A wired WAN connection is not acceptable for business reasons.

Under circumstances such as these, the implementation of a wireless LAN is a popular solution that is easily justified on economic and performance grounds.

7.4 Wireless LAN Components

The essential components of wireless LANs are the same or similar to those of wired LANs, the major change being the replacement of conventional Ethernet or Token-Ring network interface cards (NICs) with their wireless counterparts and an absence of cable connectors. Instead of connectors, there are NIC antennas (inside or outside) that transmit radio signals. An access point or control module essentially acts as a hub that connects PC devices together via the radio signals and also interconnects these devices to the wired LAN. In short, the following three components are necessary for a wireless LAN:

- *Wireless NICs.*   They may be in the form of PC (PCMCIA) Cards for notebooks or standard ISA cards for desktops.

- *Antennas for catching and relaying radio signals.*   Several different types of antennas are used. Directional antennas send network signals longer distances, such as from building to building. They are mounted on masts or flagpoles on rooftops for maximum range (up to 25 miles, or 40 kilometers). Omnidirectional antennas in covered areas are attached to access points where mobility is required. Omnidirectional antennas are also used in tandem with directional antennas for multibuilding applications where centralized access from different directions is required.

- *Access points, control modules, or hubs with control NICs.*   Please see the description of access points in Sec. 7.5.

7.5 How Does Wireless LAN Technology Work?

We shall explain the technical terms that you will find in the literature and then describe the technologies used in wireless LANs.

7.5.1 Understanding the basic terms*

The following terms appear frequently in wireless LAN literature:

Access point or hub. As in a wired 10Base-T LAN network, *access points* (or *hubs* or *control modules* in older terminology) allow multiple

---

* *Wireless LAN Encyclopedia* by Solectek Corporation. *Data Communications,* November 1994, for general discussion. *Data Communications,* September 1995, for IEEE 802.11 standard for wireless LANs.

stations to be connected with servers or with each other. In the case of wireless networking, access points have three functions:

1. They provide connectivity for mobile devices to the wired backbone of a LAN.
2. They extend the range of wireless LANs by placing access points in strategic locations. The access points themselves are connected by cable to the LAN.
3. They provide mobile users with roaming capabilities.

In a sense, an access point is like the base station of a wireless WAN. You may recall from Chap. 5 that base stations themselves are connected by a leased-line backbone to servers.

Since there are no physical ports on access points, the number of users supported by an access point is not limited by hardware (except in the case of older designs). They are more dependent on the amount of network traffic. As a rough rule of thumb, running typical office applications with medium network traffic, Solectek's AIRLAN/Access Point, for example, can support 30 to 50 wireless users. This number drops to 10 to 20 with heavy traffic applications such as large file transfers and engineering computer-aided design.

Wireless LAN bridges. *Wireless LAN bridges* connect individual LANs to other LANs within buildings or to LANs in other buildings. Since bridging products are often used in a campus environment, they are typically designed to link multiple buildings two or three city blocks apart. Some wireless bridges can link LANs up to 25 miles (40 kilometers) apart under ideal conditions. As mentioned previously, to send and receive radio signals, wireless bridges require special antennas. The type of antenna used varies with the distance covered and the physical requirements of the site.

CSMA/CA. *Carrier sense multiple access/collision detect* (CSMA/CD) was a protocol that was designed for Ethernet wired networks. *Carrier sense multiple access/collision avoidance* (CSMA/CA) is a protocol that incorporates a modified design of CSMA for wireless LAN communications. In CSMA/CA, devices await a clear channel in order to avoid collisions.

Direct sequence radio frequency. The *direct sequence radio frequency* technique sends data over several different frequencies at the same time, increasing the chances that signals will arrive at their destination. Redundant, extraneous bits, called *chips,* are added to the signal; the extent to which the signal is consequently spread is determined by the number of chips added. At any given time, components of the signal are being received concurrently at different frequencies. In order to successfully receive the signal, the receiving device must have the correct decoding algorithm to interpret the data. Direct sequence coding is the fastest and most frequently used of the spread-spectrum technologies. Throughput is typically 1.6 to 2 Mbps, with an operating range that extends to 1000 feet (300 meters), though it should be noted that performance falls off fast with increasing distance. By its nature, this technology also provides rudimentary security. NICs cost around $600 per card and $1500 per access point or control module.

Frequency hopping.   Under the *frequency hopping* schemata, the transmitter shifts frequency every few milliseconds in its transmission to a synchronized receiver. Throughput is typically 1.6 to 2 Mbps and operating ranges extend to 750 feet (230 meters), with performance degrading more slowly with distance than is the case for direct sequence coding. NICs and access points are slightly cheaper than the direct sequence variety.

Roaming.   *Roaming* is an important feature of wireless communication—it allows users to literally roam around the coverage area without being disconnected. Roaming systems employ microcellular architecture that uses strategically located access points hardwired to a LAN backbone. The handoff between access points is totally transparent to the user. Figure 7.4 shows the roaming capability in a wireless LAN.

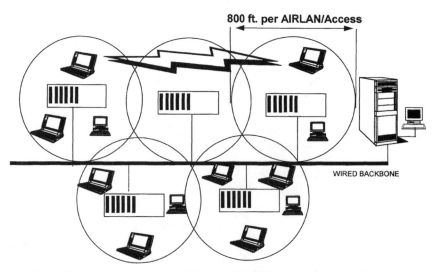

Figure 7.4   Roaming in a wireless LAN through multiple access points. (*Source: chart adopted from* The Wireless LAN Encyclopedia, *Solectek Corporation.*)

Signal security in wireless LANs. Spread-spectrum technology is basically secure, but it can easily be breached by anyone with the right knowledge and the opportunity to tamper with the hardware. It is not as secure as a wired connection. Encryption and other security software must be used to provide application-level security.

7.5.2 Technologies and frequencies

There are two transmission technologies employed by wireless LAN vendors:

- Radio-electric transmission
- Light-wave transmission

Within radio technology, there are two different frequencies and sub-technologies (Fig. 7.5):

- Licensed microwave frequency range (18 to 23 GHz)
- Unlicensed radio frequency range (902 to 928 MHz, and 2.4, 5.7 GHz)

Use of a frequency is licensed by the FCC in the United States and by appropriate regulatory bodies in other countries. Quite often, in the case of wireless LANs, the frequency is licensed to the vendors, who are required to file an application on the user's behalf stating they are a registered user of the frequency in a specific area.

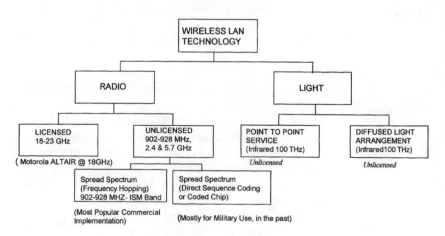

Figure 7.5   Wireless LAN technologies and frequencies.

In the second, unlicensed category, again, there are two choices:

- Spread spectrum without coded chip sequence
- Spread spectrum with coded chip sequence

In light-wave technology, there are also two choices:

- Point-to-point service
- Diffused-light arrangement (multipoint)

Both of these technologies operate in the infrared (100 terahertz) frequency spectrum and both are unlicensed.

Infrared technology is limited to use with one-room LANs, or LANs within a contiguous area like a factory shop floor. It is also used in wireless LAN bridges. Infrared technology is fast and cheap, and it resists electromagnetic interference. Its biggest drawback (albeit at the same time conferring a high level of security) is that it requires a direct line of sight—a clear path between transmitter and receiver—to operate successfully. Additionally, infrared transmissions can be interrupted by the kind of strong ambient light that is common in many corporate offices.

Laser technology, like infrared, also requires a clear line of sight to function successfully. It is very fast, but thick fog or blizzard conditions can diffuse a laser beam, disrupting data transmissions.

In the *diffused light arrangement,* signals are deliberately bounced off objects in a room, thereby obviating the need for an unobstructed path. Throughput is typically in the 1 to 4 Mbps range. The operating range is limited to 30 feet (9 meters).

Range and frequency. The possible range of a wireless transmission is directly influenced by the power output of the wireless device, the frequency at which it operates, and the presence of solid obstructions. Generally, wireless devices are able to successfully transmit signals in most office buildings. Solectek quotes AIRLAN wireless adapters as requiring a minimum of −70 dbm of signal to maintain a bit error rate of 1×10^{-8}. A signal strength below −70 dbm degrades performance. The range for Solectek's AIRLAN adapters is approximately 50,000 square feet (4500 square meters).

A wireless device's ability to penetrate solid objects is related to the frequency range at which it operates and its power output. Higher frequencies mean shorter wave lengths that have a hard time passing through obstacles like doors, walls, and floors. The low-spread spectrum frequencies (902 to 928 MHz) emit longer wave lengths that can easily penetrate solid objects (Fig. 7.6).

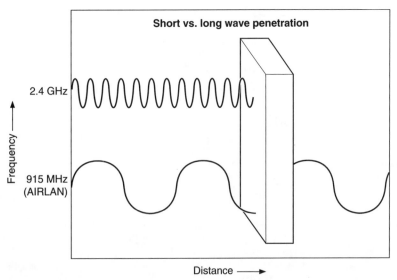

Figure 7.6 Effect of frequency on signal penetration. (*Source: Solectek documentation.*)

7.6 Wireless LAN Products

We have summarized some basic information about a few of the wireless LAN and LAN bridge products in Tables 7.2 and 7.3. Many of these products are similar, but there are differences in speed, coverage, and technology used.

7.7 Wireless LAN Applications

We have covered mobile computing applications in Chap. 2. Here, we are going to specifically mention only those applications that are suitable for wireless LANs. You may find some of the information repetitious but this is justified in the interests of keeping wireless LAN information together in one chapter.

The following sample applications indicate that wireless LANs can be justified in many different industries.

7.7.1 Health care industry

Physicians and support staff professionals are increasingly asking that a patient's information be readily available at the patient's bedside. Several hospitals have implemented infrared and spread-spectrum LANs to accommodate these wishes. You may recall that in Chap. 2 we discussed several health care applications that make effective use of

TABLE 7.2 Wireless LAN Products: LAN Adapter and Access Points

Product example	Frequency and speed	Features: coverage, S/W support
Aironet Wireless Communications, Ohio (a Telxon company)	Direct sequence spread spectrum 902–928 MHz 860 kbps	▪ Ethernet and Token-Ring cards are available ▪ Used in Telxon portable computer products
AT&T WaveLAN	Direct sequence spread spectrum 902–928 MHz Unlicensed No. of channels—N/A 2 Mbps	▪ Special NIC with built-in transceivers, outside antennas ▪ Netware, LAN Server/LAN Manager, LANtastic ▪ Up to 800 ft (243 m) with an output of 250 mW ▪ Desktop ISA and PCMCIA versions available ▪ Each WaveLAN cell is assigned a unique code; card must be set to match this
DEC Roamabout	Direct sequence spread spectrum Frequency hopping also now available 902–928 MHz and 2.4–2.4835 GHz	▪ Ethernet adapter only ▪ Complies with IEEE 802.11 standard
IBM	Frequency hopping spread spectrum 2.4–2.4835 GHz	▪ 1.6 Mbps throughput ▪ IEEE 802.11 compliant
PROXIM	902–928 MHz CSMA/CA 3 channels, 242 Kbps	▪ ISA, PCMCIA NIC versions available ▪ 3.3 million ft^2 (1 million m^2) through contiguous area; 785,000 ft^2 through walls ▪ IPX, NDIS, ODI support ▪ IEEE 802.11 compliant ▪ NetWare, MS LAN Manager, WFW, TCP/IP, LANtastic
Motorola's ALTAIR	Fixed frequency 18 GHz Ethernet transport using CSMA/CA	▪ License held and managed by vendor; assigns slightly different frequencies to different users in a specific area ▪ Control module is mounted in a high spot to get the best signal quality; user module is connected to the user PC ▪ ALTAIR II control module supports 50 users; user module can be connected to a hub for eight PC devices ▪ Not totally wireless; wired to the user module ▪ Uses scrambling technique, but security is a problem
Solectek-AIRLAN/ Internal	902–928 MHz CSMA/CA Direct sequence coding 2 Mbps	▪ ISA, MCA, PCMCIA cards available ▪ Supports Netware, MS LAN Manager, Banyan Vines, LAN Server, DEC Pathworks, LANtastic, WFW ▪ Protocols supported: IPX, NDIS, ODI, NetBEUI ▪ Coverage area: 120,000 ft^2; 60,000 ft^2 through walls; 800 ft^2 through walls, ceilings, and floors ▪ Data encryption, optional DES encryption chip
SYMBOL Technologies Laser Radio Terminal 3800	902–928 MHz CSMA/CD Six channels, 69.6 Kbps	▪ 750 ft^2 throughout contiguous area ▪ NetWare, MS LAN Manager, DEC Pathworks, TCP/IP ▪ Proprietary protocol
Telesystems ARLAN (Advanced Radio LAN)	Spread spectrum 915 MHz or 2.4 GHz range CSMA/CA 13 channels, 1.35 Mbps	▪ Heart of this wireless LAN is a Microcell controller ▪ Power output is 1 W—up to 500-ft (150-m) diameter in an office and up to 3000-ft (900-m) diameter in open factory buildings ▪ Each cell can handle up to 256 nodes ▪ NetWare, LAN Manager/LAN Server, and DEC Pathworks
Xircom	Frequency hopping spread spectrum 2.4–2.4835 GHz	▪ 1.6 Mbps ▪ PCMCIA Ethernet version ▪ Very aggressive price: less than $400 (circa 1995)

SOURCES: *Wireless Network Communications* by Bates, vendor information, *PC Week,* and other sources.

TABLE 7.3   Wireless LAN Products: LAN Bridges

Product example	Frequency and speed	Features
Solectek AIRLAN/Bridge Ultra	Spread spectrum 2.4 GHz—No FCC license required 2 Mbps	▪ Transmits data up to 25 mi (40 km) under ideal conditions; 16 mi (25 km) in rural settings; 7 mi (11 km) in urban areas ▪ Both Ethernet and Token-Ring versions available ▪ Transparent MAC layer bridging ▪ Major LAN network operating systems supported ▪ SNMP II MIB compliant
Telesystems ARLAN (Advanced Radio LAN) Bridge	Infrared light–based	▪ AirLAN 620 bridge can connect two LAN segments up to 6 mi (10 km) apart ▪ Less vulnerable to atmospheric conditions
IBM, DEC	LAN bridge products	▪ Offer comparable products

SOURCES: Vendor information, compiled by author.

wireless LANs, including applications that use pen-based computers and PDAs with easy-to-use, menu-driven touch-screen interfaces to give professionals the ability to access and enter patient information on the spot.

7.7.2   Industrial and manufacturing industries

Computer-controlled manufacturing and shop floor automation are two major application areas for handheld computers connected through wireless LANs. In factories and manufacturing plants, not only is the cost of laying cables often prohibitive, but assembly lines are frequently modified and many workers do not work in fixed locations.

7.7.3   Stock trading floor

Several leading stock exchanges, New York and Chicago included, are implementing wireless LANs for stock trading. Please refer to Chap. 2 for a more detailed description of the New York Stock Exchange's initiative.

7.7.4   Hospitality industry

Wireless LAN technology is widely used in hotels and restaurants. An order entered on a pen-based or touch-screen handheld computer is relayed directly to a base station/server which sends the order on to the kitchen. In like manner, the kitchen staff are able to instantly communicate information to a server regarding delays or other considerations. Additionally, the mobile device can be used to generate a bill and send it on to a cash register printer.

7.7.5 Retail industry

For both inventory control applications and temporary setup of sales counters, a wireless LAN is an ideal method of providing access to the main computer.

7.7.6 Distribution industry

Warehousing operations call for various mobile devices with scanners and pen-based input. These are linked to host PCs or servers that collect inventory, order filling, or shipping information, as the case may be, in real time, as the data is keyed in.

7.7.7 All industries: LAN bridging

This application of wireless LAN bridging applies to all industries, wherever a business case can be made for interconnecting LANs.

7.8 Wireless LAN Technology Evaluation Considerations

The following considerations should be taken into account when selecting a wireless LAN as a connectivity tool and when choosing a vendor:

- Suitability for the physical environment (office construction; factory, warehouse layout; types of walls and ceilings) and distances involved
- LAN operating system (NOS) support and certification by NOS supplier
- Integration with existing LANs (can the wireless LAN be grafted on to the existing LAN?)
- Ease of installation and use
- Security (is the connection secure?)
- Errors introduced by radio noise and related interference in the selected site (e.g., cellular phones and television can both cause interference)
- Interoperability based on the emerging IEEE 802.11 standard
- Vendor maintenance, installation, and technical support capability
- Total cost of hardware, software, and service charges

7.9 Design and Implementation Considerations for Wireless LANs

While we shall deal with this topic in detail in Part 5 of this book, we would like to mention here the issues relevant to wireless LANs.

1. *Integration of wireless LANs with existing LANs.* It is important to keep in mind existing LAN standards (in terms of NOS support, client operating system—Windows 3.1, Windows NT, Windows 95, OS/2 Warp—network management software, etc.) while designing a wireless LAN's configuration. The same NIC technology as in the wired LAN should always be used, if possible. Unfamiliar new technologies should not be considered without a careful assessment of the additional costs of such a move. For example, there should be strong reasons to introduce Ethernet-based wireless NICs if a majority of the existing LANs are based on Token-Ring technology. It makes sound economic sense to use as few different technologies as possible and to keep configurations simple.

It is important to remember that 60 to 70 percent of users continue to work on fixed desktop computers in most organizations. Although this percentage is decreasing, appropriate weight should be given to the technology used by fixed desktops when deciding on a particular architecture, unless the architecture being considered can be implemented transparently without adding additional support costs.

2. *Capacity planning.* It is also important to keep in mind that wireless LANs are slower than wired LANs, especially with the emergence of 100-Mbps Ethernet LANs. Because of this, data-intensive applications such as video conferencing should be discouraged on wireless LANs.

3. *Pilot wireless LANs before rollout.* It is a good idea to start with a pilot first and to be fully aware of and understand the limitations not mentioned in the glossy sales brochures. The best way to get to know these limitations is by trying pilot projects in specific physical (office, plant, or warehouse) and logical (NOS and application software) environments.

7.10   Wireless LAN Market

Wireless LANs have several niche applications. The most important is providing connectivity to portable computing devices such as notebooks, handheld terminals, and PDAs that are used, for example, in tracking the flow of goods from a warehouse to a checkout counter. The implementation of flexible LANs, for example, in the retail environment, is perhaps the second most important use to which a wireless LAN can be put. Using wireless LANs solely to replace cable is a very small market requirement.

The total size of the wireless LAN market in the United States is estimated at $350 million for 1996, growing to $700 million in 1998, according to estimates published in the trade press. The European market is expected to lag behind U.S. levels by about 12 months.

Cheaper prices can be expected as the volume of wireless LAN product shipments grows. The speeds supported by wireless LANs will also increase, but will always stay well below the levels achieved by wired LANs. The price for mobility will always be measured in both dollars and speed: that is a limitation of mobile computing that we must recognize.

Summary

In this chapter, we have discussed two aspects of wireless LAN technology. The first concerns itself with doing away with traditional LAN wiring. Mobile devices equipped with wireless network interface cards are connected to control modules in the setting up of LANs in offices, factories, and distribution centers. The second addresses the task of bridging wired LANs across metropolitan areas 7 to 10 miles (11 to 16 kilometers) wide. This implementation replaces WAN leased-line connections normally acquired from carriers for a monthly fee.

Wireless LANs are not recommended as replacements for their reliable, well-proven, high-capacity, and often-cheaper wired LAN counterparts. Rather, wireless LANs are best used as an add-on technology in niche vertical markets to provide mobility to specific categories of office or factory workers.

Wireless/Radio Networks: Mobile Computing's Information Highways*

If notebooks are the vehicles that started the mobile computing revolution, wireless networks represent the information highways on which users want to travel. Deja vu? The vehicles are there in plenty, but the highway infrastructure is only partially built. The ramps are poorly designed and drivers are being forced to change their driving habits. Since the mobile network highway is unpaved, high speeds are not possible. Services are not as good as they are on local city streets. Worst of all, the toll charges are so high that only the affluent can afford to drive on the mobile computing highway. Hopefully, this will change in the future.

CHANDER DHAWAN

About This Chapter

We discussed wireless LANs in Chap. 7. We are now ready to turn our attention to wireless WANs that truly give us the mobility we need for *anywhere, anytime* access to remote information resources. This is a subject that has been discussed in depth in several other books. The objective of our book is not to make you radio engineering specialists. Readers in search of more detailed information about radio engineering or wireless networks should refer to books by Bates, Nemzov, and others. Our aim is to provide mobile computing systems architects and integrators with a basic grounding in the technology of wireless information highways. Network specialists may well find the concise infor-

---

* For more details, please refer to the following two books: B. Bates, *Wireless Networked Communication,* McGraw-Hill, New York, 1995; M. Nemzov, *Implementing Wireless Networks,* McGraw-Hill, New York, 1994.

mation contained in the following pages useful for comparing different networks from applications and software perspectives.

Because it deals with a major subject in mobile computing, this chapter is long—perhaps too long. We did not want to break the subject down into several smaller chapters, however, because discussion of all networks logically belongs together.

We shall start with basic radio engineering concepts, explaining common terms along the way, and then we'll introduce the subject of spectrum management by regulatory bodies such as the FCC. Then we shall look at generic components of a radio network. We have adopted a chronological approach in our discussion. The first network type we shall review is the SMR that was implemented for private and public shared use for dispatch applications. After that we will focus on specialized networks, including paging. Then we will turn to circuit-switched cellular networks. Since CDPD is logically related to cellular, we discuss emerging CDPD networks in the same section. From cellular, we move on to packet data networks such as ARDIS and RAM Mobitex, and then emerging networks like ESMR, PCS, GSM, and finally satellite networks. Ending with a discussion of satellite networks is appropriate because it is in the likes of IRIDIUM that there lies the greatest potential for truly making the *anywhere, anytime* vision come true.

We end with a comparison of all the networks and review the considerations that are important in matching networks with business application characteristics.

8.1   Theory of Radio Communications

In this section, we shall discuss the key radio communications concepts that are relevant to wireless networks.

8.1.1   Electromagnetic spectrum

All wireless communication uses electromagnetic energy to convey information. Radio, light, and X rays are all different forms of electromagnetic radiation, the only differences being in wavelength and frequency. The relationship between wavelength, frequency, and different applications of electromagnetic energy is depicted in Fig. 8.1.

Relationship between frequency and wavelength. There is an inverse relationship between frequency and wavelength, as shown in Fig. 8.1. Frequency and wavelength have a significant bearing on the type of electronic circuitry and technology used for transmitting electromagnetic waves. Since electromagnetic energy is often transmitted in a

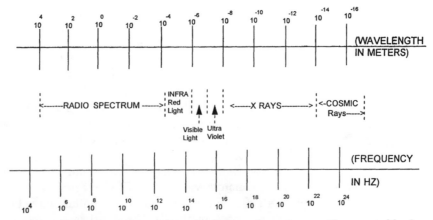

Figure 8.1 Frequency spectrum for radio, light, and cosmic rays. (Represented by frequency in hertz and wavelength in meters.)

shared medium, it is important that equipment be designed to be able to isolate users of different frequencies in the same vicinity. The reliable transmission of information is also dependent on the absence of external disturbances, as well as the proximity of other forms of electromagnetic energy.

Radio frequencies. Radio frequencies, which constitute a subset of the electromagnetic spectrum, are generated and transmitted at varying frequencies, from 10 KHz to millions or even billions of MHz. Table 8.1 shows the frequency of different types of radio energy.

Allocation of radio frequency spectrum. In every country, there is an organization that controls and regulates the allocation of frequencies used by different agencies and users—public or private. Generally, this

TABLE 8.1   Frequency of Electromagnetic Spectrum

Frequency	Band	Type of frequency
20 KHz and below		Audible
Less than 30 KHz	Very low frequency (VLF)	Radio
30–300 KHz	Low frequency (LF)	Radio
300 KHz–3 MHz	Medium frequency (MF)	AM radio—10^5
3–30 MHz	High frequency (HF)	Radio
30–300 MHz	Very high frequency (VHF)	Radio
300 MHz–3 GHz	Ultrahigh frequency (UHF)	FM radio—10^8
3–30 GHz	Superhigh frequency (SHF)	Radio
More than 30 GHz	Extremely high frequency (EHF)	Radio
100 GHz		X rays
Above 10^{22} Hz		Cosmic rays

is a federal or central governmental organization: the FCC in the United States, the Department of Communications (DOC) in the past and Industry Canada, now, in Canada, and similar bodies in Europe and Japan.

The normal practice in most countries is to submit an application for a license to use a specific frequency in a region (or an entire country) for a defined purpose. In the United States until 1993 this was the only way to acquire a broadcasting license—a long, drawn-out procedure, especially in the case of a new application or when a new band was involved.

Realizing that frequency is an important resource that has an intrinsic economic value, the FCC started auctioning specific wave bands in 1993. Since then, billions of dollars have been collected at frequency auctions by the U.S. federal government, especially where PCS-technology licenses are concerned. In most other countries, including Canada, licensing is still free and is based on the combined merits of the application, official telecommunications policies, and national interests. Of course, deregulation winds are sweeping the entire western world—and will ultimately affect developing countries as well.

What is included in a license?   A license for frequency use (in a narrow band) allows the holder to exclusive rights to use the specific bandwidth in a stipulated area. Broad-range low-level transmissions are tolerated until they interfere with narrowband licensees. Licenses are given for specific locations, areas, and periods. Transmitting equipment cannot be moved from one area to another, even in emergencies, without the prior approval of the FCC.

Bandwidth in the frequency license context.   In the traditional telecommunications literature, the term *bandwidth* refers to the raw throughput capability or capacity of a telecommunications equipment or network service in terms of bits per second. For example, a router or a communications switch may have a bandwidth of T1 (1.544 Mbps). However, in wireless network parlance, bandwidth refers to the range of frequencies assigned by the FCC or the spectrum-regulating body. Think of bandwidth in terms of the diameter of a tube or the width of a highway: the wider the tube, the greater its capacity; the wider the highway, the more lanes it has and therefore the more traffic it can carry. Likewise, the wider a frequency bandwidth, the more data it can carry. For instance, television has a very wide bandwidth (6000 KHz) because it must carry audio, video, closed captioning, and other signals. Anyone selecting a network should consider the bandwidth allocated to the particular network under consideration, since it is the bandwidth that will determine the volume of data the network is capable of carrying.

Channel. A channel represents a frequency and a bandwidth (Fig. 8.2). Quite often, network designers break the channel bandwidth into two parts—one each for the uplink and the downlink. ARDIS's public shared packet radio network in the United States, for example, uses 802–821 MHz for its uplink (terminal to network) and 851–866 MHz for its downlink (network to terminal).

Wideband and narrowband. These terms refer to the width of a band rather than the specific spectrum range allocated to a user. It is a relative term used in the context of a discussion. A bandwidth of 30 MHz allocated for packet radio networks such as ARDIS or RAM is a wideband compared to the narrowband 30-KHz bandwidth allocated to cellular networks. On the other hand, the 30-KHz bandwidth of cellular phones is relatively a wideband compared to, say, the 10 KHz allocated to another use for some other purpose.

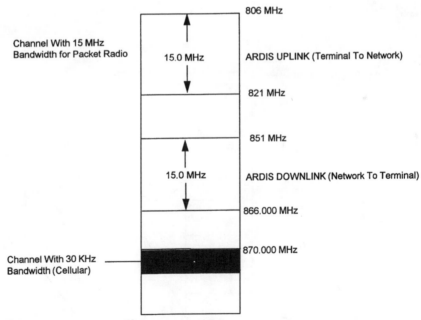

Figure 8.2 Channel—frequency with bandwidth. (*Source: Bishop Training Company.*)

8.1.2   Use of radio frequencies for different forms of wireless communication

The useful portion of the radio spectrum is from 0.2 MHz to 6 GHz. Of this, the bandwidth that is currently being used for wireless transmission is primarily in the 0.4 MHz to 2.4 GHz range. Figure 8.3 portrays the radio frequencies of different wireless communication networks in North America. While the allocation of frequencies for different types of networks tends to be broadly similar throughout the world, there are, nevertheless, some important variations from country to country, especially in Europe.

Some common usages of frequencies in the radio portion of the spectrum. Cordless phones operate in the 46 to 69 MHz frequency range. Previously, business bands were allocated the 450 to 470 MHz range for regulated narrowbands. This bandwidth is now available for equipment that collects data from portable radios. The previously unlicensed range of 902 to 928 MHz in the narrow bandwidth, originally used for garage door openers, remote control toys, and some cordless phones, is now being auctioned for PCS licenses.

New to networking—but widely used in microwave ovens—is the 2.4 to 2.4835 GHz band. It should be noted that microwave oven signals are trapped within the appliance. Leaky ovens are more of a health hazard than a source of interference to communications, though it is

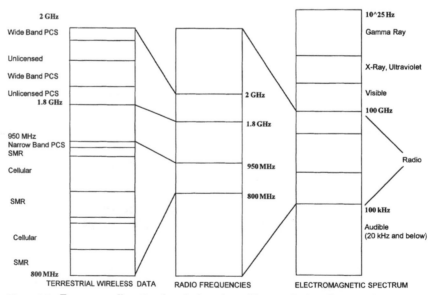

Figure 8.3   Frequency allocation for wireless data. (*Source: Bishop Training Company.*)

true some RF wireless units may not operate successfully within 15 feet or 4.5 meters of a microwave oven.

Spread-spectrum technology is commonly used for wireless LANs. Motorola has licensed narrowband 18-GHz frequencies in all major metropolitan areas in the United States for 10-Mbps point-to-point links. At the time of publication, this frequency was not regulated in Europe or other nations.

The rapidly changing scene in frequency allocation in the world. As wireless networks based on emerging technologies such as PCS are implemented during the next several years, frequency allocation bodies will continue to manage the spectrum by issuing more licenses and reallocating some of the currently unlicensed frequencies for newer uses. The spectrum usage charts are changing rapidly. You are well advised to obtain the most current information from the appropriate governing body. One source of this information is the Internet and online services such as CompuServe.

8.1.3 Modulation process: Adding information to the electromagnetic spectrum

Electromagnetic waves by themselves do not carry any information. Modulation is the process by which information is added to (or embedded in) electromagnetic waves. This is how any kind of information, whether it is the human voice or transaction data in an interactive application, is transmitted on an electromagnetic wave.

The transmitter attaches the information to a basic wave (known as the *carrier*) in such a way that it can be recovered at the other end through a reverse process called *demodulation*. Demodulation can be carried out in many ways. The simplest method is to start and stop a carrier, following a recognized pattern. Morse code uses this process to transmit telegraphic information through a series of short or long pulses, i.e., dots and dashes.

In modern telecommunications networks, information is transmitted by transforming one of the two basic characteristics of a wave, the amplitude or frequency (Fig. 8.4). Therefore, there are two basic approaches to modulation:

- Amplitude modulation
- Frequency modulation

Amplitude modulation. Amplitude modulation is used in voice communications and in most LAN transmissions. Two systems (transmit-

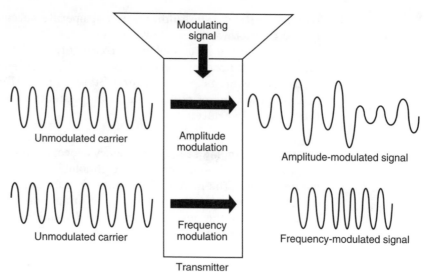

Figure 8.4 Modulation techniques.

ter and receiver) are in constant communication with each other via a carrier signal. As information (in the form of changes to the amplitude of the signal) is added to the carrier, the changing amplitude forms new signals, which are also transmitted to the receiver. The total output of the carrier signal now consists of the frequency of the original, unmodified carrier signal and the differences—called *sidebands*—between the amplitude of the original signal and the amplitudes of the new signals. Frequency is kept constant while amplitude varies.

Frequency modulation. In frequency modulation, amplitude is kept constant but frequency is varied. The modified output signal consisting of the sum of two frequencies is transmitted to the receiver.

Besides the two basic modulation techniques described above, there are also other more sophisticated techniques in use: phase modulation, pulse modulation, frequency-shift keying, phase-shift keying, and quadrature amplitude modulation.

The particular modulation technique used is probably not important to an application systems designer or systems integrator. Rather, this is a choice made by vendors of wireless modems based on engineering factors such as frequency band used, wireless network being employed, transmitter power, etc. Radio engineering specialists can be asked to assess the efficacy of a particular technique and its suitability to an application.

Analog versus digital microwave transmission. Radio microwave systems are classified as analog or digital, depending on the type of modulation technique employed. Intrinsically, all microwave transmissions are made on an analog carrier system. The use of either analog or digital modulation when inputting the information is what sets different systems apart. Even though digital microwave transmissions have been around since the 1970s, most older systems use analog technology. Only more recent systems, such as new line-of-sight systems used for full duplex transmission, are digital in nature.

There is considerable discussion in the cellular industry regarding the analog advanced mobile phone systems (AMPS) configurations prevalent in the United States today and digital AMPS (D-AMPS) proposed as a migration step towards PCS technology. New cellular systems are more often than not based on D-AMPS, except where compatibility with existing infrastructures is an issue.

8.1.4 Basic multiplexing techniques for wireless/radio: Using frequency efficiently

One of the fundamental dilemmas that a network designer has is how to use a given channel for more than one conversation (electronic), user, or data transfer (transaction or file) activity. It would be very inefficient to reserve a physical channel or an electronic switch or a wireless cell for one user only. Hardware engineers employed by network equipment manufacturers such as Motorola and Ericsson spend a great deal of time working to obtain the highest transmission speeds possible with a given design and then use *multiplexing,* the technique used to put more than one user on the same frequency in the same physical cell, as a means of sharing a channel between several users.

There are several multiplexing methods employed in radio/wireless networks. We shall only briefly touch on these techniques here. Please see Sec. 8.8 for additional information regarding the importance of multiplexing for network capacities.

Frequency division multiple access (FDMA). FDMA is the most common access method used in both satellite and terrestrial wireless transmissions. In FDMA, different frequencies are sent from different transmitters and are combined into a single composite signal for onward transmission to the receiver.

Time division multiple access (TDMA). TDMA is the second most common method of access. In TDMA, different transmitters send bursts of data sequentially at preselected intervals. TDMA operates in the temporal domain and with digital systems only, since buffering is required. TDMA uses bandwidth far more efficiently than FDMA.

Code division multiple access (CDMA). CDMA technology is relatively new and has not yet been widely implemented. It has been used by the military for some time, but has seen limited deployment in commercial applications. This access method involves a division of the transmission spectrum into codes, effectively scrambling transmitted conversations. Several transmissions can occur simultaneously within the same bandwidth, with mutual interference reduced by the degree of orthogonality of the unique codes used in each transmission.

Each receiving terminal has its own code, rejecting all others. Included in CDMA is a technique called *frequency hopping,* in which first one and then another frequency is used to carry the information. CDMA is also a more expensive technology to implement because the logic is more complex.

We use the acronyms *TDMA* and *CDMA* here in a generic telecommunications engineering sense. You will see them being used in wireless networking trade publications and technical journals in the context of technology implementations of PCS. See Sec. 8.8.1 for additional differences between these and other technologies from the viewpoint of PCS.

8.1.5 Stretching the frequency use/frequency reuse concept in wireless technology

A tremendous amount of research is being done on giving maximum numbers of users access to limited frequencies and bandwidths. Cellular networks have come up with a technique called *frequency reuse,* based on the fundamental concept of a cell in a cellular network. A *cell* is defined as a limited contiguous physical area with its own antenna and set of radios. Two callers can use the same frequency as long as there is sufficient physical distance between them in the same cell, or if they are calling from different cells. Since signal strength fades with increasing distance from the cell, the antenna and base station are able to serve the caller with higher signal strength.

This concept is illustrated in Fig. 8.5.

8.2 Components of a Wireless/ Radio Network

Different network technologies, such as private SMR, circuit-switched cellular, CDPD, packet radio (ARDIS or RAM), and PCS, all have their own unique designs and, at the same time, share several common components and characteristics. We shall describe the common aspects of these networks first and then the unique characteristics as we discuss each of the different networks in turn.

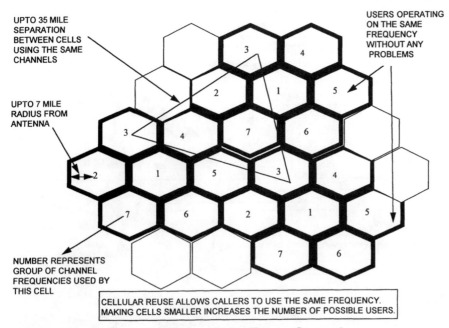

UPTO 35 MILE
SEPARATION
BETWEEN CELLS
USING THE SAME
CHANNELS

USERS OPERATING
ON THE SAME
FREQUENCY
WITHOUT ANY
PROBLEMS

UPTO 7 MILE
RADIUS FROM
ANTENNA

NUMBER REPRESENTS
GROUP OF CHANNEL
FREQUENCIES USED BY
THIS CELL

CELLULAR REUSE ALLOWS CALLERS TO USE THE SAME FREQUENCY.
MAKING CELLS SMALLER INCREASES THE NUMBER OF POSSIBLE USERS.

Figure 8.5   Frequency reuse plan. (*Source: Bishop Training Company*)

8.2.1   Base stations

This is typically the hardware/software configuration that resides on or near an antenna tower. It is the antenna that transmits or receives electromagnetic signals to or from mobile end-user devices in a specific geographic area. The height, design, and size of an antenna are all important factors. Similarly, the power output of the base station determines how far a signal will be transmitted. Base stations are connected via wired lines to wireless carriers' switching nodes.

Telephone system microwave transmissions require two separate frequencies and two systems as shown in Fig. 8.6. The first frequency is used for transmitting and the second is used for receiving. Newer systems use a single transceiver, even though they use two frequencies.

8.2.2   Repeaters

The transmission range of a microwave radio signal is finite and is dependent on the power of the base station and the height of the antenna. The path taken by the signal and obstructions along the way also affect range. Typically, line-of-sight transmission is used. Repeaters are used to extend transmissions over greater distances (Fig. 8.7). Basically, a repeater intercepts the signal from an antenna and then retransmits it. Ranges vary with the frequency used, as shown in Table 8.2.

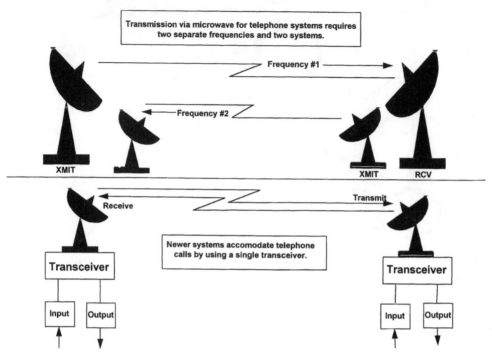

Figure 8.6   Transceivers in radio transmission. (*Source:* Wireless Networked Communications *by Bates.*)

8.2.3   Satellites as repeaters
for wireless networks

Since the mid-1960s, satellites have become increasingly important in wireless communications as repeaters in the sky at varying altitudes. There will be more discussion on satellite-based networks in Sec. 8.10.

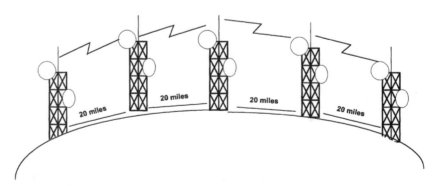

Figure 8.7   Repeaters in microwave/radio networks to increase coverage.

TABLE 8.2 Typical Microwave Radio Ranges

Frequency	Approximate range before a repeater is necessary for greater distances
2–6 GHz	30 mi/50 km
10–12 GHz	20 mi/30 km
18 GHz	7 mi/11 km
23 GHz	5 mi/8 km

SOURCE: *Wireless Networked Communications* by Bates.

8.2.4 Higher-level nodes for wireless networks

Most wireless networks use a hierarchical and distributed node design. This means that the lower-level base station nodes that service mobile devices directly are connected to higher-level nodes that act both as concentrators and switching centers. Depending on the network, there may be multiple levels in a hierarchy of nodes that is not unlike a traditional sales organization, with its districts, regions, and super regions. Overall control is vested in a *network control center* (NCC), which in turn can be compared to the head office of a business organization. The wired-line connections between higher-level nodes may be at T1 or T3 speeds, while the connections from base stations may be typically at 56-Kbps or sub-T1 speeds.

8.2.5 Connection of wireless networks to a wired infrastructure

An end-to-end network connection in a mobile computing application consists of several communications segments. Typically it is only the connection between the mobile device and the nearest network base station that is actually wireless. Anything beyond that uses wired line, because network service providers such as ARDIS, RAM, or GTE MobileNet are linked to terrestrial wired-line segments from long-distance data communications service providers. You may recall Fig. 5.6 that shows wireless and wire-line components of the network.

8.2.6 General considerations affecting performance of radio transmissions

Since radio signals are carried in free space, the following factors must be taken into consideration when assessing the likely performance of a particular radio system:

1. There are three methods of transmission used by radio engineers: point-to-point, line-of-sight, and omnidirectional. The transmission

method most suited to a specific application depends on the distances involved and the terrain being covered.

2. Noise of different kinds will always be a factor. The different methods of transmission mentioned in item one are each susceptible to various disturbances that distort signals and introduce errors. Such disturbances can result in the need to retransmit information. If the communications software/hardware does not allow for this possibility, users may find themselves having to abort and try again.

3. Range is directly related to power output. Power output/range should be a major consideration when evaluating transmitters, since they also have a bearing on the number of towers and base stations that will have to be installed in a private network.

4. Signal loss or attenuation is a factor also. Radio signals diminish in strength as they travel through certain materials (and increase in strength when they travel through conductors).

5. Heavy rain or snow can absorb some of the transmitted signal in certain frequency bands.

Figure 8.8 illustrates the effects of these factors.

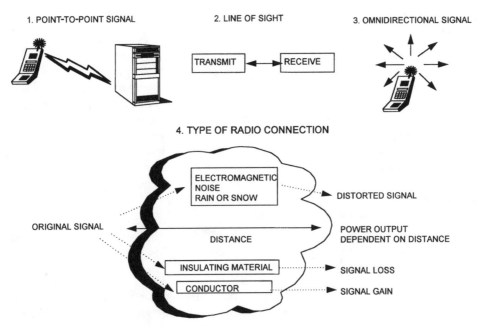

Figure 8.8   Factors affecting radio networks.

8.3 Specialized Mobile Radio (SMR) Networks

In 1974, the FCC created the Specialized Mobile Radio (SMR) service. This service, little known to the general public, has rapidly developed into one of the most exciting industries regulated by the FCC. Similar services are available in many European countries, as well as Canada and Japan. Today, SMR systems provide services in the United States to more than a million radio users through over 7000 SMR systems nationwide. According to an FCC report, the total SMR market (equipment and services combined) was well over $1 billion in 1995. By the year 2000, SMRs will be a multi-billion-dollar industry providing critical communications support to several million North American, European, and Japanese workers.

8.3.1 What is an SMR?

Licensees provide land mobile communications services in the 800 and 900 MHz bands on a commercial basis to bodies eligible to be licensed under this category. SMRs have made two-way mobile communications available to many businesses, governmental organizations, and individuals.

In simple terms, an SMR service operator is the owner of a private radio system that includes one or more base-station transmitters, one or more antennas, and other radio equipment that third parties are charged a fee to use. The third party usually, but not always, provides its own mobile radio units. The fee, together with a license from the FCC, entitles an end user to employ the SMR system to send and receive radio messages or to make and receive mobile telephone calls.

The SMR service intercepts telephone transmissions and low-power mobile radio signals and either routes them through phone lines or amplifies and retransmits them.

SMR systems consist of two types: *conventional* and *trunked* radio. Trunked systems, which constitute the majority of SMR service, are much more efficient in terms of the number of users they can support. In conventional systems, it is not possible to have more users than there are channels and a user will therefore typically be licensed for only one channel (frequency). If someone else is already using the assigned channel, the user must wait until the channel is available, even if channels on other systems in the same market are not being used. With a multichannel trunked system, the system's microprocessor seeks out open channels as needed for licensees to use. The search capability is made possible by the use of a signaling channel that makes available voice channels to user groups that wish to talk, thus allowing more users to be served per radio channel. This efficiency of use is possible because the probability of all the channels in a large system being in use at one time is lower than the probability of a single given channel being in use.

Trunked systems also have privacy benefits. Because a user can be talking on any of the channels within the trunked system, unauthorized parties have a more difficult time eavesdropping on the communications of a specific trunked SMR service user than on those of a traditional one-channel conventional SMR service user. This increased privacy is an important feature of trunked systems. Other features of trunked systems are manual or automatic selection of talk groups and multisite, multichannel capability. Since it is expensive to acquire and build a trunked system and there is a shortage of private radio spectra in major urban markets, few businesses can afford or acquire sufficient spectra for trunked radio systems without SMRs.

Many mobile radios are capable of using several different SMR systems. This feature allows operators of several SMR systems to offer a wide area and/or a roaming service to end users. One of the more advanced features offered by SMR operators is *direct inward dialing* (DID), which allows anyone to easily initiate direct telephone contact. With this option, an individual mobile telephone can be dialed with no more steps or digits than a standard phone.

SMR end users typically operate in either a dispatch mode or an interconnected mode. Many SMR services offer both modes. *Dispatch mode* involves two-way over-the-air voice communications between two or more mobile units, or between mobile units and fixed units (e.g., between the driver's office and a truck). Dispatch communications are generally short, limited to less than one minute. A common example of dispatch communications is a police dispatcher who radios a message to all patrol cars (or a specific unit) to go to the scene of a crime. Any return calls in response are also dispatch communications. A construction company with a dispatch operation in a central office communicating with several trucks would be a typical SMR customer using dispatch communications.

The *interconnected mode* allows the connection of mobile radio units with the public switched telephone network, thus making it possible for mobile radios to function as mobile telephones. It is in this area that an SMR service is similar to a cellular telephone service.

8.3.2 Common applications and an example of SMR service

SMR systems have been primarily used for dispatch applications in public safety (police, fire, and ambulance) organizations, the taxi industry, the trucking industry, and in many maintenance services–oriented industries. The following is an example of its application in the taxi industry.

ABC Limo Service sets itself up in business with several cars that the company's dispatcher needs to be able to communicate with from the dispatch office. ABC Limo decides to obtain this dispatch capability and the necessary radio equipment from XYZ SMR Service. When a customer phones ABC Limo Service and asks it to send a limo, the dispatcher transmits a radio message to XYZ's SMR station, which automatically repeats the message for pickup by any or all of ABC Limo's drivers. The drivers can talk to each other or the dispatcher back at the office via the same XYZ station. For a fee that includes the cost of the telephone service, XYZ SMR Services will also connect any of the drivers to the local phone system. Thus, if a limo driver cannot find a client's house, the client can be called direct for directions. This example is based on mobile radio voice communication, but in point of fact, data-based messaging systems are gradually replacing or supplementing these systems in many situations (see the Singapore taxi service application in Sec. 2.3.2 in Chap. 2). In a hybrid system, data-based messaging is the norm and voice is used for emergency and urgent communications only.

8.3.3 Allocated frequency band and FCC regulations for SMR

The first regulatory requirement in setting up an SMR service is to find available frequencies at a desirable site and then to apply for a license. Two distinct sets of frequencies are available for SMR operations: 800 and 900 MHz. The 800-MHz band has by far the bulk of the current installations, with the 900-MHz band—licensed much later, primarily during the late 1980s—having a small number of installations and mobile users. This is largely because the channel bandwidths for 900-MHz systems are half the bandwidths for 800-MHz systems (12.5 versus 25 KHz), and radio equipment designed for 800-MHz SMRs is not compatible with radio equipment designed for 900-MHz SMRs (and vice versa). Another reason for the incompatibility is the fact that the separation between the transmit and receive channels of a given channel pair is 45 MHz for 800-MHz systems and 39 MHz for 900-MHz systems. Additionally, the frequencies available for 800-MHz SMR systems and 900-MHz SMR systems are sufficiently far apart as to require separate antennas and other equipment for both the base stations and the mobile radios.

It is possible to get into the SMR business by purchasing an existing system, though the FCC will not permit the transfer of an SMR license to another person, corporation, or other entity, unless the licensed system is constructed and operational. This rule helps to deter the filing of applications by persons who do not intend to actually provide a service.

SMRs operate under a different set of regulations than do other commercial radio services such as radio common carriers and cellular radio operators. Over the past few years, these regulations have become extremely flexible. The most basic tenet is that SMRs are considered private carriers. By virtue of being private carriers rather than common carriers, SMRs are exempted under Section 331 of the Communications Act from rate regulation. The absence of state regulations and price regulation is considered by many to be critical to the industry's ability to achieve maximum growth and efficiency.

8.3.4   SMR licensing requirements

As a result of deregulation of the telecommunications industry, requirements are changing. The following are some of the SMR licensing rules at the time of writing.

1. In searching for available 800-MHz frequencies, the most important rule to consider is the 70-mile (112-kilometer) co-channel separation requirement. Each SMR system operating on a particular frequency is granted a 70-mile radius from its primary site to the primary site of any other system operating on the same frequency. An available frequency, therefore, is a frequency for which there are no other licensed systems within 70 miles of the proposed site. A waiver of this rule may be granted provided both cochannel licensees sign a short-spacing agreement.

2. A second major requirement for an SMR license is *loading,* which refers to the number of mobile stations served by the system. For the purposes of the loading rule, *mobiles* include mobile radios in cars and trucks, portable radios, and control stations (such as fixed units in offices). There must be a load of at least 70 mobiles per channel in order to renew a license after the initial five-year licensing term. Systems licensed after June 1, 1993, are not subject to loading standards for the purpose of channel take-backs, but to obtain additional channels under a given license, at least 70 mobiles must be loaded per channel. Preference is given to applicants seeking to expand fully loaded systems over operators seeking new licenses.

3. The FCC allows one year for the construction and operational start-up of a trunked system. This rule is intended to reduce spectrum hoarding. Once the marketing of services has begun, an operator must take care of the proper licensing of customers (end users). Each end user must have a valid license to operate mobile radios. Unlicensed end users may use a system under a temporary permit for up to 180 days, providing they have applied for a license.

4. A licensee does not own the spectrum it uses nor the license from the FCC to use that spectrum. Thus, although the FCC will allow the transfer of a license from an SMR operator to a third party to facilitate the sale of business assets (i.e., its equipment and goodwill) a licensee cannot assign its license to a third party in the event it fails to construct a system.

5. Each radio equipment manufacturer is given exclusive rights to one 20-channel trunked system nationwide. It is allowed to own and operate a system in this way so that it can demonstrate whether a facility makes economic and engineering sense. It is not allowed additional systems because of concerns about possible adverse effects on competition.

6. In the 800-MHz band, the separation between the transmit and receive frequencies of a channel pair must be 45 MHz. Each channel must have a bandwidth of 25 KHz (or 50 KHz per pair).

7. Wire-line telephone companies, once prohibited from owning or operating SMRs, are now allowed to do so. Interconnection with public switched telephone networks has always been allowed.

8.3.5 Components of an SMR system

SMR components vary from one application to the next. Whether a system is used for voice mobile applications only (as is the case with most older implementations), or is integrated for voice and data (as is the case with some of the newer installations) has a major impact on the precise components of an SMR system. Figure 8.9 shows a typical

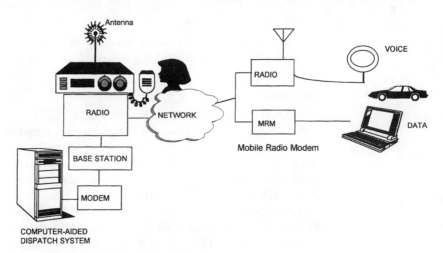

Figure 8.9 Components of an SMR system.

schematic for a simple one-site conventional system. The schematic for a multisite trunked configuration with features such as Motorola's smart-zone is more complex. You are advised to contact the vendor community for this type of information.

An SMR system may be very small, with just one tower and one base station, or it may feature a number of towers and base stations in a larger network. Many of the municipal SMR systems for public safety are of the former variety. Where the coverage area is large, there are multiple base stations. Regional SMR systems are hierarchical in topological design—similar to Mobitex or ARDIS packet radio networks.

The components involved in an SMR network are:

1. At least one base mobile station consisting of a voice radio or a data terminal, such as a notebook computer.

2. A mobile radio modem (MRM), either integrated into the base mobile station (in a more modern design), or vehicle-mounted as a separate unit. The MRM is typically $7 \times 5 \times 2$ inches ($17 \times 14 \times 5$ centimeters) in size and connects between the data terminal and the radio. MRMs operate at different speeds: 1200, 2400, 4800, 9600, and now 19,200 bps. Typically, an MRM will have an RS232 serial port and is powered by the radio.

3. Radio with antenna, installed in the vehicle.

4. Repeaters for the base station, if the distance between the mobile station and the base station is greater than the range of the base station radio transmitter/receiver.

5. Base station hardware with a radio, antenna, and a base data link controller (BDLC). This is, by far, the most expensive single component, though the cost of a mobile radio or terminal can also be high.

6. A dispatch center where dispatchers are stationed and/or a computer system where database information is stored.

8.3.6   The SMR industry today

The industry started out with many small SMR operators. In 1995, there were over 7000 licenses (trunked and conventional, with four times as many trunked systems as conventional ones), over 3000 SMR operators and an average of six to seven channels per SMR operator in the 800-MHz band in the United States. However, an interesting trend in the SMR industry during the 1980s and 1990s has been a movement toward regional and national systems. The FCC has approved waivers requested by RAM Mobile Data Communications for a 900-MHz national mobile data system, and by Millicom for a 900-MHz nationwide voice-and-data system. Motorola has implemented a national

800-MHz SMR service. This trend toward the development of regional systems has led to a significant increase in the number of rural SMRs. In response, the FCC is encouraging nationwide 900-MHz licenses and is facilitating the development of regional systems.

The SMR industry is generally considered to be both competitive and profitable, particularly in major markets. The profitability of SMR systems is best illustrated by the tremendous response to recent FCC lotteries for 900-MHz channels in the United States. It is clear from this response that, despite the higher cost of 900-MHz equipment, many people believe they can earn profits from an SMR license. The profitability is also illustrated by the waiting lists for channels in many markets.

For a summary of information on the SMR industry, see Table 8.3.

Comparisons of the SMR industry and the cellular industry. The most significant competitor faced by the SMR industry to date is the cellular radio industry. Cellular radio differs from SMRs in several significant ways. For example, cellular radio operators are common carriers and thus subject to state regulations. SMR end users, unlike cellular radio users, must be licensed. SMR services, unlike cellular radio operators, are restricted from reselling interconnection at a profit. Cellular services in any given market are provided by just two operators. In most SMR markets, there are many independent operators. Each of the cellular operators in a given market has more spectrum than all the SMR services in that same market combined.

Technologically, most cellular radio systems are more complex and costly than most SMR systems. In general terms, cellular radio technology is more spectrum-efficient for interconnection than traditional SMR technology. SMR technology, however, has some advantages, particularly for dispatch services (which cellular radio systems are not allowed to offer). SMR systems are now operating in most parts of the country while cellular radio services are not yet available in all rural markets. Finally, current SMR systems are generally smaller and less expensive to construct than cellular radio systems.

8.3.7 Typical cost of an SMR system

A five-channel 800-MHz system is generally estimated to have a start-up cost of between $60,000 and $150,000 for equipment. This cost range reflects the range of features available to an SMR system. Equipment for each additional five channels costs about $50,000. A 900-MHz system has significantly higher start-up costs (up to $100,000 more). One reason is that the use of a single antenna produces unacceptable interference between adjacent channels in a system. To handle this prob-

TABLE 8.3   SMR Networks Summary Information

	Specialized Mobile Radio (SMR) summary
Brief description	■ An SMR service provides two-way land mobile communications on a commercial basis to businesses, government, and individuals ■ A license is needed to operate the network and use mobile radios. An SMR operator may obtain a license for the user for fee ■ SMR operators provide their service for a monthly fee ■ SMR services include both voice and voice/data information exchange ■ SMR systems feature both conventional (nonshared channels) and trunked (shared channel) ■ SMR services interconnected with PSTN systems are close to cellular services in nature
Components	■ Mobile radio with mouthpiece for voice, portable radio, or mobile terminal ■ Mobile radio modem ■ Base station with antenna ■ Base-station repeaters ■ Channel assignment microprocessors assign channels to mobile user groups
Frequency bands	■ 800 MHz (since 1974) and 900 MHz (since 1990s)
Coverage	■ Licensed widely throughout urban and rural United States and other countries. Over 7000 licenses with over one million users in the United States
Capacity and speed	■ The average SMR system has 6–7 channels per system; 40–50 mobile users per channel ■ Data speeds of 1200, 2400, 4800, 9600, and 19,200 bps
Most suitable applications	■ Dispatch of police, fire, and ambulance personnel ■ Taxi dispatch ■ Truck dispatch ■ Service representative dispatch
Typical costs	■ 5-channel 800-MHz SMR system: $60,000–$150,000 ■ 5-channel 900-MHz SMR system: $150,000–$250,000 ■ Mobile radio: $800–$3000 ■ $15–$20 per month per mobile flat (current tendency is to switch to airtime usage basis) ■ $35–$45 per month per mobile for interconnection with PSTN
Availability	■ 800-MHz systems have been available for years ■ 900-MHz systems being set up now (see ESMR also)
Pros	■ Inexpensive service for specific applications, cheaper than cellular ■ Performance controlled
Cons	■ High capital costs ■ Data throughput speed limited ■ Voice-data priorities difficult to manage in older SMR systems

lem, 900-MHz system operators often employ several antennas for their 10-channel systems.

An additional cost to SMR operators is that of acquiring and setting up an antenna site, which must be above the local terrain to provide good service. The cost varies according to antenna type and geographic region. The three types of sites used most often are tall buildings, mountaintops, and antenna towers. The typical rental cost of a site, once set up, is a few hundred dollars per month. While the costs associated with antennas are significant, the main problem faced by both SMR and cellular operators is the general unavailability of premium sites.

Annual operating costs (excluding equipment costs) of a five-channel SMR system have been estimated at approximately $100,000, unless the system is operated in conjunction with other related activities that can absorb some of the overhead. Economies of scale, however, are very pronounced in the SMR industry. The minimum workforce required to operate a 5-channel system is probably the same as the minimum workforce required to operate a 20-channel system.

A typical flat-rate charge for an unlimited SMR dispatch service is $15 to $20 per month per mobile. Assuming seven channels with 70 mobiles per channel and a $17 monthly charge, income would be about $100,000 per year without interconnection to the telephone network. A growing trend among SMR systems is *airtime billing*. SMR operators may compute the amount of radio usage of each of their clients by using a commercially available software program. The advantage of airtime billing is the fact that, with end users being charged for actual time spent on the radio, there tends to be more efficient use of radio airtime. Airtime billing is particularly useful for interconnected systems.

The mobile radios used by SMR end users list from under $800 to over $3000. Motorola, the largest supplier of SMR equipment, has traditionally marketed most of its end-user equipment directly. SMR operators who use equipment from other manufacturers, including E. F. Johnson, Ericsson-General Electric, and Uniden, often market two-way mobile radios along with their service. As the SMR industry has grown, end-user equipment prices have fallen. The supply of high-quality, low-cost end-user equipment remains an important factor in the growth of the SMR industry, particularly given the declining prices of cellular radios.

8.4 Specialty Networks: Paging and Wireless Messaging

We shall discuss three specialty networks in this section:

- Paging networks
- The RadioMail network
- The SkyTel network

8.4.1 Paging networks

Paging is by far the most popular form of mobile communication. The reasons are obvious. Pagers are small, affordable, simple to use, ubiquitous, and reliable. About 16 million pagers are in use today in the United States. Almost 40 million people around the world carry them. The industry continues to grow and is forecasting substantial increases with the widespread adoption of two-way paging. One study by McLaughlin & Associates forecasts that the market for pagers in the year 2000 will be worth $21.5 billion worldwide, with the U.S. paging population around 25 million. Whether pagers will eventually be replaced by a new breed of PDAs, personal communicators, or wireless-enabled electronic organizers, or whether they will simply become functionally ever richer, providing more and more personal communications functions is a moot point: the basic concept behind paging will always be in demand.

There are two types of paging networks:

■ One-way paging networks
■ Two-way paging networks

Most of the current paging networks provide one-way paging; i.e., they allow one-way transmission of brief messages (numeric or alphanumeric) to a low-power receiver. Receivers can have their own display or can direct the display output to a laptop, PDA, or personal communicator.

In 1995, several companies (SkyTel being the first) started offering on the new narrowband PCS frequencies two-way paging that allowed users to acknowledge receipt of a message. Some companies have since enabled the transmission of other brief messages as well (see Sec. 8.4.3 on SkyTel two-way messaging and the two-way pager detailed in Chap. 14).

Figure 6.11 in Chap. 6 shows paging devices at the time of writing. It should be borne in mind that paging devices are going through considerable and rapid change. While conventional pagers are acquiring multiline LCD displays, not-so-conventional pagers are being built into wristwatches, PDAs, personal communicators, notebooks, and even desktop PCs at the office.

Paging networks are becoming functionally rich, too. Many companies are providing news services via paging. Users can now get instant information on stock prices and major sports or other news headlines.

How paging works.   Figure 8.10 shows a schematic of a typical paging network. The pager itself is a small-sized, low-power radio receiver that constantly monitors a radio frequency dedicated to pager use. It remains silent until it detects a specific ID string that tells it to display a message and alert its user. In the United States, this ID string is called a *capcode*. A capcode has nothing to do with telephone numbers

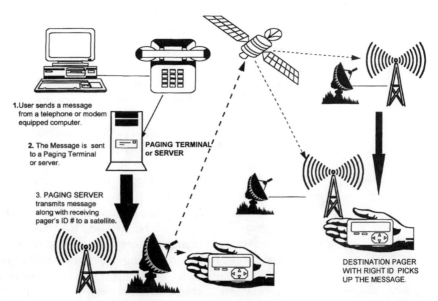

1. User sends a message from a telephone or modem equipped computer.

2. The Message is sent to a Paging Terminal or server.

PAGING TERMINAL or SERVER

3. PAGING SERVER transmits message along with receiving pager's ID # to a satellite.

DESTINATION PAGER WITH RIGHT ID PICKS UP THE MESSAGE.

Figure 8.10   How current one-way pager networks work.

or user IDs. The user ID number commonly associated with a pager is merely a key to a database that links a user ID with the capcode. The key point is that the pager company transmitter is constantly sending out pages, but a specific unit will activate only when it hears its own capcode.

Paging applications.   There are distinct types of pagers, each with a set of applications:

Numeric paging.   Used primarily to request a callback (as per the telephone number or code number sent to and indicated on the pager's display). These pagers are used in everyday business and personal life.

Alphanumeric paging.   Used for sending more detailed information as to the nature of the call or a brief message. Used in everyday business and personal life by people who want more information than just a number to call back.

Two-way paging.   Many new mobile applications are emerging where information-exchange requirements are simple and brief. Software development kits were released in 1995 and new applications will continue to emerge over the next few years.

Many e-mail applications in corporate settings (such as Lotus cc:Mail, LOTUS NOTES, and MS Mail) are being upgraded to offer, within organizations, private paging services that will contact sales and professional staff on the road. The same applications can also use public gateways such as RadioMail to reach external users.

Paging network service providers.   There are a large number of local, regional, and national paging service providers in the United States and other countries. We shall mention a few of the national providers that offer a comprehensive set of paging services today and are actively planning more advanced two-way paging services. Because the paging model for application delivery is so highly suited to business users and consumers, these future services will make simple mobile computing applications more affordable and easier to use.

   Motorola's EMBARC, MobileComm, PageNET, and SkyTel are dedicated paging service providers. ARDIS, RAM Mobile, and RadioMail provide personal messaging services with a paging option. Several other companies such as Paging Network, AT&T Wireless Services, PageMart, and AirTouch Communications have joined the PCS narrowband market. All of them have services planned for 1996 and beyond. For a summary of information on paging networks, see Table 8.4.

Paging network costs.   Paging is one of the cheapest mobile communications services available. There is generally a monthly charge, with a limit on the number of messages allowed per month. Cost of entry can be as low as $5 per month. The following costs in Table 8.5 are for a middle-of-the-road paging service.

8.4.2   RadioMail

RadioMail provides public gateway services to mobile computer users. The company offers wireless two-way electronic mail delivery, an on-demand news service, and a service that allows pagers and paging cards to receive messages from nearly any electronic mail system.

   Please refer to Chap. 14 and App. A for additional information.

8.4.3   SkyTel network

SkyTel is a subsidiary of Mobile Telecommunications Technologies Corporation. It provides nationwide wireless messaging services, including paging. SkyTel has introduced many innovative services such as SkyLink (the first wireless mailbox) and the Message Card (the first credit-card-size paging unit). SkyTel was the first U.S. paging company to introduce two-way PCS narrowband paging in the United States.

   Please refer to Chap. 14 for additional information.

TABLE 8.4   Paging/Wireless Messaging Network Summary

Brief description	■ Specialized wireless network for broadcasting a message to specific pager to call back a specific number
Components	■ Personal paging device ■ Paging computer/server at service provider's site ■ Paging transmitter ■ Use of satellite for national coverage
Frequency bands	■ PCS narrowband 901–902, 930–931, and 940–941 MHz
Coverage	■ Many local, regional or national paging networks ■ 95% of the United States is covered
Capacity and speed	■ Paging networks have different design criteria for delivering the message within specific time periods ■ 1200 bps for data transmission
Communications protocols supported	■ TAP for numeric data ■ TDP for 8-bit non-ASCII data ■ ReFlex 50 protocol developed by Motorola for two-way paging
Most suitable applications	■ Personal numeric messaging for call back ■ Alphanumeric messaging: dispatching and service ■ Two-way messaging: call dispatching with confirmation
Costs	■ $150 for pager and $10–$25 per month for messages (numeric to text to two-way) ■ ARDIS personal messaging service starts at $39 per month ■ Please note: personal messaging is richer in function than alphanumeric paging
Availability	■ Has been available for many years ■ Two-way messaging started in 1995
Security	■ Low
Pros	■ Very easy to operate for sender (from any telephone) and receiver ■ Excellent coverage: local, regional, national, international (Canada and United States) ■ Very inexpensive ■ Good building penetration ■ Many options: numeric, alphanumeric, two-way, message storage
Cons	■ Some networks are overloaded causing delays ■ No acknowledgment (though slightly more expensive two-way paging is now available) ■ Very few two-way paging applications in 1996 ■ Slow data transfer rate (1200 bps) for non-PCS narrowband paging networks

TABLE 8.5   Costs for Paging Service

Cost of pager	Numeric: $150 Alphanumeric: $250 Two-way: $400
Monthly service charge (message size 80 to 240 characters)	$10–$25 (one-way numeric to two-way) for up to 400 messages per month

8.5   Cellular Networks

Primitive radio-based cellular telephone connections were used by telephone companies as early as 1946 for the isolation of problems with terrestrial telephone hookups. This implementation was based on a simplex (one-way) operation in which only one party could talk at a time. It was not until the mid-1960s that an *improved mobile telephone service* (IMTS) was introduced which allowed direct access to dial-up telephone networks. IMTS eliminated the need for the press-to-talk requirement of a one-way connection. Higher-powered radio systems of the time (200 watts of output) provided a coverage up to 25 miles (40 kilometers). IMTS was eventually modified slightly to run in different frequency bands and was renamed *advanced mobile phone service* (AMPS). AMPS is the predominant technology implemented in the United States today.

Both IMTS and AMPS operated in different frequency bands (44 channels were assigned to public mobile services operating in 35–44, 152–158, and 454–512 MHz). Service was based on high output and line-of-sight over a broad geographical area, where frequency reuse was limited to 50 to 100 miles (80 to 160 kilometers).

Modern advanced mobile phone service (AMPS). It was only in 1978 that modern cellular service based on the 800 to 900 MHz portion of the ultrahigh frequency (UHF) band was introduced in Chicago. In 1981, the FCC finally set aside 666 radio channels for cellular use in the United States. These frequencies were assigned to two separate carriers. The lower frequencies were given to the traditional telephone companies and the higher ones were assigned to their non-telephone-company competitors.

8.5.1   Basics of cellular transmission

The fundamental concept behind a cellular network is the division of an area of coverage into a honeycomb of small cells that overlap at their outer boundaries—hence the use of the word *cellular* (Fig. 8.11). Frequencies are divided into bands, with protection zones established to prevent interference and jamming. The cellular system uses minimal power output for transmitting (the transmitter is designed to output three watts). The greater the number of users in an area, the closer the cells are, and vice versa.

Each cellular phone is assigned a unique identity called a *numeric assignment module* (NAM). Each cellular phone is also assigned a home area for traffic control. Control messages are sent on a control channel. A phone operating outside the home area is required to reregister, advising the home unit of its new location. As the cellular phone

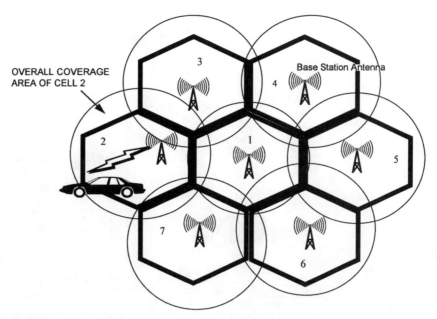

Figure 8.11   Concept of a cellular cell and its coverage area.

user moves from one area to another, it continuously sends messages to the mobile telephone switching office (MTSO) to indicate its location. All calls to a particular number are thus directed to that user's current cell location.

Initial monitoring after power on. When a user gets into a car and powers-on, the cellular phone starts monitoring the dedicated control channels to obtain signal-strength information for various channels and then tunes itself to the one specific paging channel it finds sufficiently strong in signal level (not to be confused with paging network protocols, though the concept is the same). After that, the phone essentially goes into idle mode while continuing to listen to the control information on the paging channel. If the signal falls below a certain level, the phone will go back into scan mode to look for a stronger paging channel on which to receive or send control information.

Sending and receiving calls. To make a call, the user enters the telephone number and presses the send button. The cellular phone scans and selects an available access channel and then transmits a call-set-up request. Once it has received a reply from the cellular system indicating which send channel to use, the phone tunes to the new frequency and makes the call (Figs. 8.12 and 8.13).

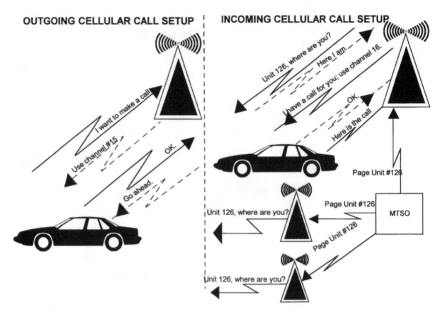

Figure 8.12 Process of cellular call setup. (*Source:* Wireless Networked Communications *by Bates.*)

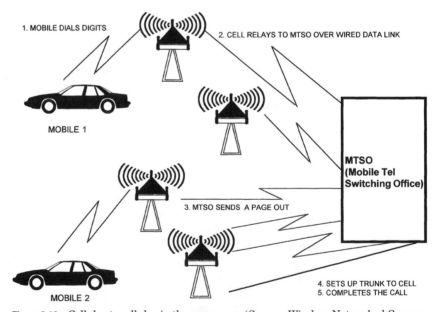

Figure 8.13 Cellular to cellular in the same area. (*Source:* Wireless Networked Communications *by Bates.*)

When the cellular system receives an incoming call, it locates the intended recipient through the paging control channel by sending a page to all cells in the area. When the phone receives a page, it is roused from idle mode. The system then instructs the phone to retune itself to a specific channel to receive the call. A microprocessor imparts to the mobile phone a frequency-agile capability that enables the phone to tune to any frequency within the operating range of the area.

The hand-off process. During the course of a conversation, a user may well move from the coverage area of one cell to that of another. When that happens, a hand-off process takes place (Fig. 8.14). As the signal strength drops below a certain level, the cell-site sends a distress call to the MTSO indicating that the signal is weakening. The MTSO then arranges to pass the call to another cell site with a stronger signal.

Cellular changing to digital. AMPS is the standard that most cellular carriers used from 1995 to 1996. However, it is an analog system that uses *frequency division multiple access* (FDMA). This has a limitation in that a frequency band can carry only one call per cell. In large metropolitan areas, the resulting congestion leads to problems during peak hours involving saturated frequencies and dropped calls. Digital cellular systems address this problem by placing several channels on

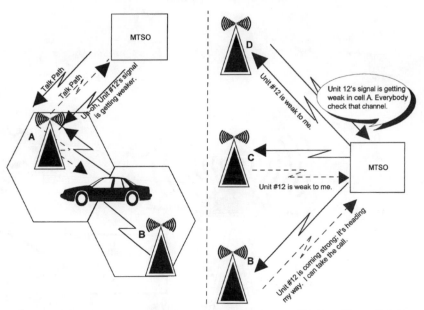

Figure 8.14 Handoff process in cellular networks. (*Source:* Wireless Networked Communications *by Bates.*)

the same frequency using a *time division multiple access* (TDMA) technique. With TDMA, the sending unit converts voice into digital data, encodes and compresses that data and transmits it. The digitally compressed voice data is transmitted in bursts, which are offset from each other by a fixed time interval, with multiple bursts placed in the same frequency in turns. The receiving unit selects the burst of data it needs, decodes and decompresses the data, and converts it into voice. Depending on the particular implementation of the TDMA standard, up to 12 conversations can be placed on any given frequency band at once, thereby increasing the overall capacity of the system.

QUALCOMM is promoting another digital scheme called *code division multiple access* (CDMA). CDMA boasts capacity improvement higher even than TDMA. However, there are not many commercial services offering either of these two technologies currently, even for cellular voice. And nobody is offering data over TDMA at all, or over CDMA until mid-1997. The reasons for TDMA and CDMA technology not gaining widespread use are many. Perhaps the most important reason is that AMPS system was developed first with digital systems following later. Other factors are cost of upgrading to CDMA/TDMA and traffic density. See Sec. 8.8 for further discussion of TDMA and CDMA.

Components of a cellular system. A cellular system consists of the following components (Fig. 8.15):

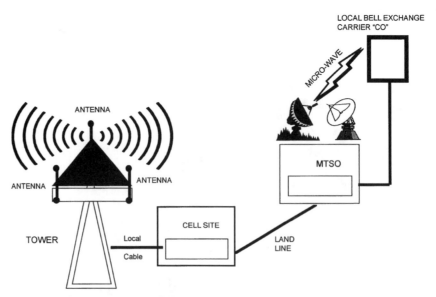

Figure 8.15 Components of a cellular network.

- A cellular handset.

- A cell site (base station) consisting of a transmitter and two receivers per channel, a controller, an antenna, and data links to the cellular office.

- The MTSO (mobile telephone switching office), which is where control and management of the various cell sites takes place and the connection is made to the local telephone (PSTN) exchange. The MTSO has dedicated circuits that link it to the cell site.

8.5.2   Circuit-switched cellular for data

Since there is now extensive cellular network coverage throughout the United Sates and the rest of the world, the opportunity exists for mobile workers to use cellular networks for data transmission. Once a notebook computer is equipped with a PCMCIA cellular modem and a cable to connect it to your cellular phone, you have a physical connection that is very much similar to a public switched telephone network. Many sales professionals and other mobile workers are using cellular data technologies already—but so far with only mixed results.

How circuit-switched cellular networks carry data traffic.   Figure 8.16 shows how the MTSO uses a modem pool and a physical communica-

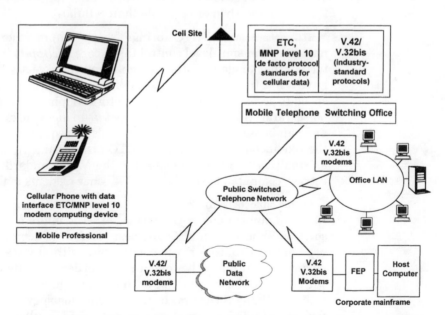

Figure 8.16   How data is carried over a cellular circuit-switched network. (*Source: Adapted from* Byte *magazine.*)

tions server to interface with mobile workers on one side, and with information servers on the other.

Problems and solutions for the data transmission over cellular networks. The analogy between a PSTN and a cellular network is a logical one only. Cellular transmission is based on a circuit-switched technology that switches between frequencies to avoid interference as new calls come online, or when hand-offs occur. Although this switching causes only momentary delays, they are nevertheless often too long for sensitive communications applications built for reliable wire-line circuits.

Thus there are a great many differences in the modus operandi of a cellular network that interfere with the reliable communication of data. Fundamentally, cellular networks are analog networks designed specifically to carry the human voice. Since the end-user device in a cellular voice conversation is an ear belonging to an intelligent being, transmission glitches and pauses caused by switches and hand-offs can be treated as annoyances only and inferences can be made regarding that which is missed.

Several vendors—AT&T in particular—have introduced specialized modems for cellular data transmission. Enhanced throughput for cellular (ETC) is one such example. ETC goes beyond the MNP-10 and V.42 error correction protocols, which, after all, were built for a PSTN that is inherently more reliable than cellular.

ETC enhancements. ETC brings many improvements to cellular modems. Transmit-level control takes care of *clipping* (a harmonic distortion that tends to flatten a signal and thus causes ETC modems to have a flat frequency response). Auto-rating adjusts the transmission speed to accommodate harmonic distortion, phase jitter, and signal-to-noise ratio. Auto-rating even allows a modem to drop to 1200 bps and can be adjusted every five seconds, if required, thereby reducing the likelihood of dropped calls. ETC also breaks up data into frames as small as 32 bytes (one-quarter the V.42 standard) and makes 20 retries, compared to the 7 that most other modems make.

AirTrue protocol. AirTrue is another noteworthy modem protocol from Air Communications of California. AirTrue offers a cellular-side only solution. Where, with ETC, the best results are achieved when ETC is used on both sides, AirTrue can operate with any standard V.42 host modem. According to the company, it has demonstrated error handling with AirTrue that is superior to the error handling obtained with ETC modems. While the claims or the test methodology have not been verified here, Table 8.6 reproduces the published results for the reader (see also Fig. 8.17).

TABLE 8.6   Error Handling Capabilities of Various Modems over Cellular Networks

Category	Connectivity, %	Interoperability, %
Modems with standard interfaces	20 to 50	24 to 55
EC^2 solutions	38 to 45	42 to 50
ETC direct connect solutions	40 to 49	45 to 53
AirTrue	90 to 96	97 to 99

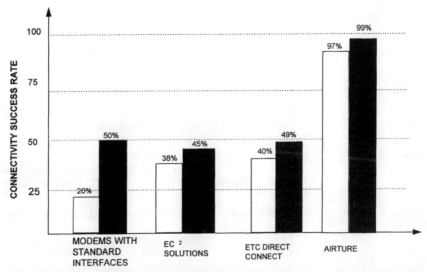

Figure 8.17   Comparison of cellular protocols for connectivity. (*Source: AirTrue white paper. Claims not verified by author.*)

Cellular network summary.   Table 8.7 summarizes the characteristics of cellular networks.

8.5.3   CDPD*

Cellular Digital Packet Data (CDPD) is a wireless wide area network (WAN) that uses idle spots in cellular channels not used by voice at any given time. CDPD technique, originally developed by IBM, allows data to hop from one idle channel to another idle channel through frequency hopping (see Fig. 8.18). The CDPD network is being deployed by several cellular carriers providing coverage to most of the nation. These cellular carriers, along with many hardware and software vendors

---

* The CDPD network overview is based on a white paper provided by GTE MobileNet, a key player in the CDPD forum in the United States. The author acknowledges with thanks the permission by GTE to include this material in the book.

TABLE 8.7   Cellular Network Summary

Brief description	■ AMPS-based analog network with the largest user base among any of the wireless networks. Predominant use is for voice, with limited data use by mobile workers for e-mail, file transfer, and sales automation
Components	■ Cellular handset ■ Cell site (base station) ■ MTSO
Frequency bands	■ 800–900 MHz
Coverage	■ Extensive, rivaled only by satellite; not available in very remote areas with low population
Capacity and speed	■ Industry is increasing capacity to meet increasing demand for voice ■ Data speed is limited to 4800–14,400 bps; effective sustainable speed is still low (4800)
Protocols supported	■ Works in an asynchronous mode similar to a PSTN dial-up connection; therefore most protocols are supported (IPX/SPX, TCP/IP, SNA)
Most suitable applications	■ Voice communication ■ E-mail, file transfer, and sales automation ■ Not suitable for applications based on interactive online transaction (OLTP)
Costs	■ Charges are for per-minute connection time intervals (25 to 50 cents per minute in United States); some networks (Bell Mobility) charge in 10-second intervals ■ A 600-character e-mail message may cost 50 to 90 cents, including setup time ■ PCMCIA modems cost $350–$500
Availability	■ 90% of United States and Canada (urban and rural areas); less extensive in Europe, but expanding ■ Planners should assume that analog cellular will be phased out and replaced by services based on D-AMPs (digital) and PCS/GSM
Security	■ Low; very easy to eavesdrop and listen to conversations, or to steal data (remember the "Squidgy" royal recording fiasco in the United Kingdom). End-to-end encryption recommended
Pros	■ Currently offer more coverage than other networks ■ No special software drivers required; applications can be used as is ■ Supports session-oriented applications ■ Compatible with remote network access (RNA) products; good LAN support ■ Cheapest network for file transfer ■ Easy to use; no special equipment other than a modem and a cellular phone
Cons	■ Cost is high for short messages ■ Unreliable transmission as compared to packet radio networks ■ Long setup (connection) time for online-transaction-type access ■ Data is normally given lower priority than voice ■ Being replaced by CDPD for short messages

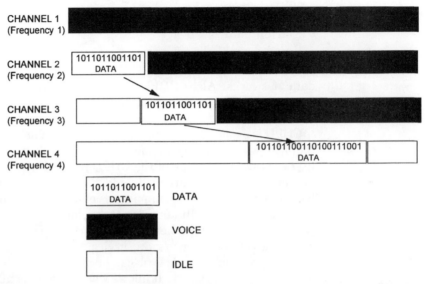

Figure 8.18 Empty spots in cellular voice channels and frequency hopping employed in CDPD.

interested in mobile computing have formed an industry group called CDPD forum. Some of the major players in CDPD forum are:

Carriers (a subset)			
Airtouch Cellular	Bell Atlantic/Nynex Mobile	GTE Mobilnet	Southwestern Bell Mobile
Ameritech Mobile Communications	BellSouth Cellular	McCaw Cellular, now AT&T Wireless	Spring Cellular

Hardware and software vendors (a subset)			
Compaq	Lotus	Novell	Racotek
Ericsson	Microsoft	Oracle	Rogers Cantel Canada
Hughes Network Systems	Motorola	Pacific Communications Sciences, Inc.	Sierra Wireless
IBM	Mobility Canada	QUALCOMM	
	Nortel		

CDPD forum objectives. The key objectives of CDPD forum are to provide:

- A seamless national wireless data network by allowing interoperability among regional networks
- Support for standard network protocols
- Security from casual eavesdropping over the airlink through sophisticated encryption techniques

- High data throughput
- Efficient network usage

CDPD technology benefits. CDPD was designed to work with current data networks. Also, CDPD uses existing network hardware from proven network technology. Many of the components in the infrastructure are commercial off-the-shelf products.

CDPD is designed to minimize the impact on network software by not requiring any changes to the higher network protocols. This allows users to take advantage of the CDPD network with little or no change to their current applications. In addition, innovative frequency-hopping techniques minimize any impact on cellular voice traffic.

Unlike current cellular voice technology, CDPD provides seamless service while visiting other cellular markets. There is no need to dial a roamer access number to gain service.

CDPD in its most basic form can be used as a wireless extension of an existing TCP/IP network. It allows workstations to talk to host computers to retrieve sales data, inventory, billing data, etc., in a similar manner as packet radio networks such as ARDIS and RAM Mobile Data.

CDPD technology. CDPD differs significantly from data communications over a traditional cellular connection, called a *circuit-switched connection*. Using circuit-switched cellular technology, once a connection has been established, it owns that link and the users are paying to use it, whether or not data is actually flowing. Further, any line noise, dropped connections, or other impairments to which modems are susceptible are visible to the modem. Often connections have to be reestablished at additional cost.

CDPD is a *packet-switching network*. Data transmissions are broken down into packets of data. When there is no data to transmit there are no packets and hence no communication charges.

CDPD is not an overhaul to the existing cellular infrastructure or a new network requiring millions of dollars in capital expenditure. The technology required to make CDPD a reality is an overlay to the existing cellular infrastructure (see Fig. 8.19). The bolded blocks indicate what components have been added to the cellular infrastructure to provide CDPD capability. CDPD uses the same cellular frequencies and has the same coverage as cellular telephony. CDPD cohabitates with cellular voice traffic by using idle time (the time not used by a cellular phone) to transmit data. For example, a typical cell site may be equipped to handle up to 30 simultaneous voice calls. At any given point in time, the full capacity is probably not being used. During these periods, packets of data are transmitted over the unused voice channels. When a voice call is assigned to a channel currently transmitting

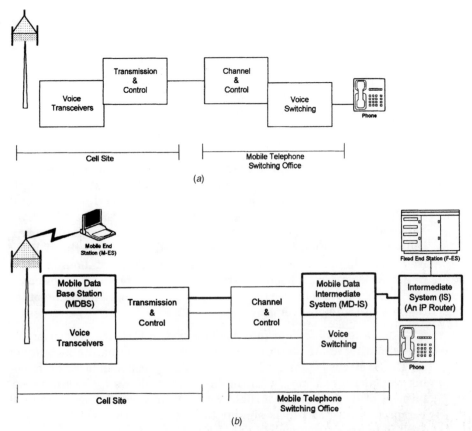

Figure 8.19   CDPD overlay on existing cellular infrastructure: (*a*) existing cellular architecture; (*b*) CDPD overlay. (Note that the bolded blocks are the components needed for CDPD operation.) (*Source: GTE MobileNet.*)

data, the voice call will have priority, and the data session will "hop" to a new idle voice channel. As a result, CDPD carriers will incur lower infrastructure costs which they could use as a competitive advantage in future.

CDPD network components.   The components of CDPD actually function like familiar networking components. Figure 8.20 shows the CDPD nomenclature with the familiar networking names. A brief description of the functionality of each follows.

M-ES.   The *mobile end system* (M-ES) is typically a laptop or hand-held computer with a CDPD modem installed. However, the M-ES can also be a nonpersonal device such as a telemetry device. The M-ES software can either be native IP or a package that supports the Hayes AT command set. The network software is discussed in Chap. 12.

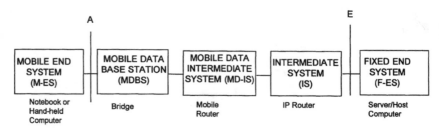

Figure 8.20   Components of a CDPD network. (*Source: GTE MobileNet.*)

MDBS. The *mobile data base station* (MDBS) is primarily responsible for radio frequency management. For example, as data is transmitted while traveling, the mobile end system may travel through several cellular cells. It is the responsibility of the MDBS to manage the channel handoffs. The connection between the M-ES and the MDBS is typically called the *airlink* or *A* interface. The MDBS informs the M-ES of frequency changes much like the collision avoidance standards found in network protocols. From a network transport protocol perspective, the MDBS acts as a bridge that simply relays messages from the router to the M-ES and vise versa.

MD-IS. The *mobile data intermediate system* (MD-IS) is responsible for validating an M-ES on the network and managing the mobility of an M-ES (discussed later). The MD-IS is also responsible for exchanging the encryption keys with the M-ES that allows for secure transmission of data over the airlink. From a network transport protocol perspective, the MD-IS is similar to a router with special functions built in to handle the mobility of M-ESs.

IS. The *intermediate system* (IS) is an IP router that sends datagrams to the outside world. Hosts connected to the IS think they are receiving data from a common IP network.

F-ES. The *fixed end system* (F-ES) is the customer's host. It is the final destination of the message sent from an M-ES. This connection between the F-ES and the IS is typically called the *external* or *E* interface. It is important to note that an M-ES actually functions as an F-ES when two mobile devices send messages to one another.

CDPD software. The CDPD software architecture is purposely designed around standard network protocols. Figure 8.21 shows the OSI layers and how these layers are mapped to the CDPD environment. Because of the uniqueness of CDPD, the network layer comprises a portion of the CDPD-specific software layers and the API. As mentioned above, CDPD must be able to manage the mobility of an M-ES and, therefore, must control part of the network layer.

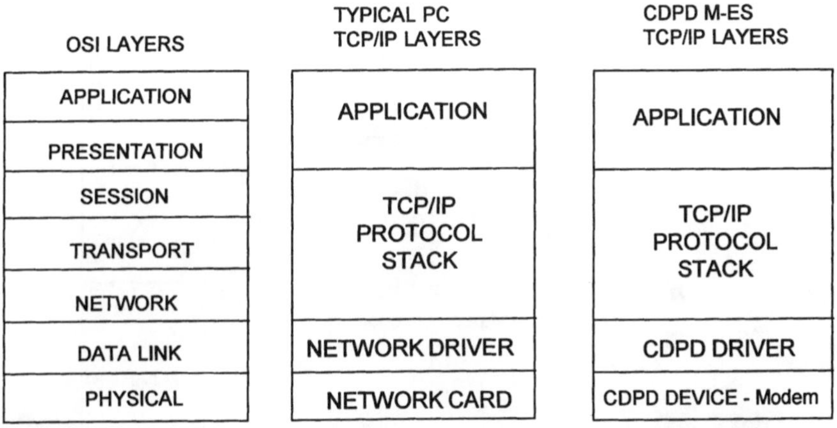

| | TYPICAL PC | CDPD M-ES |
OSI LAYERS	TCP/IP LAYERS	TCP/IP LAYERS
APPLICATION	APPLICATION	APPLICATION
PRESENTATION		
SESSION	TCP/IP PROTOCOL STACK	TCP/IP PROTOCOL STACK
TRANSPORT		
NETWORK		
DATA LINK	NETWORK DRIVER	CDPD DRIVER
PHYSICAL	NETWORK CARD	CDPD DEVICE - Modem

Figure 8.21 Matching of CDPD software layers with OSI and TCP/IP. (*Source: GTE MobileNet.*)

Comparing the CDPD layer to a common architecture may help demonstrate how the CDPD network operates (see Fig. 8.21). The standard OSI layers are on the left. In the middle is the typical PC TCP/IP stack with the LAN device driver at the data/network layer. Above the LAN device driver is an off-the-shelf TCP/IP protocol stack that provides services to applications.

Mapping the PC stack to the M-ES stack, shows that very few changes are needed to operate over the CDPD network. In place of the LAN device driver is the CDPD device driver. The application still sees a network card. This is one of the keys to the transparency CDPD offers. The application does not need to concern itself with the fact that the network card is actually a wireless communications device.

Mobility management. A unique feature of CDPD is its method of handling M-ESs moving from one CDPD location to another. For example, a salesperson may use the CDPD network in Los Angeles on one day and access the network again in New York on the next day. Standard networks are not equipped to handle this movement. In a standard network, moving a device to a different geographical location requires reconfiguring the network software.

The CDPD network is designed to work with standard off-the-shelf TCP/IP products; therefore, it has to manage mobility, yet keep this movement transparent to TCP/IP products. To achieve this, the architects of CDPD created a protocol sublayer named *mobile network location protocol* (MNLP). The MNLP sublayer manages the movement of an M-ES from one location to another.

So how does the CDPD network know where to send a message if the destination is changing from day to day? The method is not too differ-

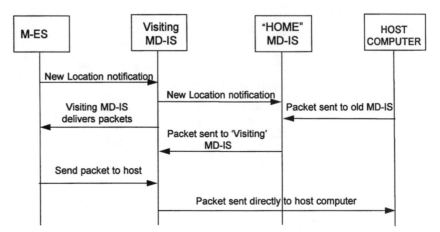

Figure 8.22   Mobility management in CDPD. (*Source: GTE MobileNet.*)

ent from the forwarding of letters in the post office system (Fig. 8.22). For example, assume someone has just moved to a new city and needs to receive mail at the new address. Many senders still send letters to the old address. If the post office has a change of address form, however, it knows where to forward the mail. Letters arrive at the old post office with the old address on the envelope. The post office sees the old address and sticks on a yellow sticker indicating the new address and forwards the letter to the new post office. Upon arrival in the new post office, the letter is loaded on the postal vehicle for delivery to the new address.

Continuing the analogy, letters sent in the other direction go directly to the original sender. Now look at how an F-ES transmits to an M-ES that has moved since the last transmission (Fig. 8.22). The M-ES registers with the MD-IS in the new city. This visited MD-IS notes the visiting status and notifies the home MD-IS of the new location through the CDPD infrastructure network. When the F-ES sends a message to the home M-ES, it is forwarded to the visited MD-IS for delivery. The visited MD-IS receives the message and forwards the message to the M-ES. Upon receipt, the M-ES replies to the F-ES by sending a message to the visited MD-IS. The visited MD-IS sends the message directly to the F-ES rather than send it to the home MD-IS since the home MD-IS need not be aware of traffic going from the M-ES to the F-ES.

Remember that this process is completely transparent to the M-ES and the F-ES. As far as the F-ES is concerned, the M-ES has not moved and expects the M-ES to receive the message at its home location. It is this transparency to the standard IP application that makes this movement of an M-ES from market to market possible.

Issues for application software developers. Ignoring for now the differences between typical desktop machines and mobile computers themselves, two sets of issues confront software developers writing applications for mobile users. The first set of issues arises due to the different communication medium itself. These problems are low throughput and delay, dropped packets, duplicated packets, packets arriving at the destination out of order, and cost. The second set of issues arise due to mobility itself. These issues are changing resources, abnormal disconnections, and deferred I/O.

Problems due to radio as a communications medium. The biggest problems facing software developers writing for new wireless data communication networks are: (1) delays, (2) dropped packets, (3) duplicated or out-of-order packets, and (4) cost.

Low throughput and delay. *Throughput* is the amount of information that can be transmitted per unit of time. *Delay* is the time required for a packet to travel from the transmitter to the intended receiver. In many cases, low throughput has the same effect as delay. Because of limited bandwidth on the airlink and the need to relay packets several times before they reach their destination, delay is usually the most significant technical problem to be overcome by application writers. Keep in mind that delays may not be consistent. Because of variable network traffic, some packets may travel quickly while others encounter delays of many seconds.

Mobile applications must allow plenty of time for packets to arrive at their destination. Communication protocols that require an acknowledgment for each packet should be discouraged. Protocols using sliding windows or a metering mechanism should be used instead. Note that many protocols implementing sliding windows are configured by default for a small window size—it should be increased.

Another approach to dealing with delay is to combine multiple functions in one packet. For example a request testing a file's existence could be combined with an OPEN FILE request.

Dropped packets. Dropped packets are especially frequent in times of congestion. Intermediate devices use this tactic to cope with links that are so unreliable packet buffers become overloaded. Even when each component of a communication link guarantees delivery, the system may not always succeed due to component failures.

Most connection-oriented protocols such as SPX and TCP include methods of recovering from dropped packets; however, applications using datagram protocols such IPX or UDP need to implement their own recovery mechanisms. Most introductory technical books on networking include descriptions of methods to recover from dropped packets (as well as duplicated and out-of-order packets), with the use of sequence numbers and e-transmission algorithms.

Duplicated and out-of-order packets. It is more unusual but not rare to encounter duplicated or out-of-order packets. Most mechanisms that handle dropped packets also handle these problems.

Cost considerations in CDPD networks. Unlike land-line networks, CDPD networks charge for their use by kilocharacters sent and not by the time. Therefore it is important to write applications that make the most out of each packet transmitted or received. This requires a different approach to writing applications. As an example, let's look at a specific mobile-aware application. An application at a remote weather-sensing station reports weather observances every 30 minutes. This application continuously sends status to a host computer over the CDPD network. Typically, to verify the data has been successfully received, the host would send an acknowledgment back to the remote device if it receives the data successfully.

Using CDPD, the acknowledgment packet would incur its own charge, increasing packet charges. A more efficient method may be to use an unacknowledged protocol. When the remote application sends the status packet to the host, the remote assumes it was received successfully and no acknowledgment of receipt is sent by the host. For an application that sends data constantly to the host, acknowledgment is unnecessary and, in the CDPD world, more economical.

Abrupt disconnections as a result of mobility. Wireless communications links are frequently disconnected without intent. Applications must always allow for this possibility and recover gracefully. Error recovery routines should be thoroughly tested. Complex transactions should, if possible, be broken into smaller ones. Developers may want to explore checkpoint and recover/resume techniques for some applications. Also, if possible, applications should revert to using deferred I/O when immediate connectivity is not available.

Deferred I/O. Even with wireless communications, there are times when I/O should not be performed. Some new operating systems (PenPoint, for example) allow for deferred I/O. Most developers are aware of print queuing as one form of deferred I/O. Another example is deferred database record updates using a mobile copy of a central database. Upon reconnection to a central database, the central database and the mobile copy are synchronized, usually leaving it up to the user to resolve any inconsistencies.

One use of wireless communication may be that databases are accessed in real time. Updates need not be deferred. UPS's package tracking system is one example of a mobility-enabled database appli-

CDPD forum report card coverage report
January 1995–October 1995

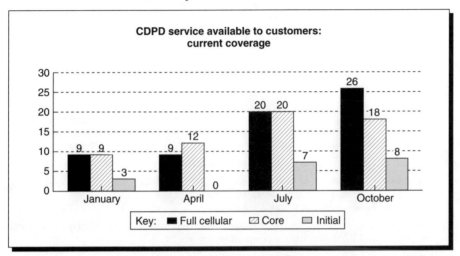

For the third quarter, there was a significant decrease in the amount of core coverage. However, it seems this decrease is due to a shift in a number of MSAs from core coverage to full coverage.

Also, initial coverage increased, indicating new areas are being introduced to CDPD technology.

Figure 8.23   CDPD coverage score card from CDPD Forum.

cation. However, throughput considerations may still dictate a hybrid system of remote and direct access, especially for existing applications.

Deferred I/O, when coupled with the propensity of mobile users to connect to different peripherals or entirely different environments at different times, creates some interesting design challenges for application software.

CDPD coverage status. See the CDPD Forum's Report Card as of December 1995 (Fig. 8.23). Also, see comments in Table 8.8.

CDPD costs. See the brief description in Table 8.8.

8.5.4 Comparison of CDPD and circuit-switched cellular data*

We shall review these differences from design, coverage, cost, and interoperability considerations.

---

* This comparison is based on a CompuServe paper filed by Kevin J. Surace of Air Communications Inc. of California. The author acknowledges with thanks the permission to edit the information and include it in the book.

TABLE 8.8 CDPD Network Summary

Brief description	▪ CDPD is a wireless mobile packet data network, created as an overlay on the existing cellular network. It uses same frequency as cellular and utilizes holes in channels not being used for voice transmission at any given time.
Components	▪ Mobile end station (M-ES), mobile data base station (MDBS), mobile data intermediate system (MD-IS), and fixed-end system (F-ES).
Frequency bands	▪ Same frequency as cellular.
Coverage	▪ Currently in metropolitan areas, expected to match cellular coverage by 1996 year-end.
Capacity and speed	▪ Presently enough for the current set of users; carriers will upgrade based on demand. 19,200 bps for data. However, because of voice priority and packetization, it is difficult to assume that you can effectively achieve 19,200 as speed from user perspective.
Protocols supported	▪ TCP/IP. Other protocols supported through third-party gateways.
Most suitable applications	▪ Short OLTP transaction-based messaging, such as telemetry, credit authorization, e-mail-based messaging, public safety, transportation, etc. ▪ Long file transfer is *not* suitable for CDPD.
Cost	▪ Prices under constant change at present time; expect them to be slightly below ARDIS and RAM initially for competitive reasons.
Availability	▪ It is available now but in large metropolitan areas only.
Security	▪ Higher than cellular but some applications may need end-to-end encryption.
Pros	▪ Potential of full nationwide coverage in future (1997) with seamless roaming within metropolitan areas. ▪ Link speed of 19,200 bps can give good response times for OLTP-type short-message-length applications, if carriers have enough channel capacity. ▪ Direct TCP/IP support means fewer application changes. ▪ Support from RNA industry is coming quickly.
Cons	▪ Voice priority means applications that are not mobile-aware could suffer from poor response time and disconnection. ▪ Current level of coverage is not adequate for national applications. Limited coverage in rural areas. ▪ Interoperability between regional CDPD carriers has not been proven in production application environments. ▪ Cannot transmit facsimiles directly to fax machines. ▪ Emerging network end-to-end systems-integration expertise is scarce.

Design differences. CDPD is a packet technology designed to send small packets (up to 2000 bytes) for small bursts of time in between voice calls, using much of the same infrastructure as cellular networks.

CDPD is designed as a TCP/IP network. Instead of using phone numbers, it uses addresses for everyone on the network.

CDPD employs a modulation technique known as *gaussian minimum shift keying* (GMSK) to modulate the carrier in a half-duplex mode (data can be sent in only one direction at a time). It also uses a forward error correction technique known as *Reed Solomon* coding. Due to network and protocol (including the forward error correction) requirements, the raw modulation rate of 19,200 baud may be reduced to 9600 bps of actual user data. Since a CDPD channel is shared by multiple users, actual CDPD channel control by a single user may be low and effective speed lower than 9600 bps.

Coverage. Currently, CDPD is available in fewer metropolitan and rural areas than is cellular. However, this will change in the next few years. From a long-term planning point of view, it is fair to say that with minimal additional cost to cellular carriers, coverage of CDPD can be extended to any area where a cellular circuit-switched network is available.

Cost. CDPD is billed on a per-packet and per-byte basis. Circuit-switched cellular is billed on a per-minute basis. The cost of use of both technologies (CDPD and cellular) can be high or low depending upon need and how the network is used. The ultimate cost of each service to the user greatly depends upon usage patterns and the type of data sent. In general, as one approaches two-way paging usage (high message count with very little data in each message), CDPD decreases in cost. Circuit-switched cellular is more cost-efficient for larger amounts of data transferred in fewer connections, as, for example, with file transfers.

In order to show the different costs, several examples will be given. These examples do not include monthly access charges, which can range from $20 to $65 for either service. While both CDPD and circuit cellular prices vary nationwide and by pricing plan and usage, these examples are based on average 1995 nationwide prices for these services. The number of messages and message lengths have been selected to stress a point of comparison, rather than convey the suitability of a specific application on a particular network. Readers should do their own calculations with their application characteristics and current CDPD and cellular circuit-switched prices in their region.

Example 1: Trucking/Messaging application. A trucking company's vehicles each make 35 deliveries a day. At every stop, an address is sent showing the next stop. The average message length is 150 characters. Assuming 22 workdays a month, that makes a total of 770 messages per truck per month.

- The average CDPD airtime/packet usage cost per user would be $46 per month.
- The average circuit cellular airtime usage cost per user would be $270 per month.

Example 2: Sales automation application. A sales company has a number of salespeople equipped with wireless communications to check inventory, enter orders, check for e-mail, and send faxes to customers. The average salesperson sends three 20-KB faxes per day and accesses the order management system once a day to review inventory, during which 20 KB of data flows, twice a day to retrieve e-mail at an average 10 KB a time, and four times a day to enter a 1-KB order.

- The average CDPD airtime/packet usage cost per user would be $850 per month.

- The average circuit cellular airtime usage cost per user would be $92 per month.

Example 3: Insurance adjuster application. An insurance company decides to send digital photographs of accident claims directly to headquarters. An average of four 250-KB photos are sent each day.

- The average CDPD airtime/packet usage cost per user would be $8140 per month.

- The average circuit cellular airtime usage cost per user would be $128 per month.

Example 4: Field service application. Field service technicians go online six times a day to query a parts database, review parts lists, check delivery availability, and obtain address statuses. Each inquiry is 500 bytes in each direction, for a 1-KB total data exchange.

- The average CDPD airtime/packet usage cost per user would be $50 per month.

- The average circuit cellular airtime usage cost per user would be $46 per month.

Interoperability. Besides cost, interoperability is a major consideration in choosing an appropriate system. Table 8.9 illustrates the basic connectivity differences between circuit-switched and packet services. As the table illustrates, circuit-switched services allow access to any modem or fax machine. CDPD networks can communicate only with other network addresses, such as mailboxes or gateway addresses.

Type of communications. The third item to consider is the most likely form of data communications that will take place. If it is likely that only two-way paging or short messages (less than half of a page) will be sent out and received, then CDPD is the least costly and most reliable choice. It should be borne in mind, however, that CDPD is a *store-and-forward* packet data system, which means it can take several seconds for a data

TABLE 8.9 Connectivity Differences Between Circuit-Switched and Packet Services

Function	Circuit-switched cellular	CDPD
Dial phone numbers and modems	Yes	No
Call fax machines	Yes	No
Compatible with LANs and user software	Yes	Limited, but sometimes by way of a direct service or gateway
Gateway always needed	No	Yes
Talk only to other addresses	No	Yes

packet to be received by another address. Where time-sensitive applications are a priority, this may be a fact worth remembering.

Circuit-switched data is a real-time full-duplex system, meaning that a key stroke at one end of the connection is almost instantly echoed at the other end of the connection. Therefore, for high-data-content applications such as remote PC DBMS queries, file transfers, and faxes, e-mail (with attachments), LAN connections, online services, remote desktop access, BBS services, and the freedom to dial up any modem, circuit-switched cellular is the right choice. Essentially, circuit cellular data makes it possible to do remotely whatever can be done from a desktop.

Summary of CDPD versus circuit-switched cellular. We can summarize the differences between the two systems with the following list:

CDPD

Packet data system designed for short bursty messages

System is essentially a store-and-forward wireless messaging system

Cannot dial phone numbers, but can send messages to addresses

Faxes limited to text only (and only through a fax service gateway)

Cost-effective for large numbers of short messages, expensive for faxes and large files

Circuit cellular

Analogous to the standard land-line phone system and modem

Dial virtually any modem in the world and connect

Complete faxing capability to any fax machine or fax modem

Cost-effective for large content, high-information-flow applications

Expensive for short messages

8.6   Packet Radio Data Networks

Packet radio data networks are wireless networks specially designed to carry data traffic only. There are two major nationwide packet radio networks in the United States: ARDIS and RAM Mobile Data. While they have many common elements in terms of suitable applications and support by third parties, we shall discuss each of these networks separately. Differences and commonality will emerge out of this discussion.

8.6.1   ARDIS packet radio network

Network description. *ARDIS* which stands for *Advanced Radio Data Information Service,* is the first and largest wireless data network provider in the United States. ARDIS was formed in 1990 as a joint venture between IBM and Motorola. The network was designed and built by Motorola for IBM as a specialized private wireless network for IBM's customer services application. It was used primarily to dispatch service technicians, to query databases for parts availability, and to determine call statuses. In 1995, IBM sold its interest in the network to Motorola. In Canada, Bell Mobility ARDIS is co-owned by Bell Canada Enterprises (BCE) Mobile and Motorola.

Network infrastructure.   The network infrastructure consists of the following (Figs. 8.24 and 8.25):

- 1600 base stations throughout the United States connected to local switching nodes, which, in turn, are connected to a single national switching node through a high-performance, nationwide telecommunications backbone

- Information nodes, from PCs to mainframes, connected to ARDIS via radio modems, dial-up, dedicated leased lines, and a variety of value-added networks

- Variety of mobile devices including selected models of notebooks, PDAs and handheld computers from Apple, Compaq, Dell, Enloc, ETE, Hewlett, IBM, Itronix, Kenwood, Motorola, Norand, Psion, Symbol Technologies, Telxon, and Toshiba (list supplied by ARDIS; we believe, other compatible devices which will accept ARDIS network compatible modems e.g., INFOTAC or PCMCIA type II from Motorola/IBM, and appropriate software drivers will work as well)

- A national support center located in Lincolnshire, Illinois

Network coverage.   ARDIS provides national coverage with seamless roaming, which allows users to travel anywhere there is coverage and

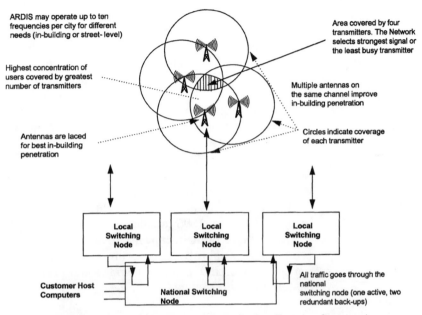

ARDIS may operate up to ten frequencies per city for different needs (in-building or street- level)

Area covered by four transmitters. The Network selects strongest signal or the least busy transmitter

Highest concentration of users covered by greatest number of transmitters

Multiple antennas on the same channel improve in-building penetration

Antennas are laced for best in-building penetration

Circles indicate coverage of each transmitter

Local Switching Node

Local Switching Node

Local Switching Node

Customer Host Computers

National Switching Node

All traffic goes through the national switching node (one active, two redundant back-ups)

Figure 8.24   ARDIS network schematic. (*Source: Bishop Training Company.*)

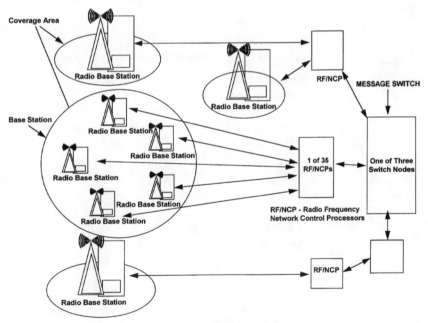

Coverage Area

Radio Base Station

Radio Base Station

RF/NCP

MESSAGE SWITCH

Base Station

Radio Base Station

Radio Base Station

Radio Base Station

Radio Base Station

Radio Base Station

1 of 35 RF/NCPs

One of Three Switch Nodes

RF/NCP - Radio Frequency Network Control Processors

Radio Base Station

RF/NCP

Figure 8.25   The ARDIS wireless data network. (*Source:* Using Wireless Communications in Business *by Andy Seybold.*)

be registered as soon as they power-up their modems. ARDIS covers the top 410 metropolitan areas (10,700 cities and towns) of the United States, Puerto Rico, and the Virgin Islands—which includes more than 90 percent of all businesses and 80 percent of the total population.

According to the vendor, ARDIS is optimized for in-building penetration and provides the industry's best in-building coverage, ensuring continued wireless connectivity inside tall buildings. In fact, OTIS elevator service representatives use the ARDIS network to communicate with their offices.

Network capacity. ARDIS currently handles 55 million messages monthly sent by more than 60,000 mobile customers. Almost 12,000 of these are IBM service engineers, with 14,000 Sears field services staff being added from 1995 to 1996.

ARDIS has a built-in capacity to add new users in certain areas. The company is constantly monitoring market requirements and increasing capacity both in proactive mode and in reactive mode to meet large customers' requirements.

Key features of ARDIS

- The greatest national coverage and largest subscriber base in the United States among the packet data networks.

- Transparent roaming.

- Deep in-building penetration of signals. This is important for many sales automation and service representative applications.

- Guaranteed delivery of messages in less than 14 seconds, with an average message (composed of one ARDIS packet of 240 characters) being delivered in less than 5 seconds. Being less hierarchical than Mobitex, ARDIS has fewer intermediate nodes in the path of data traffic for users who roam outside their local areas. This feature, along with RDLAP 19,200-bps support in more areas in the future could give ARDIS users a significant response-time advantage.

- Guaranteed network availability of 99 percent, with an actual up time of 99.9 percent, and a target of 6-sigma up time. The ARDIS network has been designed with built-in redundancies of most of its components, including the base stations, switching nodes, and wide area links that form the backbone.

Application suitability. Like Ram Mobile Data and CDPD (once it is fully deployed and can guarantee performance in terms of response time), ARDIS is best suited for short-transaction-based mobile-aware applications. OLTP applications with small message lengths are ideally suited. Because it was originally conceived as a dispatch application for IBM, the ARDIS network is well suited to customer service (sales and maintenance) applications in small and large high-rise buildings. Apart from vertical applications, ARDIS has started marketing horizontal applications bundled with Motorola-supplied hardware, such as the Motorola Marco and Envoy PDAs with built-in transceivers and type II PCMCIA cards.

ARDIS also offers an e-mail service called the ARDIS personal messaging (APM) service, and RadioMail, along with PDA hardware. RadioMail offers such business services as global e-mail, faxing, immediate stock market information, news, and support information.

Many custom applications have been and are being written for the ARDIS network. Application architects and designers should avoid using ARDIS for long file transfers and session-oriented applications.

Large custom application examples.   IBM and Sears are the two largest users of the ARDIS network.

IBM customer engineer dispatching.   IBM was the first and continues to be the largest user of the ARDIS network. IBM implemented a large application for dispatching engineers to IBM customer sites to solve hardware and software problems. As many as 7000 customer engineers are equipped with mobile devices that are connected through the ARDIS network to an IBM mainframe where a parts database and service record statuses are kept. Initially, the engineers used a device called a brick that was equipped with a Motorola INFOTAC modem. In 1995, IBM started a pilot project with IBM 750 and 755 ThinkPads equipped with PCMCIA wireless modems (Fig. 8.26).

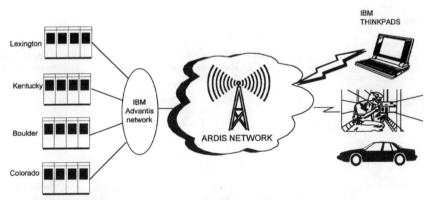

Figure 8.26 IBM's use of ARDIS network for its customer service network. (*Source: Schematic drawn by author from information in* www.raleigh.ibm.com.)

The business case for IBM was simple, it appears. Often, the contractual requirements for the service of mission-critical computer infrastructures are such that full and immediate contact with a customer engineer is the only solution. Such a complete and fast response can be achieved only through a wireless network. A typical scenario would have a customer engineer retrieving the service-call details, checking the equipment's history, checking for parts availability, and putting the parts on hold before even arriving at the actual scene. Once the engineer arrives at the site, the shipping of necessary parts is confirmed. The customer engineer has all the service manuals

TABLE 8.10   ARDIS Services

ARDIS personal messaging services
ARDIS personal messaging service is a wireless two-way data messaging service

Two-way messaging	The ARDIS personal messaging two-way service provides real-time wireless messaging among ARDIS personal messaging subscribers. Messages reach subscribers even if their devices are turned off. The network stores messages for 72 hours and send them to subscribers as soon as the devices are activated. If a subscriber wants to know exactly when a sent message was received, a wireless acknowledgment can be sent within seconds of a message being viewed.
Paging	Subscribers can send and receive pages from any telephone. Both touch-tone numeric pages and operator-assisted alphanumeric pages can be sent.
Faxing	Wireless fax messages can be sent to any fax machine in the country from a mobile device.

RadioMail
RadioMail is a wireless gateway to the Internet, as well as to a host of information and business services accessible only via wire-line connections

Two-way messaging	Messaging via RadioMail is interactive and virtually instantaneous. Subscribers can communicate with other RadioMail subscribers or anyone with an Internet e-mail address, including users on Prodigy, CompuServe, America On Line, corporate e-mail systems, and other online services.
Information services	Bulletin services like the latest news, sports scores, and business news are also accessible via RadioMail.
Call-in service	The RadioMail call-in service allows anyone to reach a subscriber's wireless device by simply phoning RadioMail, providing the device ID number, and dictating the message to be sent.
Directory service	RadioMail offers subscribers a RadioMail address directory. If a subscriber needs to know a RadioMail address, an inquiry message can be sent to this service. A message will be sent back with the requested ID. If desired, a subscriber can be unlisted, as with telephone directory services.
Faxing	Messages can be sent to any fax machine in the world from RadioMail-registered wireless devices.
Message charges	Message unit charges on RadioMail are calculated on the number of message units sent and received by each registered device, including fax messages sent and call-in service messages received.

installed in a laptop, which obviates the need to locate and search large, heavy books. IBM has earned a speedy payback on its investment in this technology—fast, efficient, and high-quality service are de rigueur in today's computer services business. IBM uses this wireless infrastructure to great competitive advantage in its services business.

Sears service representative dispatching. See the description of this application in Chap. 2.

ARDIS messaging services. ARDIS offers a personal messaging service and RadioMail business services through its network (Table 8.10).

TABLE 8.11   Communications Interfaces Supported by Different Third-Party Products

Platform	Server/ Client	Vendor name	Product name	OSI model layer	ARDIS connectivity
OS/2	Server	Nettech	RFgate	Network	Synch./X.25 PTP Asynch. to X.25PTP
OS/2	Server	Nettech	RFLink	Transport	Synch./X.25 PTP Asynch. to X.25 PTP RF Host/MG
DOS, Windows*	Client	Nettech	RFMlib	Network	Native Mode Version 1.2
DOS, Windows*	Client	Nettech	RFLink	Transport	Native Mode Version 1.2
OS/2	Server	Motorola— MSP	AirMobile Server SDK Suite	Network Transport	Synch./X.25 PTP Asynch. to X.25 PTP RF Host/MG
DOS, Windows*	Client	Motorola— MSP	AirMobile SDK Suite for DOS/Windows	Network Transport	Native Mode Version 1.2
Novell Netware	Server	Aironet	FieldNet Gateway NLM	Transport Network	Synch./X.25 PTP
DOS, Windows Version 1.2	Client	Aironet	FieldNet Client ODI Network	Transport Version 1.2	Native Mode
SCO UNIX	Server	Complex Architectures	Enterprise Messaging Services	Transport Network	Synch./X.25 PTP Asynch. to X.25 PTP RF Host/MG
DOS, Windows, Macintosh System 7	Client	Complex Architectures	EMS	Transport Network	Native Mode Version 1.2

SOURCE: ARDIS software developer's binder.

TABLE 8.12   Prices for ARDIS Services

Packet-based services	
Regular message service	
Data is sent over the ARDIS network in 240-character blocks. Each block is called a message unit (physical message). Sending physical messages via ARDIS incurs two charges: (1) a basic physical message charge; (2) a per-character charge	
Peak	$0.06 per packet
Off-peak	$0.04 per packet
Charge for characters sent	$0.03 per 100 bytes
Short message service	
Per physical message comprising eight characters or less, transmitted or received by a radio terminal device, regardless of time. (Characters-sent charges do not apply.)$0.03 per message	
Other miscellaneous charges	
Registration fee	$65.00 per user
Monthly minimum air time charge	$20.00
Monthly port rental	$495.00

Personal messaging services plan		
Bronze Pack: 100 message units;	ARDIS Personal Messaging (APM)	$39.00
36 cents for each additional unit	RadioMail	$39.00
*(Motorola Envoy and Marco only)	Both APM and RadioMail (Executive Pack)	$49.00
Silver Pack: 250 message units;	ARDIS Personal Messaging (APM)	$69.00
28 cents foreach additional unit	RadioMail	$59.00
*(For Motorola Envoy and Marco only)	Both APM and RadioMail (Executive Pack)	$79.00
Gold Pack: 425 message units;	ARDIS Personal Messaging (APM)	$99.00
23 cents for each additional unit	RadioMail	$89.00
*(For Motorola Envoy and Marco only)	Both APM and RadioMail (Executive Pack)	$109.00
Platinum Pack: 650 message units;	ARDIS Personal Messaging (APM)	$139.00
21 cents for each additional unit	RadioMail	$129.00
*(For Motorola Envoy and Marco only)	Both APM and RadioMail (Executive Pack)	$139.00

SOURCE: ARDIS price sheets (1995).

Interoperability and communications interfaces supported. ARDIS supports X.25, asynchronous, bisynchronous, IBM SNA directly. TCP/IP, IPX/SPX protocols are supported through third-party gateways.

The ARDIS network can interoperate with Rockwell (American Mobile Satellite), paging networks (through RadioMail gateways). Interoperability with RAM Mobile Data and CDPD has been tested in a laboratory environment, but service is not yet offered commercially.

Table 8.11 details the communications connectivity support for the ARDIS network of several third parties.

Costs for ARDIS network services. Costs for wireless data networks have been steadily decreasing for several years. This trend will continue, with an estimated 25 to 30 percent reduction taking place every

TABLE 8.13 ARDIS Network Summary

Brief description	▪ National terrestrial, trunked packet data radio network for data applications only; currently there is no voice.
Components	▪ Multiple transmitters in base stations connected to a local switching nodes. Local switching nodes connected to one national switching node. Mobile devices connected to ARDIS via supported modems. Typically, information servers at host computers are connected to the national switching node.
Frequency bands	▪ ARDIS operates in the 806 MHz to 821 MHz range for uplinks and in the 851 MHz to 866 MHz range for downlinks. 25 KHz channels are used.
Coverage	▪ National; top 410+ MSAs with 10,700 cities: 90% of U.S. urban businesses and 80% of the total population are covered. ▪ ARDIS has subsidiaries or affiliated networks in the United Kingdom, Canada, Germany, Australia, Malaysia, Singapore, and Thailand.
Capacity and speed	▪ ARDIS operates one to ten 25-KHz channels per coverage area. ▪ Currently (1995–1996) supporting 44,000 users nationwide—more than RAM. ▪ Total capacity is far from full in some areas. ▪ Gradually increasing capacity and coverage as demand increases. ▪ Currently supports 4800 bps (MDC 4800) in most areas; RDLAP at 19,200 bps is available in 33 areas in the United States and some parts of Canada. RDLAP service is being expanded, based on customer demand and business economics.
Protocols supported	▪ X.25, asynchronous, bisynchronous natively: TCP/IP and IBM's SNA (LU 2 and LU 6.2) supported through third-party gateways.
Most suitable applications	▪ Short OLTP transaction-based and messaging, such as credit authorization, sales automation, public safety, transportation truck tracking, e-mail, etc. ▪ Long file transfer is NOT suitable for packet network.
Cost	▪ $39 per month for Bronze Pack (100 messages) to $139 per month for Platinum Pack (650 messages) for ARDIS personal messaging service. ▪ $0.06 per packet plus $0.03 per 100 bytes for packet-based nonmessaging applications with no e-mail.
Availability	▪ Available throughout urban United States, Canada, United Kingdom, and selected countries in Asia.
Security	▪ Higher than cellular but some applications also need end-to-end encryption.
Pros	▪ Good nationwide coverage with seamless roaming within metropolitan areas ▪ Deep building penetration ▪ RDLAP at 19,200 bps gives good response times for OLTP-type applications ▪ Network is less hierarchical than RAM Mobile's Mobitex. This may leads to fewer hops and better response times for some roaming users ▪ Increasing support by hardware, systems, and application software vendors
Cons	▪ Limited throughput—not suitable for file transfer ▪ Not suitable for session-oriented applications relying on continuous synchronous connections ▪ Limited coverage in rural areas ▪ Limited support by remote access hardware vendors such as Shiva ▪ Cannot transmit facsimiles directly to fax machines ▪ End-to-end systems-integration expertise is scarce

year for the foreseeable future. The actual cost of using the ARDIS network depends on the services used. There is per-packet pricing for online transaction-based applications and bundled monthly charges for various categories of messaging services. The following costs based on 1995 prices are provided as rough guidelines to help with preliminary cost estimates (Table 8.12). The reader should contact the vendor for the current prices.

Summary of the ARDIS network. Table 8.13 presents a summary of the features of the ARDIS network.

8.6.2 RAM Mobitex packet radio network

Network description. Mobitex is a mobile packet radio data network that provides remote access to data and two-way messaging for mobile computing applications. It is the underlying network for the nationwide RAM Mobile Data network in the United States, United Kingdom, and Australia. The technology on which Mobitex networks are based is similar to that used in ARDIS and cellular telephone systems. In 1995, there were public Mobitex networks in operation in fourteen countries on four continents. A recent new Mobitex network in Singapore is the first in Asia. Many more new networks in Europe, Asia, and South America can be expected in 1996. (See Sec. 9.8 on Mobitex's international coverage.)

All Mobitex networks use the same protocols and operate under the same specifications. Mobitex is a worldwide standard, introduced and controlled by Ericsson of Sweden and administered by the Mobitex Operators Association (MOA). RAM Mobile Data in the United States and Roger Cantel in Canada are two such Mobitex operators. Under the MOA's leadership, European Mobitex operators are now introducing *international roaming,* a service that will allow Mobitex users to continue to send and receive messages and access data outside their home countries. The Mobitex specification administered by the MOA also ensures that all operators follow the same standard.

The elements of Mobitex: base stations and cells. Like mobile telephone systems, the Mobitex network is based on radio cells, with radio links replacing the wiring that connect telephones or data terminals to networks. Intelligent base stations provide the links by allocating channels to active terminals within limited geographic areas. The size of a radio cell is determined by such factors as the number of channels available, the maximum number of units that must be served, and the output (transmission) power of both the terminals and the base station.

Base stations serve several essential functions in a mobile communications system. First and foremost, they function as connection points so that calls or messages can be switched from one base station or local exchange in the telecommunications network to another.

A second function of base stations is the ability—in conjunction with mobile user devices—to hand off calls as users change locations. This is the function that makes roaming possible. This concept is similar to circuit-switched cellular networks as described in Sec. 8.5.

Acting as network nodes is a third function of base stations. By fulfilling this role, base stations enable the forwarding of traffic statistics to the network NCC for subscriber billing. This function also makes possible the downloading of control data and software to base stations for the purpose of altering operating characteristics.

Packet switching for efficiency. Virtually all Mobitex's components employ *digital technology,* which means that information is transferred across the network as data, not analog voice signals. The critical system parameter for data communications is the method of transmission employed. Two transmission techniques are available: circuit switching and packet switching.

In *circuit-switched* networks, such as those used for mobile telephone systems, a physical connection must be maintained between the sending and receiving nodes for the duration of a call, connection, or session. Setting up the connection also takes some time, which is a disadvantage when the user needs to transfer only a small amount of data.

In *packet-switched* networks, data is divided into small packets that can be transmitted individually as traffic permits. Setup time is eliminated and network connections are instantaneous. In action, packet switching makes more efficient use of channel capacity, typically allowing 10 to 50 times more subscribers over a radio channel, compared with circuit switching. This point is very important, because the radio spectrum is a scarce resource. Mobitex is a packet-switching network.

Mobitex network architecture. The Mobitex network is based on the OSI network model of the ISO standard. It provides the bottom three layers of the OSI model: physical, data link, and network. It supports most common transport protocols such as X.25 and TCP/IP through third-party interfaces that use Mobitex's MTP/1 network protocol. Table 8.14 illustrates the correspondence between OSI and Mobitex.

A Mobitex network is organized as a hierarchy, with nodes at three levels: base stations (BRS), area exchanges (MOX), and main exchanges (MHX). At the top level there is the NCC (Fig. 8.27).

TABLE 8.14   OSI and Mobitex Correspondence

OSI layer	Mobitex protocol	Protocol function
Not restricted to any specific layer	MCP/1 (Mobitex compression protocol)	Data compression for faster transport in fewer packets
Transport layer	MTP/1 (Mobitex transport protocol)	Guaranteed packet delivery over Mobitex
Network layer	MPAK (Mobitex packet protocol)	Data transmitted over the network in a series of Mobitex Packets (MPAKS)
Data-link layer	MASC (Mobitex asynchronous protocol)	Machine interface that allows packets to be transferred over the Mobitex data link
Physical layer	ROSI (Radio Open System Interface)	Used by radio units to communicate with base stations

As we have seen, coverage is provided by overlapping radio cells, each served by an intelligent base station. Mobitex users communicate with the closest base station. Because intelligence is distributed throughout the network, data packets need only be forwarded to the lowest network node common to the sender and receiver. The base station can handle all local traffic between mobile terminals. Only billing information is passed to the higher levels.

The area and main exchanges handle switching and routing in the Mobitex network and provide connection points for fixed terminals. A number of interfaces (including several ISO and IBM standards) are supported, but X.24—the ISO and CCITT international standard for public packet-switched data networks—has become the most widely

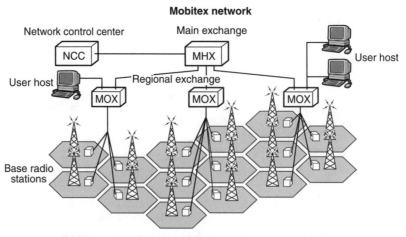

Figure 8.27 RAM's mobile data network from Ericssson. (*Courtesy of Ericsson's Mobile Data News.*)

used standard for Mobitex, though TCP/IP support is also becoming common. Subscriber information for billing of services is also processed at this level before being forwarded to the NCC.

Mobitex connectivity hardware and software. Ericsson and its business partners have developed a number of connectivity products (see Chap. 13) based on several Mobitex proprietary and some industry-standard interfaces available on all major network platforms. These products include radio modems that allow mobile users to access the network, software drivers, and MCSS (switches/gateways) that provide links to other data networks and computer systems.

The Mobidem line of radio modems supports several data protocols, including an extended AT command set. This makes them compatible with existing communications software that runs on standard PCs. Mobidem models include the M1000 series, which are portable, battery-operated units; the M2000 series, which now includes a radio modem on a PCMCIA card; and the new M4060, intended for vehicle mounting or fixed installations.

On host systems (commonly called *the fixed side*) many earlier Mobitex applications used the X.25 protocol, the international standard for packet-switched data networks. Now TCP/IP support is becoming more common. Gateways are available, however, for most major network standards, including IBM SNA (Systems Network Architecture).

These connectivity solutions make Mobitex applications adaptable to different customer configurations so that any standard PC or handheld computer with Mobitex communications software drivers can be used to access the Mobitex network. Mobile communication server switches/gateways installed at the area-exchange level or on a customer's premises allow mobile users to access mainframe-hosted databases or global messaging systems such as the Internet.

Key features of Mobitex. Transparent roaming throughout the country or even internationally in Europe, store-and-forward capability, fault tolerance, shortest path, interoperability, and security are the key features of Mobitex, according to Ericsson and Ram Mobile Data. We shall describe these features briefly.

Transparent roaming. The RAM Mobile Data network provides nationwide, seamless, transparent roaming within the United States. It is *transparent* because no manual intervention is required by a subscriber or mobile user once a radio modem has automatically registered with the network. It is *seamless* because the networks know which base station a mobile is linked to at any given moment. Frequency-agile RAM-compatible radio modems are able to operate on all 200-RAM-system frequencies, roaming from one to another automatically, switching to the best currently available channel as needed, and thus allowing RAM mobile users to stay connected as they move around within RAM's coverage area. International roaming is also now available in certain markets.

Store and forward. RAM offers store-and-forward capabilities within its networks to ensure that messages are delivered. If a mobile user turns the modem off or drives through a tunnel, crosses a steel bridge, or leaves the RAM mobile data coverage area, the message is retained within the NCC's store-and-forward facility. The RAM network will try to retransmit the message automatically, provided other Mobitex OSI layers or third-party-developed software have been programmed to handle the packet in transit and have not timed out.

Reliability and shortest path. According to RAM Mobile Data, its network engineering design features 99.99 percent reliability, ensuring transmissions with a bit-error rate equivalent to that of wired-line modems. Numerous fault-tolerant components are employed in the system, including link-level data checking between adjacent network nodes and forward-error correction. Base stations and other equipment have uninterruptible power supply (UPS) and/or generator backups. Messages are routed through the shortest path when they traverse multiple MOX nodes.

All facilities and communications links are remotely monitored by network management staff located at RAM's NCC in Woodbridge, New Jersey.

Interoperability and interfaces supported. Ericsson provides lower-level APIs and other technical specifications to third parties who have developed various higher-level application interfaces to multiple network transport protocols and information servers. These software interfaces are discussed in Chap. 11. Chapter 13 lists several third-party products that support the RAM network. Briefly, the following hardware and software protocols are supported:

- *Modems:* Ericsson, GE, and Motorola radio modems used with RAM's service.
- *Protocols:* Mobitex supports SNA 3270, LU2 and LU3, X.25, TCP/IP, and asynchronous protocol, as well as the MTP/1 transport protocol.

Security. Unlike the vulnerability of cellular voice communications, it is generally more difficult to tap into and decipher wireless data networks such as the RAM Mobile Data network with its frequency-agile modems moving from one channel to another. However, no network is considered absolutely secure. Users should investigate, therefore, the appropriateness of end-to-end encryption for their applications. Certainly it is required for public safety and banking transactions.

Mobitex coverage worldwide. In the United States, the RAM Mobile Data network (owned by BellSouth) operates multiuser two-way shared public Mobitex networks in the top 266 MSAs—covering over 7700 cities, as well as airports and major transportation corridors. This covers 92 percent of the U.S. urban business population, according to RAM Mobile Data. RAM's networks have 1058 base stations—along with local switches and network control centers—that send and receive data messages between portable and fixed mobile devices in the field, host computers, servers, and information services. RAM has up to 30 channels available in each major MSA.

There were 16 public and 4 private (for a total of 20) Mobitex networks installed in 15 countries and 5 continents at the end of 1995.

RAM Mobile's capacity. According to the vendor, RAM Mobile Data's networks in the United States can handle more than one million subscribers (mobile users). This appears to us to be a theoretical capacity for certain components, and we were unable to verify the precise method used in making this estimate. In 1995, RAM had approximately 20,000 users. As demand for its services has increased, RAM has, over the past few years, increased its capacity by adding more channels at each base station, as well as by subdividing cells (a technique similar to the one used in cellular network design). In 1995, the network also increased the number of base stations by 150 to provide additional geographic coverage and increased in-building penetration. The organization's marketing and network-capacity design goal is to offer no waiting for channel availability.

RAM's market positioning. RAM Mobile Data is focusing its marketing approach on a few horizontal groups and on three vertical industry groups: financial, manufacturing, and distribution/transportation. Functional applications within these industries include executive management, financial management, and field (sales and service) management.

In order to provide complete solutions, RAM has entered into business partnerships with many hardware vendors, systems integrators, and application developers who are developing new applications or customize existing applications for RAM Mobitex networks. Some examples are:

Horizontal

- E-mail: Lotus cc:MAIL, LOTUS NOTES, Microsoft Mail, Novell's Group Wise, and other mail packages that can be accessed via the RAM Mobile network

- Field sales automation: packages such as SSI SalesLink, Thinque
- Field service automation: packages such as ASTEA, RF Dispatch, SCA

Vertical Applications

- Brokerage: Dow Jones Telerate, Telescan, Davidge, Mobile Trader
- Transportation/Dispatch: Ascar, AII, RoadShow

Application suitability.   Most suitable applications for Mobitex networks are those that send short transaction-based messages. The dispatching of service representatives, POS credit authorizations, sales orders, customer inquiries, inventory queries, and public safety are typical.

Major customer implementations of the RAM Mobile Data network.   Four interesting high-profile customer implementations of the RAM Mobile Data network are described as follows:

Food Fair supermarket in Winston-Salem, North Carolina, now offers to its customers the convenience of using credit cards for the purchase of groceries. Food Fair uses RAM's wireless service to reach MasterCard's automated point-of-sale program (MAPP) processing system for credit card authorization. By sidestepping the installation of dedicated telephone lines for automated authorization terminals, Food Fair was able to offer the new service without any disruption to its business. Market research shows that customers prefer stores that accept credit cards and that they spend about 30 percent more per visit than they would if they had to pay cash.

GE Customer Service dispatchers use the RAM Mobile Data network to send work assignments and consumer histories to service personnel on the road. The service personnel, for their part, query order status, pricing, and parts information while on site with customers. According to the company (as reported in RAM's press release), GE has increased its service calls by an average of two per day per service representative, which means that 257,502 more calls are being made per year with no increase in the number of service representatives. In addition, use of the network has increased overall responsiveness to customers and has reduced order processing times.

British Airways sought to speed up passenger check-ins for its international flights, obtain better control over baggage/cargo handling, and decrease aircraft service time. By using RAM Mobile Data network, BA is reducing passenger lineups at check-in counters and customs desks by bringing wirelessly connected portable terminals with the same passenger, flight, and fare information as the traditional countertop reservation computers directly to the passengers waiting in line.

The airline uses RAM's services to monitor maintenance and repair operations on the ground and to track all luggage and expedite the loading of food, fuel, and cargo. The availability of the Mobitex network in many European countries eased the international implementation of this application for British Airways.

Conrail manages and tracks its fleet of 2000 locomotives and 65,000 freight cars by sending and amending work orders via the RAM network while trains and work crews are en route. Communications that used to take up to 48 hours are now completed in a few minutes.

Costs. As noted previously, costs for wireless data networks have decreased dramatically and will continue to do so. The following figures are provided for RAM pricing as a rough guideline for a preliminary cost estimate. Contact the vendor for the most current information.

Entry level (up to 100 KB) (35 cents added per additional KB)	$25/month
Mobile user (up to 200 KB) (33 cents added per additional KB)	$66/month
Mobile professional (up to 275 KB) (31 cents added per additional KB)	$85/month
Power user (up to 500 KB) (27 cents added per additional KB)	$135/month

Summary of RAM Mobile Data's Mobitex network.   Table 8.15 gives a summary of features of the RAM Network.

8.7   Enhanced Specialized Mobile Radio (ESMR) Networks

Several ESMR networks are being installed for specialized services. We shall decribe Nextel and Geotek here.

8.7.1   Nextel network description

The specialized mobile radio networks described in Sec. 8.3 employ older technology and consequently do not use the frequency band as efficiently as they might. Though numerous, SMRs are very small in capacity, as measured by the number of channels and mobile users supported. SMRs also concentrate on voice communications primarily. Enhanced specialized mobile radio networks are essentially SMR net-

TABLE 8.15   RAM Mobile Data's Mobitex network summary

Brief description	■ National terrestrial, trunked packet data radio network for data applications only ■ Currently no voice, but that might change
Components	■ Hierarchical network base stations, message exchanges (MOXs), regional exchanges, and a network control center
Frequency bands	■ RAM Mobile Data operates in the 896 to 901 MHz range and the 935 to 940 MHz range. Each channel width is 12.5 KHz
Coverage	■ National: top 266 MSAs; 7700 cities; 92% of U.S. urban business population ■ Mobitex networks available in 15 countries in 5 continents
Capacity and speed	■ Current (1995–1996) capacity is estimated at over one million mobile users ■ Currently supports 8000 bps; possible to increase this to 16,000 bps with newer base station hardware and modems
Protocols supported	■ X.25, asynchronous, TCP/IP and IBM's SNA (LU 2 and LU 3) through third-party gateways
Most suitable applications	■ Short OLTP transaction-based messaging, such as credit authorizations, sales automation, public safety, transportation truck tracking, e-mail, etc.
Costs	■ $25/month for entry level user (100 KB) to $135/month for power user (500 KB) ■ Approximately 30 cents per KB
Availability	■ Available now in most urban areas in the United States and other countries
Security	■ Higher than cellular but some applications may require end-to-end encryption
Pros	■ Good nationwide coverage within metropolitan areas ■ Capacity for new users ■ Performance acceptable for OLTP short-message-length transactions ■ Increasing support by major hardware and software vendors, including application developers ■ Store-and-forward capability is good if integrated with network transport software
Cons	■ Limited throughput—not suitable for file transfer ■ Limited coverage in rural areas ■ No voice capability ■ Cannot transmit facsimiles directly to a fax machine ■ Older modems are bulky ■ End-to-end systems integration expertise is scarce

works that operate in the same frequency band, but with enhanced capabilities that address the shortcomings of SMRs. ESMRs differ from SMRs in the following ways:

■ ESMRs use more advanced digital technology.

■ ESMRs use cellular techniques such as channel reuse.

■ ESMRs are regional and even national with wider geographical coverage.

Enhanced SMRs are only now emerging in the United States as a result of a policy, recently adopted by the FCC, which encourages consolidation and the establishment of regional and even national SMRs which can use the scarce spectra in the 800 and 900 MHz bands more efficiently. (Most ESMRs use the new SMR spectrum allocation in the 900-MHz band.) The FCC calculates that more mobile users can be supported by larger networks with a greater number of channels, a calculation based on the mathematics of queuing theory familiar to telecommunications engineers. In simple terms, more than twice as many users can be supported on a single 10-channel system than on two 5-channel systems.

As a result of the FCC's new policy, a start-up company named Fleet-Call has aggregated the assets of several fleet dispatchers (taxi companies, construction companies, and others in similar businesses) and plans to implement a completely new digital wireless network called Nextel. FleetCall also has an equity share in Clearnet in Ontario, which is setting up a similar network in Canada. Nextel and Clearnet combined could provide coast-to-coast continental ESMR service in the next few years (after 1996). The planned Nextel network includes the frequencies owned by OneComm and DialPage. See Table 8.16 for a summary of features of the Nextel network.

Application suitability.   The Nextel ESMR network will offer the following services:

- Dispatching with two-way radio
- Two-way alphanumeric paging with acknowledgment and other data applications
- Cellular telephone capabilities
- Session or packet-data transmission without a separate modem

In a sense, ESMR is a combination of cellular, paging, packet radio, and SMR in one network. It follows that ESMR's market emphasis is on combining services for customers who use these features. Initially, FleetCall is starting off with analog (eventually to be converted to digital) dispatching networks.

Network equipment and schematic

All-in-one mobile radio equipment.   Nextel will use its own brand-name version of a Motorola-designed mobile radio similar to Motorola's Lingo-brand radio. The proposed all-in-one radio will come in several models. In addition to its basic walkie-talkie capability, the following additional configurations will be available:

- Fully operational cellular phone
- Limited-access phone with preset numbers
- Dashboard-mounted radio with handheld microphone

TABLE 8.16 Enhanced Specialized Mobile Radio (ESMR): Nextel Summary

Brief description	■ ESMR service is an enhancement of SMR service ■ Combines dispatching, paging, and cellar voice services into one network ■ Currently based on analog technologies, but will eventually become a digital network
Components	■ Components similar to SMR but based on new digital switches ■ Engineering work in progress ■ Exact makeup of components still to be known
Frequency bands	■ Mostly 900 MHz
Coverage	■ Currently limited to a few sites (21 in 1995) but the intention is to have national coverage. Expert estimates are that full coverage is at least 3–4 years away (1999)
Capacity and speed	■ Since the network is not fully operational, this information is only partially available ■ Circuit-switched data speed is expected to be 7200 bps with a 4800-bps effective rate after error correction ■ Packet data speed is expected to be greater than 19,200 bps
Protocols supported	■ PI, frame relay, dial-in remote and X.25
Most suitable applications	■ Same applications as SMR—primarily dispatching ■ Paging ■ Cellular-like mobile telephone service ■ Short-messaging applications
Costs	■ Mobile equipment costs $750–$1000 per unit in the United States ■ Activation fee varies according to the service plan selected ■ Approximately $50 per month per user for dispatch, paging, and phone ■ Cellular phone service at 15–45 cents per minute ■ Paging: $5 per month for numeric; $10 per month for alphanumeric
Availability	■ Only conventional (SMR equivalent) services currently available, and only in selected cities ■ Full coast-to-coast digital services, including data, are 3–5 years away
Security	■ Information not available at time of writing. Our assessment is that it will be similar to SMR. We suggest higher levels of security, including encryption, be added on
Pros	■ The proposed full functioning all-digital network will provide high-quality, reliable service ■ Two-way radio dispatching, paging, and telephone in a single unit ■ A short message service (SMS) with acknowledgment by the mobile unit will be an entirely new service ■ A modem is not required for data communication: functionality is in the switch ■ A single company (Nextel) will manage a coast-to-coast network. In conjunction with Clearnet in Canada, the potential exists for the provision of a full-continental service ■ Both packet-switched and circuit-switched technologies will be available
Cons	■ Only conventional SMR services on an analog network are available today. The digital network currently exists only on paper. Full coast-to-coast deployment is still 3–5 years away ■ Many customers find all-in-one phones to be unduly cumbersome

Network schematic. Figure 8.28 shows a schematic of the proposed Nextel ESMR network infrastructure that will be accessible by any mobile user with a handheld or vehicle-mounted communicating device. The network switch will direct the call to the appropriate host.

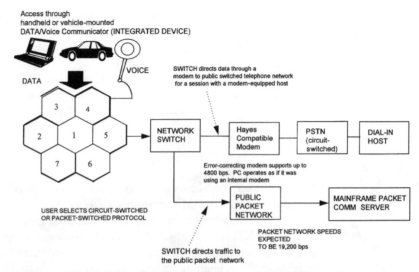

Access through
handheld or vehicle-mounted
DATA/Voice Communicator (INTEGRATED DEVICE)

VOICE

DATA

SWITCH directs data through a
modem to public switched telephone network
for a session with a modem-equipped host

NETWORK
SWITCH

Hayes
Compatible
Modem

PSTN
(circuit-
switched)

DIAL-IN
HOST

Error-correcting modem supports up to
4800 bps. PC operates as if it was
using an internal modem

USER SELECTS CIRCUIT-SWITCHED
OR PACKET-SWITCHED PROTOCOL

PUBLIC
PACKET
NETWORK

MAINFRAME PACKET
COMM SERVER

PACKET NETWORK SPEEDS
EXPECTED
TO BE 19,200 bps

SWITCH directs traffic to
the public packet network

Figure 8.28 Planned Nextel data network. (*Source: Bishop Training Company.*)

Communications software interfaces supported.   Other than TCP/IP, it is
not clear what other transport software Nextel will support. It should
be noted that the data capability has only very recently become available. Since most of the data plans are still on the drawing board, we
should wait and see what will actually be delivered.

Network coverage.   By the end of 1995, Nextel/OneComm was partially
deployed in 21 areas. More areas will be covered in 1996 and 1997. The
following cities/states/areas in the United States were included in the
combined network in 1995:

Athens, Georgia

Atlanta, Georgia

Birmingham, Georgia

California

Charlotte, North Carolina

Chicago, Illinois/Milwaukee,
Wisconsin

Denver/Colorado Springs,
Colorado

Greensborough/Winston-Salem,
North Carolina

Greenville/Spartanburg, North
Carolina

Kansas City, Missouri

New York, New York/Newark, New
Jersey/Hartford, Connecticut

Oklahoma City, Oklahoma

Portland, Oregon

Raleigh/Durham, North Carolina

Seattle, Washington

St. Louis, Missouri

Tulsa, Oklahoma

Wichita, Kansas

Detroit, Michigan/Toledo, Ohio

Philadelphia, Pennsylvania

Washington, D.C./Baltimore,
Maryland

Cost considerations.   Nextel has announced several different pricing plans. Service costs are expected to be approximately $50 per month per user for dispatch, paging, and phone services. Costs for data-based services are not currently available.

Cellular telephone service (connection to the PSTN) is expected to be around 15 to 45 cents per minute. Paging charges are expected to be $5 per month for a numeric service and $10 per month for an alphanumeric service.

8.7.2   Geotek

Geotek, in partnership with Mitsubishi, is also building a digital wireless network using SMR channels. The Geotek network uses a proprietary technology based on a frequency-hopping multiple-access principle in the 900-MHz band (Fig. 8.29). This technology uses the channel bandwidth very efficiently, with as many as 30 simultaneous users per channel. Geotek has been allocated up to 10 channels in different markets.

Geotek plans to offer services similar to those proposed by Nextel:

- Paging and messaging
- Circuit-switched and packet-switched data
- Private point-to-point communications
- Cellular telephone services

The company has already completed trials in Philadelphia. High-power transmitters capable of covering larger geographical areas are

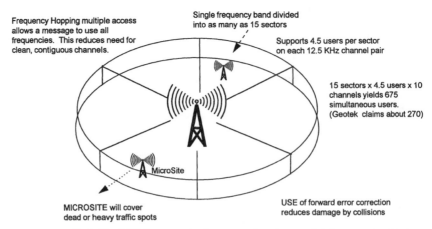

Frequency Hopping multiple access allows a message to use all frequencies. This reduces need for clean, contiguous channels.

Single frequency band divided into as many as 15 sectors

Supports 4.5 users per sector on each 12.5 KHz channel pair

15 sectors x 4.5 users x 10 channels yields 675 simultaneous users. (Geotek claims about 270)

MicroSite

MICROSITE will cover dead or heavy traffic spots

USE of forward error correction reduces damage by collisions

Figure 8.29   Geotek's ESMR network (frequency-hopping multiple access). (*Source: Bishop Training Company.*)

being used so as to lower infrastructure costs, which in turn will lead to lower service charges.

8.8   PCS/PCN

Personal communication services (PCS), or *personal communication network* (PCN), is a brand-new telecommunications technology that promises to completely revolutionize the way we use telephones and wireless communications. The fundamental concept behind PCS is the assignment of a personal, unique, cradle-to-grave worldwide communications number: the *numeric assignment module* (NAM). This number is not tied to a specific wire-pair or jack in a house or office (a fundamental limitation of the wired telephone world) and therefore does not have attached to it any of the limitations of current analog technologies, where roaming across cellular companies is far from transparent. Instead it is encoded in a microchip installed in a personal communicator—a personal communicator that, in addition to being a mobile telephone with worldwide roaming capabilities based on a second number, a *mobile assignment number* (MAN), also receives e-mail, faxes, and multimedia communications. Supporting this personal device is an intelligent digital communications network that keeps track of where you are and forwards any voice, data, fax, or multimedia messages to you by interfacing with both legacy telephone switching networks and the new digital cellular networks.

Such a system will truly make *anytime, anywhere* communication ubiquitous. It is the vision of PCS—and the ultimate hope of thousands of companies, entrepreneurs, and technology enthusiasts.

8.8.1   Theory behind the PCS vision

Let us attempt to understand how PCS will work. First of all, we have a device—a device that for want of a better title we call the *personal communicator*. This personal communicator, when placed in a charger or smart PCS box connected to the traditional wired telephone infrastructure (which will continue to exist in all its glory), acts as a cordless telephone. Now, however, the traditional infrastructure is supplemented by a *wireless local loop* that allows the personal communicator to become a cellular telephone when not in a zone of cordless phone operation (Fig. 8.30). With this new hybrid wire/wireless infrastructure, the personal communicator assumes the dual functionality of both a wire-based cordless phone and a network-based cellular phone. When a conventional call comes in, it travels to the personal communicator cradled in the intelligent PCS box, If the user picks up the personal communicator, it acts like a cordless phone, until the user

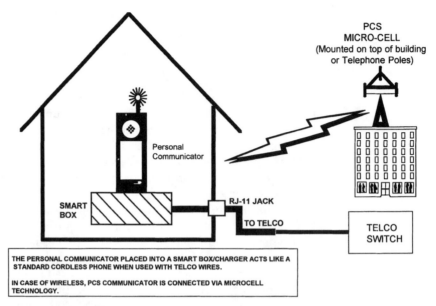

THE PERSONAL COMMUNICATOR PLACED INTO A SMART BOX/CHARGER ACTS LIKE A
STANDARD CORDLESS PHONE WHEN USED WITH TELCO WIRES.

IN CASE OF WIRELESS, PCS COMMUNICATOR IS CONNECTED VIA MICROCELL
TECHNOLOGY.

Figure 8.30 How PCS will work with wired and wireless networks. (*Source:* Wireless Networked Communications *by Bates.*)

moves out of the zone of cordless phone operation, at which point the personal communicator transmits to one of thousands of ubiquitous microcells mounted atop light poles, telephone poles, tall buildings, etc., in locations called *telepoints*. The personal communicator is wireless to the microcell, but the microcells themselves are hardwired via new fiber or existing coaxial cables to an upgraded telephone infrastructure.

The personal communicator continues to be in wireless communication with the first microcell it finds until the user reaches the boundary of that microcell's coverage area. As the user moves out of one microcell's coverage area and into the coverage area of another, there is a handoff to the next microcell. The typical range of current engineering designs of microcells is 100 to 200 ft (30 to 60 m), a distance that will doubtless increase in the future.

The microcell telepoints are connected to a central device, with a cell serving 20 to 30 concurrent calls and a central controller managing 32 to 64 cells. These capacity points will improve in the future to 96 microcells with each microcell itself capable of handling 20 to 25 calls. With a little bit of simple arithmetic, you can figure out that the future capacity of a controller could be in the range of 1920 to 2400 calls.

As the user moves around, the intelligent network keeps track and forwards calls to the user's current location. PCS makes this possible with a capability called *Signaling System 7* (SS7) that tracks the movement of personal communicators. Thus, in the true spirit of mobile computing, users are able to receive or make calls anywhere to anywhere, anytime.

This scenario continues to function even when a user moves out to another city or into a different telephone company's coverage area. Let us assume that a user travels on a business trip from New York to San Francisco. As the user emerges from the plane, the personal communicator broadcasts an alert message to the local wireless node. The local PCS controller passes the registration request to the local central office (CO). When the CO fails to recognize the user's NAM or MAN, it queries a centralized database server to identify the user and the user's access privileges. While this inquiry is taking place, a message is sent to the personal communicator to confirm physical contact. As soon as clearance is received from the central database, logical communication can take place, if the user so wishes.

Let us now consider what happens if an individual in New York calls our business traveler in San Francisco. The call is, of course, sent first to the local CO in New York. When the user does not respond to a message sent out in the New York area, an SS7 inquiry is sent to the network: "Where is the user with the following NAM and MAN located?" The remote CO in San Francisco responds to the New York CO with the relevant information. The call is then passed on to San Francisco and an end-to-end contact is established.

This is a brief outline of the PCS scenario envisioned in the United States and Canada. Bring in GSM, the European equivalent of PCS, and interconnection between the two, and the scenario can be extended to Europe—and so on, until ultimately the scenario is global in nature.

8.8.2 Components of a PCS network

The following important components are essential to a PCS network (Fig. 8.31):

- Personal communicator
- PCS interface box
- Microcells (telepoints)
- PCS controller
- MTSO
- SS7 network and SS7 Database

Personal communicator. Several names have been suggested for this device by different vendors and the trade press: *personal communica-*

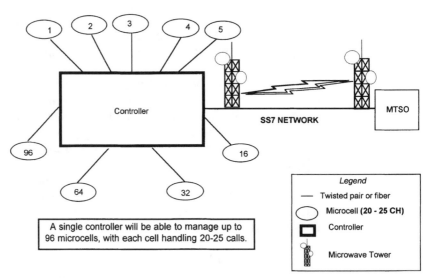

Figure 8.31 Components of a PCS network. (*Source:* Wireless Networked Communications *by Bates.*)

tor will suffice here. The exact shape and function of personal communicators and the services available through them will develop as the technology develops and users voice their preferences. However it eventually evolves, it appears certain that the personal communicator will in one way or another be an amalgam of cellular telephone, pager, fax, and PDA, no doubt with enhancements. Above all, it will be convenient, easy to use, and portable.

PCS interface box. The *PCS interface box* will interface the personal communicator with the legacy wired telephone infrastructure and will probably look much like the base unit of a portable telephone, with built-in charger and a few more controls.

Microcells. A *microcell* is the smaller, low-powered PCS equivalent of a cellular base station. Economics and size will be preeminent in the design of microcells. To cover an entire landscape, many small, inexpensive microcells will be needed. Since the geographic range of a microcell will be minimal, the number of concurrent calls that a microcell can handle will be far fewer than is the case with conventional cellular base stations.

PCS controller. Current designs are capable of handling 32 to 64 cells; future designs will handle up to 96 or even 128 cells.

MTSO. The *MTSO* is the next hierarchical switching box in a PCS network.

SS7 network and database. The heart of an intelligent PCS system is an SS7 network in which transport functions are handled independently of control functions. In an overall PCS network, signaling acquires great importance because of the mobility of users and the need to track movement from one cell to another.

Figure 8.32 shows the hierarchy of the various components of a PCS network.

Future innovations to PCS: The smart card. The PCS companies are considering use of a smart phone card that will contain information about your privileges in a different region or country. This information is downloaded into the network and you can roam around anywhere with your new PCS communicator.

8.8.3   Technologies and standards behind PCS

There are several technological approaches that are being considered for PCS. The cost involved in implementing new standards, backward compatibility with and migration from existing analog cellular net-

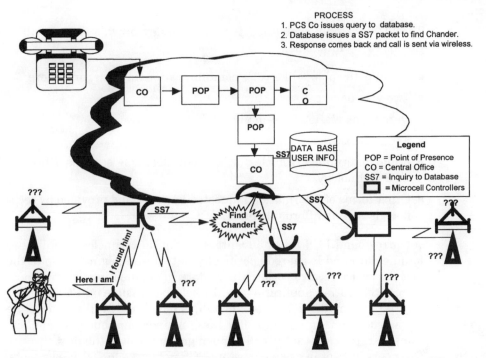

Figure 8.32 How local PCS company issues a query to database to find a user. (*Courtesy of* Wireless Networked Communications *by Bates.*)

works, and internetworking requirements are some of the pointers that network providers should keep in mind when considering technology for a PCS network. Although integration of voice and data with PCS networks is a future possibility, technology has not advanced sufficiently in this area that we can give planning advice to practitioners. Instead, we are providing basic introductory information about the technologies being considered. Mobile Computing professionals should obtain the most current information at the time and evaluate the data-transmission capabilities of a specific PCS implementation. There are four technologies being considered:

- Time Division Multiple Access (TDMA)
- Enhanced TDMA (E-TDMA)
- Code Division Multiple Access (CDMA)
- Narrowband Advanced Mobile Phone Service (N-AMPS).

Table 8.17 compares these technologies.

8.8.4 PCS functions and mobile computing applications

The PCS personal communicator may one day provide some or all of the following functions:

- Cellular-telephone-like voice communication anytime, anywhere
- Two-way paging
- Data transmission for e-mail and limited file transfer
- Facsimile traffic
- Radio-determined vehicle- or personal-location services (through GPS technology)
- Other forms of information requested by users and provided by technologists at an affordable price

Narrowband PCS takes advantage of digital technology's ability to handle data and is geared specifically to two-way paging. Despite the fact that voice receives priority over data in engineering designs, most mobile computing applications will likely be possible on PCS eventually. What remains to be seen, though, is how long it will be before the infrastructure is in place to offer these nonpersonal business applications. Our guess is that it will happen gradually over the next three to five years, except in cases of custom PCS networks for selected clients, where it will happen sooner.

TABLE 8.17   Comparison of Multiplexing Technologies Being Considered for PCS

Capability	Technology			
	TDMA or D-AMPS	Enhanced TDMA	CDMA	N-AMPS
Basic description	TDMA splits the frequency spectrum into time slots	E-TDMA uses a more robust allocation of time slots (see reuse factor)	With CDMA, a direct-sequence code is digitally applied to a signal. Multiple signals can be transmitted on the same channel, since each has a unique code	N-AMPS splits a 30-KHz cellular channel into three 10-KHz channels
Bandwidth	12.5 KHz	12.5 KHz	12.5 KHz	10 KHz
Radio channel	30 KHz	30 KHz	1.25 MHz	30 KHz
Calls/channel	1	3	36–38	
Calls/cell	19	57	360–380	
Frequency reuse gain factor	3 times	10–15 times	20 times	3–5 times
Digital	Yes	Yes	Yes	No
Services offered	Voice Data Paging	Voice Data Paging E-mail	Voice Data Paging Radio determination	Voice Data Paging
Costs	Low	High	High	Low
Quality	Fair	Good to excellent	Good to excellent	Good
Examples of implementors	Ericsson D-AMPS 1900	Hughes Network systems	QUALCOMM satellite carrier; Telezone in Canada	Not available
Other comments	Simultaneous voice and data	Simultaneous voice and data	Improved hand-off Fewer base stations	Stopgap technology

SOURCE: *Wireless Networked Communications* by Bates, modified and upgraded by the author.

8.8.5   Where is PCS now?

The basic technology for PCS exists today. Prototypes of personal communicators are available and are being exhibited in trade shows. Microcell hardware and control software is available from equipment suppliers such as Ericsson and Northern Telecom. Narrowband PCS has started already. Broadband PCS for voice is a bigger challenge, but it will start soon. The two streams will take their own course. In the case of narrowband mobile computing applications, the key is the establishment of sufficient infrastructure to persuade third parties to start developing applications for the emerging networks. Of the several PCS network trials started in 1995 and scheduled for 1996, the following are noteworthy:

Destineer (a subsidiary of Mtel in the United States) started the Pioneers Preference narrowband PCS network in 1995 using a 50-KHz frequency. A Pioneers Preference license is offered to innovative companies who promote state-of-the-art emerging radio technologies. Destineer is offering a software developers kit for extending mainstream applications to its two-way wireless infrastructure.

MobileComm (a subsidiary of BellSouth Corporation) will launch its narrowband Personal Communications Services (NB-PCS) network in the United States at the 1996 Olympics in Atlanta. MobileComm will provide event results over its PCS network to members of the news media covering the Olympics. The BellSouth system uses a 50-KHz outbound channel with a 12.5-KHz return path for two-way traffic. Olympic results will be sent to a gateway and then via satellite to MobileComm's PCS base station. From there, the information will be sent to portable devices. Nationwide installation will start in 1996.

PageMart and several other companies also started PCS narrowband operations in 1995.

American Personal Communications (APC) and *Ericsson* demonstrated the first PCS voice call on July 26, 1995, on Capitol Hill, Washington D.C. The demonstration was made possible with a PCS 1900 network supplied by Ericsson.

Pacific Bell Mobile Service is building a wireless network across California and Nevada using the Ericsson's PCS 1900 technology.

AT&T has invested $1.7 billion in 21 PCS broadband licenses. In August 1995, AT&T Wireless Services started PCS trials in Atlanta, Georgia, with Ericsson's D-AMPS 1900 technology. The tests, focusing on technology verification, included a limited number of base stations and one mobile switching center.

8.8.6   Integration with current cellular technology

Since there is a huge existing infrastructure of cellular networks, wireless service providers are trying to integrate the new PCS services with older D-AMPS networks. CMS-8800 is a technology from Ericsson that provides a dual-band 800/1900 MHz system. It started with D-AMPS and is now migrating toward the D-AMPS IS-136 standard. This will allow the D-AMPS infrastructure to evolve toward digital PCS in a step-by-step manner. CMS 8800 technology supports seamless internetworking and infrastructure sharing with D-AMPS networks as well as analog AMPS networks.

8.8.7 PCS cost considerations

Along with better coverage, clearer communications, devices that consume less power, and data transmission speeds that will rival ISDN, PCS promises services that will be 40 percent cheaper than cellular is today.

8.8.8 Comparison with other networks

PCS provides the following advantages over analog cellular:

- Better voice quality
- Better security because digital voice and data can be encrypted
- More efficient use of the spectrum than analog cellular
- Data capability that is inherent in base technology

Table 8.18 demonstrates the technical differences between cellular and PCS technologies. Table 8.19 summarizes the features of the personal communication system.

TABLE 8.18 Comparison of Cellular and PCS Technologies

Technical and business factors	Cellular	PCS
Antenna site	Higher, with more space required for the basestation site	Smaller footprint.
Number of base stations	Fewer base station sites	Many more sites required (as many as 25 times more).
Cost of base stations	Equipment is more expensive (economies of scale: fewer units always cost more)—$500,000 to $1 million per site	Much cheaper—($20,000–$50,000).
Power output	3 to 15 W	0.1 to 0.5 W
Right of way issues	No major advantage to existing network providers	Telephone companies own right of way on poles—the likely site for most microcells.
Service fees	Higher charges (30–60 cents/minute)	Lower prices expected (10 to 20 cents/minute).
Bandwidth	800 to 900 MHz almost fully used up	1850 to 1990 MHz, with additional spectrum being allocated by the FCC
	Only two operators in each area	Multiple operators—more competition

SOURCE: *Wireless Networked Communications* by Bates, modified and upgraded by the author.

TABLE 8.19   Personal Communication System (PCS) Summary

Brief description	▪ Brand-new completely digital network with a true *anywhere, anytime* capability and a universal personal number associated with a personal communicator
Components	▪ Personal communicator ▪ Microcell base station ▪ Microcell controller and mobile telephone switching office (MTSO) ▪ SS7 network and personal service database
Frequency bands	▪ For broadband PCS, two 60–MHz bands have been allocated in the 1850 to 1990 MHz range (1850 to 1910 MHz and 1930 to 1990 MHz) ▪ For narrowband PCS, 901 to 930 MHz
Coverage	▪ Currently only trials in a few cities. Ultimately a PCS network will cover the entire United States and Canada. This is expected to take many years (upwards of five, in our estimate)
Capacity and speed	▪ Number of channels in broadband for cellular voice communication vary with the multiplexing technology used—up to 40 channels in 1.25 MHz with CDNA ▪ In broadband, network speeds will start at 19,200 bps and could ultimately reach ISDN speeds (2×64 kbps) ▪ Narrowband PCS is primarily for paging, which will be at 25,600 bps
Communications protocols supported	▪ Not known at this stage
Most suitable applications	▪ Personal-communications-based services initially ▪ Paging ▪ E-mail, facsimiles
Costs	▪ No firm cost figures were available at the time of writing ▪ Expected to be cheaper than cellular; (at maturity, may be 40% cheaper than current cellular rates)
Availability	▪ Not available currently except for pilots. Late 1996 and 1997 are the most realistic estimates for availability large metropolitan areas. Nationwide coverage for data to match ARDIS and RAM may take be seen before the turn of the century
Security	▪ Much higher than cellular: voice and data can be encrypted
Pros	▪ Fully digital network with voice/data integration ▪ Universal, portable, unique telephone number usable anywhere in the world ▪ Per-minute charges will be less than cellular ▪ More efficient use of the spectrum ▪ Considerable potential for voice and data integration
Cons	▪ Trials started only in 1995—PCS networks will not be available in 1996 for serious large-scale mobile computing applications ▪ Only the raw network exists for data—no software interfaces have been defined so far ▪ No network software drivers available ▪ No APIs available

8.9   Global System for Mobile Communications (GSM)

Please refer to Chap. 9 for a discussion of GSM, which is essentially a European standard for digital cellular voice and data.

8.10   Satellite-Based Wireless Infrastructure Networks

This is the final wireless network category that we shall discuss in this chapter.

8.10.1   Satellite networks: The universal wireless networks

Wireless industry visionaries and missionaries have long used the words *anywhere, anytime, ubiquitous,* and *global* to describe a future in which the ability to communicate electronically will not be dependent upon location. They have excited the imaginations of users, fully aware all the while that there are vast territories of the world where it is too difficult to install terrestrial networks and/or too costly to justify them on an economic basis. It is to the reach of satellite-based communications that these visionaries look when they extend the wireless scenario to remote areas, mountains, lakes, nations, continents, and, ultimately, the globe.

The prospect of being able to connect a home-office computer to a remote hydroelectric site, say, or to collect data from an isolated lake in Alaska, or to have a police car in northernmost Canada access a law enforcement database in Ottawa—or even to have a sales executive pick up e-mail from a ski chalet—is very exciting indeed.

Satellite-based communications make all this possible (Fig. 8.33). Although a proven technology, however, that has been around since the 1960s (unlike PCS, which is a fundamentally new concept), its implementation in wireless applications has been relatively slow, despite its unique advantages. Chief among these advantages is the fact that it takes only a minimal number of satellites (as few as three in one implementation) to cover vast areas that can include anything from forests, mountains, and valleys, to lakes, seas, and deserts. One reason for the slow implementation has been the high cost of using satellites. Now that is changing: not only do existing satellites have capacity to spare (as in the case of Canada's Telsat), but developing countries like India and China are also beginning to offer much cheaper satellite launchings.

Costs of satellite-based communications.   Satellite-based services use relatively inexpensive very small aperture terminals (VSAT) to provide

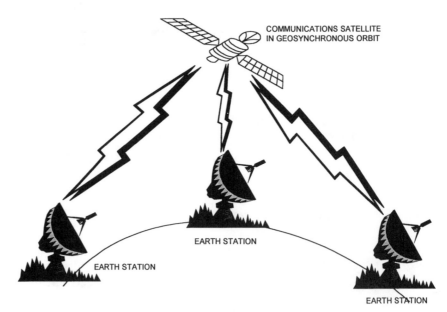

COMMUNICATIONS SATELLITE
IN GEOSYNCHRONOUS ORBIT

EARTH STATION

EARTH STATION

EARTH STATION

Figure 8.33   Satellite-based wireless communications network.

global connectivity. In the United States, 1996 lease and service charges are $400 to $500 per month, with costs dropping to $250 to $300 per month per site for 100 or more sites. Compare this with a PCS installation of about $2500 to $3000 per month (assuming that a copper infrastructure from the local CO is already in place). VSAT is also competitive when compared with multidrop 9600-bps or 56-Kbps connections. Comparisons from countries like Canada, Europe, and Japan show similar trends, though actual figures vary from country to country. Table 8.20 illustrates these comparisons. For certain mobile computing applications, a satellite-based wireless service may well be economical and may meet performance goals. Readers should certainly evaluate this as a network option, depending on the application.

TABLE 8.20   Comparison of Satellite-Based Communications

Service	Setup cost	Cost/month	Bandwidth
VSAT	$300	$300	128 Kbps
T-1	$3,000	$3,500	1.544 Mbps
T-3	$6,200	$4,500	44.92 Mbps
Frame relay	$3,500	$2,800	356 Kbps
9.6-kbps lease	$1,200	$400	9.6 Kbps
56-kbps lease	$1,200	$750	56 Kbps

SOURCE: *Implementing Wireless Networks* by Nemzov.

The race to control the wireless sky. There are several satellite-based worldwide infrastructure networks that are already operational, being implemented, or in the planning stages. Over a dozen suppliers are competing for frequency allocations to offer voice, data, paging, and radio-based position-sensing services. These companies are considering one of the following three orbital approaches for launching their satellite based services:

- Low-earth orbit (LEO)
- Mid-earth orbit (MEO)
- Geosynchronous orbit (GEO)

8.10.2   LEO: IRIDIUM

This satellite system has been proposed by Motorola. An application to construct, launch, and operate an LEO global mobile satellite system called IRIDIUM was filed with the FCC in 1990. The IRIDIUM concept involves launching a constellation of satellites specifically for mobile services, operating in the 1.6100 to 1.6265 GHz bands. Motorola has proposed a configuration of 66 radio-linked satellites traveling in polar orbits at a 420-nautical miles (778-kilometers) altitude to provide global coverage. Implementation was originally targeted to start in 1996, with services being offered in 1998, but this schedule may well slip because of the politics and the size, complexity, and scope of such a large project.

The IRIDIUM project is expected to cost a consortium of 17 investors from around the world over $3.4 billion. The China Great Wall Industry Corporation, IRIDIUM Africa, IRIDIUM Canada, IRIDIUM India, IRIDIUM South America, Lockheed, Motorola, Nippon IRIDIUM Corporation, Raytheon, and Sprint are among those who will participate in the operation of the system, and/or in providing major components for it. While the costs are high, so is the potential for satellite-based wireless services.

Components of the IRIDIUM satellite network. The IRIDIUM proposal is based on the use of mobile or handheld portable transceivers with low-profile antennas. The following components will be fundamental to the system:

- A GSM-based IRIDIUM subscriber unit (ISU) from Motorola
- The constellation of 66 satellites designed by Motorola
- Terrestrial radio gateways
- Antennas by Lockheed and Raytheon

The ISU, which looks like a one-piece telephone, first attempts to access a terrestrial cellular system before looking to the satellite for a connection. This arrangement minimizes demand on the satellite system's resources on the one hand, while making possible global roaming and seamless coverage, on the other.

Services provided by IRIDIUM.   The following services are proposed:

- Mobile voice communications (digital) 4800 bps from anywhere in the world to anywhere in the world
- Data communications: two-way messaging (e-mail) applications at 2400 bps
- Two-way facsimiles at 2400 bps
- Global two-way alphanumeric paging
- Radio determination: location of fleets, aircraft, and marine vehicles.

Benefits of IRIDIUM's services.   Many of the following benefits can be derived from any satellite service, though some are unique to IRIDIUM:

- Global service based on the 66-satellite constellation
- Uninterrupted 24 × 7 (24 hours a day, 7 days a week) service
- Disaster relief
- Low-power design for the ISU, made possible by the proximity of the satellite
- Efficient use of the spectrum: 5 × reuse in the United States and 180 × reuse throughout the world
- International technical and business cooperation: IRIDIUM is a consortium of international PTTs and other technology partners

Figure 8.34 shows a schematic of the IRIDIUM network and its use by various devices. Please note that many of these devices are conceptual at this stage. Table 8.21 summarizes IRIDIUM's system facts and features* and Table 8.22 estimates the costs for various services.

Availability and coverage.   The IRIDIUM consortium hopes to start commercial service in North America in 1998 and to spread to the rest of the world in the years following.

---

* Iridium Systems Overview, supplied by Iridium Corporation Canada.

TABLE 8.21   IRIDIUM System Facts

Space segment	
Number of satellites	66 interconnected
Number of orbital planes	6
Orbital height	420 nmi (778 km)
Inclination of orbital planes	86.4°
Orbital period	100′ 28″
Coverage	5.9 million mi² (9.5 million km²) per satellite
Satellite weight (WET)	1543 lb (700 kg)
Spot beams per satellite	48
Link margin	16 dB
Lifetime	5–8 yr
Frequency bands	
L-band service links	1616–1626.5 MHz, L-band
Intersatellite links	23.18–23.38 GHz, Ka-band
Gateway/TT&C links:	
Downlinks	19.4–19.6 GHz, Ka-band
Uplink	29.1–29.3 GHz, Ka-band
Switching equipment	
Siemens GSM-D900	
Signaling	
GW to ISC	PCM transmission and SS7-ISUP or MFCR2
IRIDIUM Telephone	Frequency division/time division (FDMA/TDMA), quanternary phase shift keying (QPSK)
Transmission rates	
Voice	Full-duplex, 2.4 Kbps
Data/facsimile	2,400 Bd
Launch	
McDonnell Douglas Delta II	5 IRIDIUM satellites per launch
Khrunichev Proton	7 IRIDIUM satellites per launch
China Great Wall Long March IIC	2 IRIDIUM satellites per launch

SOURCE: IRIDIUM Systems Overview brochure supplied by Iridium Canada.

TABLE 8.22   Estimated Costs for IRIDIUM Services

Equipment	Cost
Basic handset (ISU telephone)	$2,000
Pager	$200–$300
Radio determination system	$200
Data add-on module	$1,000–$1,500
Charges	
Per minute, or part thereof	$2–$3
1-page fax at 2400 bps	$4–$6 for 1.25 min; 30 s to dial and ring through, 45 s for the handshake and transmission
E-mail	$4–$6 for a 600-character message

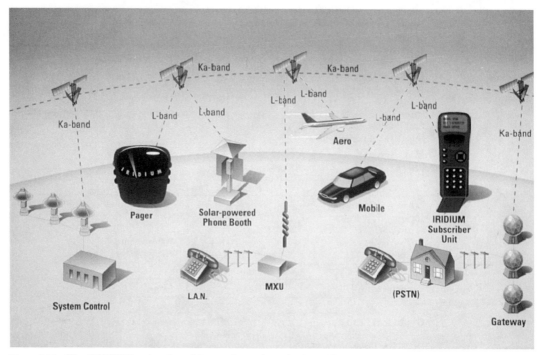

Figure 8.34 The IRIDIUM network and its use by various devices. (*Courtesy of Iridium Canada.*)

8.10.3 QUALCOMM's satellite-based OmniTRACS network

QUALCOMM, a company based in San Diego, California, currently offers the OmniTRACS network using a geostationary satellite system that sends and receives data from moving trucks on interstate highways in the United States, Canada, South America, Europe, and Japan (Fig. 8.35). OmniTRACS provides the following services:

- A two-way messaging service between trucks and their dispatch centers (called QTRACS)
- An automatic vehicle location (AVL) service
- A vehicle information system (VIS) that provides vehicle sensor telemetry (speed, engine rpm, other diagnostics) information for service management purposes
- Other wireless data applications

For a fixed monthly fee, QUALCOMM provides its 100,000 users in the transportation industry with a bundled service that includes an in-vehicle terminal, a modem-like in-vehicle satellite radio unit, com-

COMMUNICATIONS SATELLITE
IN GEOSYNCHRONOUS ORBIT

QTRACKS and VIS Applications

Figure 8.35   QUALCOMM's satellite-based service for truck fleet service.

munications and application software, satellite network services, and custom integration services.

Figures 8.35 and 8.36 show the in-vehicle hardware and use of Omni-TRACS software in a transportation-industry context. The OmniTRACS system consists of three major components:

- Mobile hardware that includes a ruggedized, portable keyboard

- A communications terminal housing the electronics

- A continuous tracking antenna in a sealed dome

The communications terminal provides the intelligence for sending and receiving text messages to and from the vehicle. At the heart of the system is the Viterbi decoder and an Intel microprocessor. The Omni-TRACS antenna is mounted outside and is based on a proprietary Ku-band design.

8.10.4   Other satellite systems

Another significant player is Inmarsat's Project 2, or Inmarsat-P, which is already widely used for voice and data purposes. Inmarsat could use this 11-GEOS (22,300-mile or 35,887-kilometer altitude) satellite network to offer services similar to IRIDIUM.

Figure 8.36   OmniTRACS system's in-vehicle hardware.

American Mobile Satellite Corporation (AMSC) has been providing services similar to QUALCOMM's OmniTRACS since 1992. AMSC's Skycell, whose satellite "footprint" covers the United States, Canada, Cuba, Puerto Rico, and the Virgin Islands, is a fill-in service providing voice and data communications services in areas where there is no terrestrial cellular server available.

Telesat Canada, a satellite services company based in Ottawa, Ontario, started offering mobile data services in Canada from the ANIK E satellite network in 1994.

By 1999, a consortium called the TRW Space and Electronic Group expects to have launched into a mid-earth orbit (1800-mile or 2897-kilometer altitude) a 12-satellite network called Odyssey. TRW believes that an MEO is more economical than an LEO because it requires fewer satellites, and the satellites stay in orbit longer. Odyssey will cover the entire earth with only two satellites, each providing 2300 channels for voice, data, radio-location, and messaging services.

For 1998, Loral QUALCOMM Satellite Services, a joint venture of Loral Corporation and QUALCOMM, has proposed Globalstar, a network of fifty-six 600-pound (272-kilogram) satellites—including eight spares—in eight 750-nautical-mile (1389-kilometer) LEO orbital planes.

Two other proposed LEOs are Constellation Communications' Aries system and Ellipsat Corporation's 12-satellite Ellipso system.

8.11   Comparison of the Features
of Wireless Networks

Table 8.23 lists the different features of various wireless networks for comparison purposes.

8.12   Choosing the Right Network
for Mobile Applications

The selection of a suitable network for a mobile application is an important topic, but one that should be considered as a part of an overall plan. We shall discuss this subject in Chap. 15.

8.13   Wireless Network Design Issues

We shall discuss these considerations in Chap. 16 as a discrete part of the overall technical design of a mobile computing solution.

Summary

We have covered a lot of material in this chapter. We explained basic electromagnetic theory and the terms tossed around by network suppliers. Armed with this knowledge, readers who perhaps were not before, should now be in a position to engage in meaningful discussions of the composition of various networks, their coverages, their cost implications, and their suitability for business applications.

One of the messages we hope we have conveyed in this chapter is that private wireless networks are no longer economically viable, except, perhaps, in contained metropolitan areas, where traffic flow and packet charges for ARDIS, RAM Mobile Data, and CDPD are high. Special financial arrangements with public shared network providers can always be negotiated, on the other hand, because of their economies of scale.

There is no doubt that network options will continue to increase. CDPD will no doubt be used extensively in certain types of applications, regardless of negative comments from the press and proponents of other networks. The CDPD industry will find ways to overcome its shortcomings. PCS in North America and GSM in Europe will play an increasingly important role in data transmissions once the industry has satisfied the needs of voice communication (which is where all the money is right now). The demand for ever

TABLE 8.23 Comparison of Different Networks

	SMR	Paging	Cellular (circuit switch)	ARDIS	RAM Mobile Data	CDPD	ESMR	PCS	GSM	Satellite
Frequency band (MHz)	Mostly 800, some 900	930–931	800–900	806–821, 851–866	896–901, 935–940	800–900	Mostly 900	901–930 1850–1990	849–890, 935–960 1.7–1.8	Varies (IRIDIUM: 1000.61–1000.625)
Private/Public	Private shared (mostly)	Public	Public	Public shared	Public shared	Public shared	Public shared	Public shared	Public shared	Public shared
Voice/Data	Yes	Very few	Both	Data	Data	Data	Data	Both	Voice now, data later	Both (including video)
One-way/ two-way messaging	Mostly one-way	Both	Yes	Yes	Yes	Yes	One-way now Two-way later	Yes	Yes	One-way mostly
Availability (now/future)	Yes, since 1980	Available	Yes	Yes, since 1990	Yes, since 1992	Limited in 1995	Limited in 1995	Trials in 1995	In Europe, but not in the U.S.	Some now, more in the future
Coverage	Excellent: 95%	Excellent: 95%	Very good: 90%	Good: better than RAM	Good	Limited, but expected to grow	Limited	None for data in 1996	Limited, but now growing	100% in United States and Canada
Building penetration	Fair	Excellent	Not optimized, but improving	Very good	Good	Same as cellular	Fair	Not known	Not known	Limited

		Very low (512 bps)	2,400–9,600 bps rarely 14.4 Kbps	4,800–19,200 bps	8,000 bps–16,000 bps–	19,200 bps	4,800 bps	>19,200 bps	9,600 bps for SMS	2,400 bps
Speed		Very low (512 bps)	2,400–9,600 bps rarely 14.4 Kbps	4,800–19,200 bps	8,000 bps–16,000 bps–	19,200 bps	4,800 bps	>19,200 bps	9,600 bps for SMS	2,400 bps
Data volume	Low	Very low	High	Medium	Medium	Medium	Low medium	Not known	Low	Low
Complexity of use		Very simple	Simple	Medium	Medium	Simple: modem	Medium	Not known	Not known	Simple (trucks)
Hardware cost per user	$800–$3,000 per mobile	Low ($100–$250)	PCMCIA modem only ($350–$500)	Medium ($500–$700)	Medium ($500–$700)	Higher now ($1,000)	Not available	Not known	Not known	$4500 (Omni-TRACS)
Fees	$15–$20 dispatch, $45 interconnect	$10–$25 per month	25–50 cents per minute Low for volume	See Sec. 8.6.1	See Sec. 8.6.2	Lower than ARDIS/RAM ($0.20/KB)	Not available	Not known	$1.20–$1.40 per minute	$0.05 per message plus $0.02/char.
Reliability	High	Good	OK for batch data; poor for interactive	Very high (redundancy)	Very high (redundancy)	No field data; better than cellular	No field data	Better than cellular	No field data available	Excellent
Unique features	For private network	Cheap for short messages	Good for long-file-transfer-type applications	Good for in-building OLTP	Good for short OLTP applications	Cheaper, low data priority	Good for dispatch applications	Voice first, data later (U.S. only)	Voice first, data later (Europe)	Good for trucking; ubiquitous

Notes: RAM offers 8000 bps now. Chart shows cost estimates only for future services.

increasing voice communication capabilities continues to grow—and is easy to fulfill. Wireless data transmission is more difficult, and every customer's situation is unique. It will be a few more years before data transmissions occupy the attention of PCS vendors. In the meantime, ESMR and narrowband PCS will continue to nibble at the data market through SMS.

Increasing competition as more and more networks come into being will continue to force network costs down by 25 percent to 30 percent every year for at least the next three years—a good thing for both the vendor community and the user community.

Finally, decide on a network only after a thorough analysis of the market. Also, it pays to thoroughly understand business application requirements and to match them against the present and future capabilities of the networks under considerations.

9

The International Wireless Network Scene

With the globalization of business, physical trade barriers between countries are coming down rapidly. Many countries in Europe, Japan, and Asia are moving forward by establishing national wireless infrastructures so that they can be players in the future economic order, where information flow within and across nations will play an increasingly important role. International wired and wireless networks will become important tools for global competitiveness. CHANDER DHAWAN

About This Chapter

In Chap. 8 we discussed wireless networks, including SMR, cellular, CDPD, packet radio, paging, ESMR, PCS, and satellite-based networks. In the course of this examination, however, we focused largely on the United States. International wireless networks constitute a whole different scene again—a unique and important scene with which we need to familiarize ourselves in the context of the global economy. In this chapter, we shall study the global system for mobile communications (GSM) as an international standard before briefly reviewing wireless networks in different countries.

9.1 The Global Nature of Wireless Networks*

While the United States has always been in the forefront when it comes to adopting wireless data network technology, international

---

* For more detailed information, refer to R. Schneiderman, *Wireless Personal Communications: The Future of Talk,* IEEE Press, 1994.

markets in Eastern and Western Europe, the Pacific Rim, and Russia are warming up, too, as are markets in developing countries such as India and China. As well, the cellular voice market, which has always moved ahead of the cellular data market, is now seeing significant growth in Europe, albeit a late start—so much so that by the end of 1995, almost every European country had at least one cellular system installed. Optimism over the European market is not based just on the growth of the cellular voice market, however, but also on the rapid progress made in establishing standards and services for mobile data applications. In fact, Europe has made greater progress toward a standardization of the digital cellular market than has the United States. While North Americans are still struggling with the design and implementation of a digital PCS standard that is backward-compatible with AMPS infrastructure, the Europeans have established GSM technology as the de facto standard for digital cellular. Moreover, they are succeeding in persuading other countries to join the standard and as a result, GSM networks are being implemented at a faster rate in Europe than their counterpart, PCS technology, in the United States. It is important to realize, then, that the future potential size of the wireless data market in Europe is quite likely equal to the U.S. market.

While we do not have accurate estimates of the current total mobile-data market in Europe, statistics indicate that in 1994 there were 16 million mobile users of radio paging and cellular telephones, as well as private and public mobile radio. Based on industry growth rates, we estimate the figure for 1996 to be around 25 million.

Similar general comments can be made about Japan as well. Other Asian markets have also started to grow. Paging is becoming so popular in India that Motorola has set up a large manufacturing facility there for local and export markets. We shall make more specific comments about these markets later in this chapter.

As we have mentioned, the most significant development in Europe is GSM, which we described briefly in Chap. 8. In this chapter, we shall examine it in greater detail. Although GSM services are initially being maintained within the borders of individual countries, roaming agreements between service providers will eventually enable international travelers to access their home networks from abroad using a GSM phone and a PC PCMCIA Card. While there are only a few roaming agreements in place currently, over 100 GSM operators have agreed to maintain conformity with the international standard. Ultimately, there will be cross-border communication between GSM countries in Europe and PCS countries such as the United States, Canada, and Japan.

9.2   Canada: A PCS Country

The Canadian telecommunications market has always stayed in concert with the American market, though usually with an approximate 12-month time lag in reaching the same level of penetration.

The two major mobile data operators in Canada, Bell Mobility and Cantel, have been providing mobile data services since the early 1990s. Bell Mobility, owned by Bell Canada Enterprises (BCE), provides wireless services that include paging and packet data through Bell-Ardis. Roger-Cantel offers similar services through paging and Mobitex networks. Both companies provide national coverage in predominantly urban areas, based primarily on user demand. Additionally, paging services are also provided by a number of smaller companies.

The Canadian market is moving towards a PCS implementation on the same lines as the U.S. market. Because of the proximity of urban areas along the Canadian/U.S. border, there is an agreement between the two countries to use the same frequencies for similar applications, thus paving the way for continentwide access to digital cellular voice and mobile data services.

There are several new entrants in the Canadian mobile data market. Along with the wireless network incumbents, Bell Mobility and Roger Cantel, 16 other companies (Clearnet, LanSer, MicroCell, Telezone, and several others) are vying for the emerging market. Industry Canada, a Canadian federal government department responsible for radio communications, is fostering additional competition by limiting the current cellular carriers to certain frequencies.

The results of applications submitted for PCS licenses during 1995 were announced by Industry Canada in December 1995. Besides Bell and Cantel, MicroCell and Clearnet were the early first-round winners. Total investment during the next five years in the new services is expected to be more than $2 billion.

LanSer of Montreal was given a license for an experimental broadband in the 1.9-GHz range. The company had filed an expression of interest with Industry Canada to obtain a 30-MHz license so that it could begin offering a PCS service in major urban areas of Canada by the end of 1996. Working with LanSer were Ericsson, which was going to provide the phones; Harris Farinon, which was going to provide a direct microwave link system, and Canadian Marconi, which would have provided smart antennas to eliminate static interference. LanSer was not successful in the initial licenses awarded by Industry Canada, however.

Telezone, a Toronto-based company, has acquired Beeper People, one of Canada's oldest paging organizations. The company has tested various PCS technologies and found Q-CDMA to be its best choice. It reports that with Q-CDMA it will be able offer a data service initially at 14.4 Kbps—and eventually at 76.8 Kbps. According to Telezone, Q-CDMA requires fewer base stations than other PCS technologies such as TDMA, costs less to maintain, and can carry more traffic, giving it a higher reuse factor.

Clearnet is the largest operator of SMR wireless communications in Canada. In 1995, the company offered dispatch services to over 53,000 users in its service territory, which includes all of Canada's major population areas and traffic corridors. Clearnet is currently building a new all-digital ESMR wireless network linked with cellular and land-line systems. It will offer fully integrated enhanced dispatch (group calling), mobile telephone, text messaging with acknowledgment (two-way paging), and mobile data services—all on one network and all accessible to users through one handset. In addition, the company plans to augment its ESMR services with other innovative wireless products and applications, including PCS (for which it was granted a license in 1995).

On the research, development, and manufacturing side, Northern Telecom (Nortel), a Canadian multinational company with headquarters in Ontario, and Bell Northern Research (BNR), one of the most advanced telecommunications research organizations in the world, are designing architectures and software that will enable PC users to access and control a wide range of voice, data, and video communications applications.

Nortel is also a major developer and supplier of PCS infrastructures. It was the first company to market a certified, unlicensed PCS communicator called the Companion Portable Telephone. This is a wireless PBX system designed for operation in the 1920 to 1930 MHz band, the isochronous (voice) side of the recently allocated unlicensed PCS band. Nortel has already won several international contracts to supply digital cellular network hardware to European customers, its PCS 1900/ DCS 1800 (GSM) product being chief among its premiere offerings for PCS/GSM infrastructures.

According to BIS Strategic Decisions' *Canadian Wireless Market Forecast,* a market research company based in Norwell, Massachusetts, there were 2.7 million wireless subscribers in Canada in 1994. Of these, approximately 800,000 used pagers and 33,000 used data. By the year 2000, it is expected that this number will grow to 11.8 million subscribers: 650,000 using data, 4.6 million using paging, and 6.5 million using cellular phones.

9.3 Europe's GSM Technology

The GSM has become *the* international standard for digital cellular service. By the end of 1995, more than 74 countries had signed a GSM memorandum of understanding to adopt it as a standard. Almost all major European countries have committed to it. The commercial operation of GSM networks started in mid-1991 in European countries. According to Ericsson, there were over 50 networks and 6 million GSM subscribers by the end of the third quarter of 1995, and there will doubtless be many more by the time this book goes to press in 1996. Even reminding ourselves that there were some 40 million AMPS and 1.5 million D-AMPS subscribers reported at the same time in North America, we should nevertheless be aware of the singular fact: GSM is Europe's answer to North America's PCS. We shall talk about the differences between the two later in this section.

9.3.1 Basic objectives of GSM

The development of GSM started in 1982 when the Conference of European Posts and Telegraphs (CEPT) formed a study group called Groupe Spaciale Mobile (from which the letters GSM were originally derived). The group's mandate was to study and develop a pan-European cellular system in the 900-MHz range, using spectra that had been previously allocated. At that time, there was a plethora of incompatible analog cellular systems scattered throughout Europe. Some of the basic criteria for the group's proposed system were:

- Good subjective speech quality
- Low terminal and service cost
- Support for international roaming
- Ability to support handheld terminals
- Support for a range of new services and facilities
- Spectral efficiency
- ISDN compatibility

In 1989, the responsibility for GSM was transferred to the European Telecommunication Standards Institute (ETSI), and the Phase I recommendations were published in 1990. At that time, the United Kingdom requested a specification based on GSM, but for higher user-densities with low-power mobile stations operating at 1.8 GHz. The specifications for this system, called the Digital Cellular System (DCS1800), were published in 1991.

9.3.2 Evolution of GSM services

The full implementation of GSM will take place in three phases. The first phase introduced basic voice services and emergency calling features. The second phase, which started in 1995, is the addition of advanced features such as call waiting, call forwarding, caller information services, and improvements in the subscriber identity module (SIM). The third phase will involve full roaming capabilities, with appropriate business agreements between the participants. In Europe, digital cellular phones that work digitally in GSM mode are already being manufactured and marketed, in contrast to the United States, where cellular carriers can only offer dual-mode, analog/digital cellular phones.

Besides voice, various data services are currently supported, with user bit-rates up to 9600 bps. In a variety of ways, specially equipped GSM terminals can also connect with PSTN, ISDN, packet-switched, and CSPD networks, using synchronous and asynchronous transmission. Supported as well are the group 3 facsimile services, videotex, and teletex. Other GSM services include *cell broadcast,* where traffic reports and the like can be broadcast to users in particular cells.

A *short message service* (SMS)—similar to two-way paging services in narrowband PCSs in the United States—allows users to send and receive point-to-point alphanumeric messages up to 150 characters in length.

9.3.3   Global scope of the GSM

The GSM was created to provide a modern digital cellular service that would enable people to travel at will without encountering problems with voice and data wireless applications. Its acceptance as an international standard and the creation of an infrastructure based on this standard is fundamental to the success of the enterprise. Although the actual number of roaming agreements is minimal as yet, and what few exist are limited to voice traffic only, a technical infrastructure, which will make roaming possible as soon as business and political leaders can agree, is nevertheless well on its way to being established. And despite the fact that the United States has opted for PCS as a base technology for digital cellular, it is entirely possible that cross-border GSM services will one day be extended even to non-GSM countries like the United States, Canada, and Japan.

GSM's data capability.   Although GSM networks have initially been able to offer only voice services, the fact that the technology is digital in nature means that data is an integral capability. By the end of 1996, it is expected that more than 50 GSM network operators in Europe will be offering data and fax services, with many networks also offering X.25 service.

9.3.4 Frequency bands and channels

In Europe, spectra in the 849 to 890 MHz, 935 to 960 MHz, and 1.7 to 1.8 GHz frequency bands have been reserved for GSM operation. GSM radio links are made up of two channels: a circuit-switched Bm and a packet-switched Dm channel. The Bm channel delivers 9600 bps of synchronous and asynchronous traffic, and the Dm channel carries signaling information at a slow 382 bps. Some service providers are using the Dm channel for an SMS up to 160 characters.

Speech in GSM is digitally coded at an efficient 13 Kbps with *forward error correction,* a so-called full-rate speech coding. One of the most important Phase II additions will be the introduction of a half-rate speech encoding operating at around 7 Kbps, an effective doubling of the capacity of a network. Forward error control and equalization contribute to the ability of GSM radio signals to withstand interference and multipath fading.

9.3.5 Components of a GSM network

GSM networks are set up with the same components as conventional analog cellular networks: antennas, base stations, terrestrial wired backbones, and gateways to public telephone systems (Fig. 9.1). When

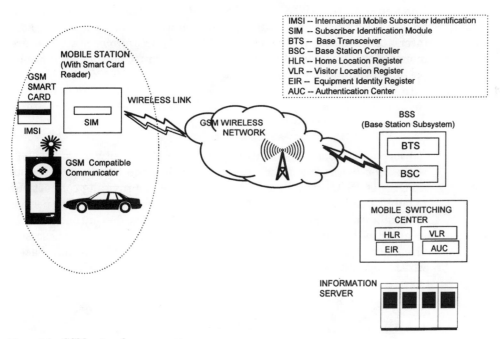

Figure 9.1 GSM network components.

a mobile user makes a call with a GSM phone, the signal is picked up by an antenna and the call is handed off from cell to cell as the user moves from one area to another.

The functional components of a GSM system can be broadly divided into the *mobile user device* (or *mobile station*), the *base station subsystem,* and the *network subsystem.* Each subsystem is comprised of functional entities which communicate through various interfaces using specified protocols.

Mobile user device (mobile station). A GSM mobile user device has two distinct parts. There is the actual hardware, which comprises the mobile user equipment itself, and then there is the subscriber information, which includes a unique identifier called the *international mobile subscriber identity* (IMSI), which is stored in a subscriber identity module (SIM) implemented as a smart card. By inserting the SIM card in any GSM mobile equipment, the user is able to make and receive calls at that terminal and receive other subscribed services. The charges are automatically billed to the user's home network. By encoding the subscriber's information on a discrete card in this way, personal mobility is greatly enhanced for GSM users.

Base station subsystem. The base station subsystem comprises a *base transceiver station* (BTS) and a *base station controller* (BCS). The BTS houses the radio transceivers that define a cell and handle the radio interface protocols with mobile stations. Since numerous BTSs are mounted in unprotected environments, they must be rugged, reliable, portable, and inexpensive.

In addition to handovers, the base station controller manages the radio resources for one or more BTSs, including setup, teardown, frequency hopping, etc.

Network subsystem. The central component of the network subsystem is the *mobile services switching center* (MSC), the GSM equivalent of a PSTN or ISDN switching node. In addition to providing connections to public fixed networks, the MSC provides all the functionality needed to handle mobile subscribers, including registration, authentication, location updating, inter-MSC handovers, and call routing. These services are provided in conjunction with four intelligent databases, which, together with the MSC, form the complete network subsystem.

The four databases are:

- A *home location register* (HLR) that contains all the administrative information pertaining to each subscriber registered in the corresponding GSM network, along with the current location of the subscriber. The HLR assists in routing incoming calls to mobile users and is typically the SS7 address of the visited MSC. There is logically one HLR for each GSM network, although the register may on occasion be implemented as a distributed database.

- A *visitor location register* (VLR) that contains sufficient HLR administrative information to maintain call control and provision of subscribed services for each mobile in the area controlled by the VLR. Although the VLR can be implemented as an independent unit, to date all manufacturers of switching equipment have implemented the VLR together with the MSC so that the area controlled by the MSC corresponds to the area controlled by the VLR. This proximity of the VLR to the MSC speeds up access to information that the MSC requires during a call.

- An *equipment identity register* (EIR) that contains a list of all the valid mobile equipment in use on the network. Each piece of equipment is identified by an *international mobile equipment identity code* (IMEI). The IMEIs of mobile equipment that has been reported stolen or of equipment that is not type-approved are marked in the database as invalid, and consequently will not function.

- An *authentication center* (AuC) that comprises a secure database containing a copy of the secret key stored in each subscriber's SIM card. It is used for authentication and ciphering on the radio channel.

9.3.6   Typical costs of GSM service

Currently, GSM services are not cheap. The German company Mannesmann Mobilfunk charges a peak rate of DM1.21 per minute (US$0.82, at the time of writing) for domestic calls and DM2.07 (US$1.40) per minute for international GSM calls originating in Germany. No figures were available for data services.

9.3.7   GSM in the United States

While many American companies such as Motorola, Ameritech, Bell Atlantic, AT&T Wireless (ex-McCaw Cellular), Northern Telecom, and MCI have substantial stakes in Europe, only one provider is offering digital cellular service in the United States based on the GSM standard: Nextel is currently providing a GSM-based service in southern California.

The drawback to the implementation of the GSM standard in the United States is its lack of compatibility with existing analog infrastructures. With analog cellular so well established throughout the continent, widespread adoption of a fully digital standard that does not support analog cellular in some shape or form is inconceivable. Therefore, GSM networks in the United States will be implemented by new entrants rather than existing cellular network providers.

Table 9.1 gives a summary of the features of the GSM network.

9.4 Europe: A GSM Continent

The following is a country-by-country status report on European wireless networks.

9.4.1 England

The U.K. market represents one of the largest wireless markets in Europe, with more than 2 million subscribers at the end of 1995.

TABLE 9.1   GSM Network Summary

Brief description	■ A digital cellular communications network technology that has been established as a worldwide standard except in North America and Japan. Corresponds to PCS in services offered, but its standards are different
Components	■ Mobile user equipment with international mobile subscriber identity (IMSI) in a SIM module implemented as a smart card ■ Base station subsystem ■ Mobile services switching center (MSC) ■ SS7 network and four databases: home location, visitor, authentication, and security
Frequency bands	■ 849–890 MHz, 935–960 MHz, and 1.7–1.8 GHz frequency bands have been reserved for GSM operation
Coverage	■ Over 50 networks with 6 million subscribers reported for the third quarter of 1995
Capacity and speed	■ Greater than AMPS-based cellular and growing rapidly: 22 million subscribers can be expected by the year 2000 ■ 9600 bps for data
Communications protocols supported	■ X.25 and ISDN LAPD
Most suitable applications	■ Short message service (SMS) similar to North America's narrowband PCS two-way paging ■ Facsimile
Costs	■ 50% higher than analog cellular in the United States
Availability	■ Now, ahead of PCS implementation in the United States
Security	■ Very good
Pros	■ International standard ■ Low-cost all-digital mobile user set ■ Purely digital
Cons	■ No backward compatibility with analog ■ No dual-standard user sets available in Europe ■ Higher service fees than PCS despite advanced infrastructure development

Originally, the British government licensed four companies to provide a PCS-style CT-2 telepoint (microcell) service shortly after the publication of *Phones on the Move,* a report by the U.K. Department of Trade and Industry. Unfortunately, the four companies subsequently found themselves encountering high operational costs and a lack of interest from potential users. After the resulting industry rationalization, Hutchison Personal Communications emerged as the only company offering telepoint service.

Hutchison invested over $100 million in more than 8000 base stations in train stations and along highways in the United Kingdom in the course of setting up its Rabbit network. The service was met with apathy. Hutchison tried desperately to promote the service by selling phones directly at drastically reduced prices. By the end of 1993 the company had withdrawn from the venture. There does not appear to be any demand for a telepoint service in England, or for that matter, anywhere in Europe.

After the telepoint license debacle, the Department of Trade and Industry issued three full, nationwide PCS licenses to Mercury Personal Communications (initially a joint venture between Cable and Wireless of the United Kingdom and US West), Unitel, and Microtel. Unitel soon abandoned its plans for PCS and merged with Microtel, which is now owned by Hutchison.

Vodafone and Cellnet (a subsidiary of British Telecom) are now the two major digital cellular providers in the United Kingdom. Vodafone launched a regional GSM service in late-1991, and relaunched it two years later in 1993 along with an SMS, after initially encountering a lukewarm reception. Cellnet started a GSM service in December 1994.

RAM Mobile Data's Mobitex-based packet radio network for data exclusively is also available in the United Kingdom. The range of applications offered is similar to those of RAM Mobile Data in the United States.

9.4.2 Germany

Germany has the fastest growing GSM market in Europe, with an estimated 2 million subscribers at the end of 1994 (1996 numbers are estimated at approximately 3 million).

Initially the German government treated the cellular market as an extension of the traditional telephone market and granted only one monopolistic (though regulated) license to Deutsche Bundepost Telekom (DBT), Germany's national public telephone provider. Recently, however, the government launched its *Telekom 2000* initiative with the objective of modernizing the telecommunications market in eastern Germany and linking it more completely with western Germany. To this end, the government issued two additional licenses. One went to DBT and the other went to Mannesmann Mobilfunk Gmbh, a privately held consortium of European and U.S. telecommunications organizations (PacTel holds a 26-percent share). There is one difference between the two licenses: Mannesmann is restricted to GSM technology only while DBT is allowed to operate both analog and digital cellular services.

DBT and Mannesmann Mobilfunk Gmbh are currently the two major players with almost equal shares of the cellular market. Mobilfunk started D2 private service in June 1992 and DBT started D1 private service soon after, in July 1992 (after starting an analog cellular service earlier). Both services have grown steadily. A third license was awarded to E-plus in February 1993 for a national service based on the DCS 1800 standard and using Nokia equipment. The E-plus network had 30,000 subscribers by the end of 1994.

As in the United Kingdom, Germany has other wireless data services such as Mobitex-based packet radio networks, paging networks, and a large number of public mobile radio (PMR) networks and public access mobile radio (PAMR) networks (trunked PMR networks).

9.4.3   France

As France's national telephone carrier, France Télécom is an important player on the French telecommunications scene. In addition to its Radiocomm 2000 analog cellular service, it operates Itineris, a GSM network that the company first implemented in 1992 in Paris and Lyon with equipment from Alcatel, Matra, and Ericsson. The country's second digital cellular carrier is SFR, a private firm supplied by Alcatel, Siemens, and TRT, the French Phillips subsidiary.

GSM service is gaining increasing popularity in France and is expected to spread throughout the country. The market research company, EMCI, estimates that by the end of 1996 there will be 900,000 subscribers.

In France, strategic alliances are necessary in order to enter markets, which are controlled by an organization called the PTT. Companies in the United States are taking up the challenge. BellSouth has a 12.5 percent interest in France Télécom Mobiles Data, a subsidiary of France Télécom that offers mobile data services. The consortium expects every city with a population of more than 100,000 to have access to mobile data services by 1997. US West has a 9 percent interest in La Lyonnaise Communication.

TDR, a sister company of SFR, launched a Mobitex-based mobile data network in 1994 with equipment supplied and supported by Ericsson. The network, started in Paris and expected to be available to 60 percent of France's population of 55 million by the middle of 1996, is being used for several mobile data applications, including electronic point of sale (EPOS).

9.4.4 Other European countries

Eastern European countries and the former Soviet republics are also emerging as markets ripe for wireless networks. Many governments have adopted wireless networking as an efficient way of providing telephone services in nonserviced areas, where the costs of laying cables and constructing central office infrastructures are high and installation is a lengthy process. Wireless local loop is a very attractive technology in these regions.

Hughes Network Systems, along with GTE Spacenet International, SEL Alcatel, and San Francisco/Moscow Teleport (SFMT) have installed the world's first large-scale residential wireless telephone system in Tatarstan, a city of 3.6 million people, 500 miles (800 kilometers) east of Moscow. Hughes supplied E-TDMA digital cellular telephone technology; SFMT was the general contractor; GTE Spacenet built the earth station; and SEL Alcatel provided the digital base station. US West is involved in a cellular project in Moscow.

Tables 9.2 and 9.3 summarize the status of wireless networks in various European countries.

9.5 Japan

Analog cellular was introduced in Japan in 1979, almost four years before its debut in the United States. Nippon Telegraph & Telephone Company (NTT) is Japan's government-owned public corporation. It held a monopolistic position in the cellular market until 1985, when the country enacted the Telecommunications Business Law, which effectively abolished the monopolies held by NTT. In 1986, the Ministry of Posts and Telecommunications (MPT) issued regional licenses to two new cellular service providers: Nippon Ido Tsushin and Daini Dendon Incorporated (DDI). Nippon Ido is limited to the Tokyo-Nagoya area. DDI is licensed to operate in the remaining urban areas of the country.

Japan's digital mobile communications system is called Japan Digital Cellular (JDC). It is a TDMA-based system in the 800-MHz and 1.5-GHz bands—similar to the American TDMA network, but with one major difference: it is not dual mode (analog and digital). Instead, customers will be able to choose among four providers of digital cellular service.

Japan's telecommunications policy was a significant irritant in United States/Japanese economic and trade discussions in 1995. While it is true

TABLE 9.2 Status of Wireless Networks in Europe (End of 1993)

	Analog cellular		Digital mobile telephone		Paging		Mobile data		PMR/PAMR	
	Operator	No. of users	Operator	No. of users	Operator	No. of users	Operator	No. of users	Operator	No. of users
Austria	Austrian PTT	220,000	Austrian PTT	0	Australia PTT	90,000	None	0	PMR Networks	175,000
Belgium	Belgacom Mobile N.V.	66,000	Belgacom	0	Belgacom	197,000	None	0	PMR	100,000
Cyprus	Cyprus Telecom	15,200	License	0	None	0	None	0	None	0
Denmark	Tele Denmark Mobile	252,300	Tele Denmark, Sonofon	64,000	Tele Denmark	60,300	None	0	PMR	150,000
Finland	Telecom Finland	460,000	Telecom Finland	20,300	Telecom Finland	42,000	Telecom Finland	800	Telecom Finland + PMR	100,000
France	France Telecom, SFR	470,000	France Telecom, SFR PCN License	91,000	France Telecom	358,000	France Telecom	0	Telecom France, PMR	500,000
Germany	Deutsche Telekom	800,000	DTK, Mannesman	975,000	DTK	464,000	Dete Mobil	300	DTK, PMR	1,450,000
Greece	None	0	Panafon < Stel Helles	21,500	OTE	20,000	None	0	PMR	20,000
Iceland	PTT Iceland	17,400	License awarded	0	PTT Iceland	5,600	None	0	None	0
Ireland	Eircell	55,300	Eircell, 2d license	300	PTT Ireland	5,600	None	0	PMR	15,000
Italy	SIP	1,177,000	SIP, 2d license	6,200	SIP	217,000	None	0	PMR	250,000
Luxembourg	Luxembourg PTT	500	Luxembourg PTT	4,400	Lux. PTT	6,600	None	0	PMR	5,000
Netherlands	PTT	213,000	Lic. awarded, 2d planned	0	PP	381,000	RAM	0	PMR, PAMR, Traxys	200,000
Norway	Tele-Mobil	363,600	Tele-Mobil, Netcom	8,000 8,000	Norwegian Telecom	123,100	Norwegian Telecom	400	PMR Networks	75,000
Portugal	TMN	31,000	TMN, Telecel	63,700	Tele Portugal	41,000	None	0	PMR	50,000
Spain	Telefonica	258,000	Two lic. awarded	0	Telefonica	120,000	None	0	PMR	200,000
Sweden	Telia, Comvik	795,000	Telia, Comvik, Nordictel	49,000	Telia, Comvik	179,000	Telia	10,000	PMR, Telia (PAMR)	270,000
Switzerland	PTT	249,000	PTT	9,000	PTT, Aircell	75,800	None	0	PMR	150,000
United Kingdom	Cellnet, Vodafon	1,957,000	Vodafone, Mercury, Cellnet, Microtel	42,100	BT, Vodapage, Aircall Hutchison	758,000	RAM, Cognito, Paknet, Hutchison	3,000	PMR, PAMR, others	953,000
TOTAL		7,402,000		1,354,500		3,151,000		14,500		4,463,000

SOURCE: Carse, Phillip and Gerry Garrard, "Continental Drift Towards Wireless Data," *Data Communications*, March 1994.

TABLE 9.3   Status of GSM-Based Data Services (Early 1995)

Country	Operator	Number of subscribers	Service launch dates				Short message services	
			Data	Fax	ISDN	X.25	Mobile terminated	Mobile originated
Australia	Optus Comm Pvt	50,000	Mid-1995	Mid-1995	4 qtr. 1995	N/A	10/93	1995
	Telstra Corp	N/A	Mid-1995	Mid-1995	N/A	N/A	4/94	4/94
	Vodafone Pty.	N/A	11/94	11/94	6/95	N/A	10/94	Mid-95
Belgium	Belgacom Mobile N.V.	67,000	2/95	2/95	2/95	2/95	4 qtr. 1995	4 qtr. 1995
Denmark	Danish Mobile	100,000	5/94	5/94	1995	1995	12/94	12/94
	Tele Danmark Mobile	150,000	1/95	1/95	1995	1995	10/94	4 qtr. 1995
Finland	Telecom Finland Mobile Service	70,000	3/95	3/95	N/A	1995	9/94	4/95
	Alands Mobiltelefon	500	1/95	1/95	N/A	N/A	1/95	1/95
	Oy Radiolinja	55,000	5/94	9/94	1994	5/94	1993	1/95
Greece	Panafon, Athens	85,700	12/94	12/94	N/A	N/A	12/94	3/95
Hungary	Pannon GSM, Budapest	N/A	12/94	12/94	N/A	N/A	N/A	N/A
	Westel 900 GSM Mobile RT	65,000	4/95	4/95	N/A	N/A	3/94	4/95
Germany	Deutsche Telekom Mobilfunk	>900,000	1/95	1/95	3 qtr. 1995	1/95	9/94	2 qtr. 1995
	E-Plus Mobilfunk	30,000	3/95	3/95	N/A	N/A	5/94	5/94
	Mannesmann Mobilfunk Gmbh	900,000	10/94	10/94	1/95	1/95	3/95	Mid-1995
Luxembourg	P&T Luxembourg	13,000	6/95	6/95	N/A	N/A	5/95	5/95
Netherlands	PTT Telecom, The Hague	90,00	4/95	7/95	N/A	N/A	8/94	8/94
New Zealand	BellSouth New Zealand	N/A	10/94	10/94	In trials	2 qtr. 1995	10/94	4/95
Norway	Netcom GSM, Oslo	71,000	10/94	10/94	1/95	10/94	3/95	3/95
	Telenor Mobil	80,000	5/94	5/94	N/A	N/A	2/95	N/A
South Africa	Mobile Telephone Network	120,000	4/94	4/94	1995	1995	3/94	3/94
	Vodacom Group	160,000	5/94	5/94	1995	6/95	1/95	Mid-1995
Sweden	Comviq GSM	200,000	10/94	10/94	Mid-1996	N/A	4/94	Mid-1995
	Europolitan AB, Sweden	70,000	1/95	1/95	N/A	N/A	4/94	4/94
	Telia Mobitel	240,000	5/94	5/94	N/A	N/A	5/94	4/95
United Kingdom	Hutchison	100,000	6/95	6/95	1995	1995	4/94	4/94
	Telecom Securicor Radio	20,000	12/94	12/94	2 qtr. 1995	12/94	N/A	N/A
	Vodata, Newburgh, United Kingdom	118,000	10/94	10/94	Late 1995	Late 1995	N/A	N/A

SOURCE: Carse, Phillip and Gerry Garrard, "Continental Drift Towards Wireless Data," *Data Communications*, March 1994.

an agreement to relax the trade barriers inherent in the Japanese market was eventually signed, only time will show just how effective the agreement really is and to what extent U.S. companies will be able to penetrate the Japanese wireless market. Before this agreement, no foreign manufacturer was allowed to produce cellular phones in Japan. So far since the agreement was signed, Motorola has supplied a token number of cellular phones to NTT and DDI. The company hopes to increase its market share significantly.

NTT has selected Motorola, AT&T, Ericsson, and six Japanese manufacturers to develop its digital cellular network. Fujitsu, Matsushita, Mitsubishi, and Motorola are supplying the digital phones. Motorola, NEC, and Ericsson have supplied infrastructure technology to DDI. Nippon Ido has selected AT&T, NEC, Fujitsu, and Nokia as its equipment suppliers. Ericsson and Toshiba have formed a joint venture to develop their digital telephone business in Japan.

DDI has introduced a two-way PCS-type personal handy-phone (PHP), or personal handy-communications system (PHS), which looks like most cordless phones. The PHP operates in the 1.9-GHz band and started operating in 1993. PHP base stations are similar to PCS microcells in the United States and cost only $25,000 to $30,000. However, because their coverage is only 330 to 650 feet (100 to 200 meters), more of them are required. Japan hopes to see the PHP widely accepted and expects to have sold 40 million units by 2010.

9.6   Asia/Africa

Hong Kong has the largest per capita cellular phone count in the world, with Motorola's CT-2 being the most popular model. Paging services are also very strong. The government licensed Hutchison Telephone, Pacific Link Communications, and Hong Kong Telecom CSL to start cellular service in 1983. Subsequently, the huge demand necessitated a fourth license. The four carriers are now planning their conversion to digital. It is expected that they will use U.S.-style dual-mode telephone sets supplied by Japanese and U.S. manufacturers (who dominate the market in Hong Kong).

Motorola and Ericsson are major equipment suppliers to the People's Republic of China, which has the potential to be a huge market. Two cellular phone systems have already been constructed there under a joint venture between Novatel of Canada and Fonic of Hong Kong. Novatel-Fonic is also producing phones, pagers, and CT-2 phones in China.

In Taiwan, the public-sector-controlled Computer and Communications Laboratory in Hschichu is developing a DSECT-based handheld phone and PBX. Estimates as to the possible number of DSECT users in the country by the year 2000 run as high as 42 million. However, these numbers seem to be very high indeed to us.

In South Korea, state-run mobile telecommunications carrier Korea Mobile Telecommunications Corporation introduced mobile phones in 1984. In 1992, it had 340,000 subscribers, a figure that is expected to grow to five million by the year 2000. The Korean Ministry of Communications has selected QUALCOMM's CDMA digital technology for the Korean cellular telephone system, with commercial trials expected to start in 1996. The equipment will be manufactured by Korean electronics companies such as Goldstar, Hyundai, Maxon, and Samsung under license from QUALCOMM. It is expected that these manufacturers will become alternate suppliers of CDMA equipment in the United States and other countries implementing CDMA technology.

India is another sleeping giant that is just waking up as it starts opening its markets to foreign participation. One interesting aspect of installing cellular network infrastructures in countries like India is that it often turns out to be more economical to provide cellular network services instead of traditional wired-line networks in rural areas. In fact, there is a huge potential for wireless networks in India for both cellular applications and inexpensive paging applications. Motorola has already set up a plant in India for manufacturing pagers. India is always interested in joint proposals from foreign companies who would like to manufacture components in India. Such undertakings can serve a dual purpose: they can satisfy local demand, and they can provide low-cost products for export to Europe and North America.

9.7 South America

While Latin America has generally been slow to adopt cellular technologies, several nations are now eager to embrace PCS in order to make telecommunications services more readily available to consumers and businesses.

In 1995, at a conference in New York on investment opportunities in Latin America, speakers from five of the six countries that made presentations raised the subject of PCS in discussing wireless plans. Chile has one of the most liberalized telecommunications markets in South America. PCS licenses have been awarded and the licensees are free to use whatever technology they choose. Service is expected to start in late 1996.

Because PCS operators in Chile will face stiff competition from established cellular services, prospective PCS operators have already begun to advertise deep discounts on usage rates. Such tactics may be short-lived, however, since the PCS industry, like any other, must eventually provide a suitable return on investment.

Chilean telecommunications operator CTC Cellular set up a Mobitex network in 1994 to serve the six million people in the Santiago/Valparaiso metropolitan area. Several public safety organizations and oil companies were among the first users. Point of sale and credit card validation are other applications. CTC has been awarded a private license and must seek a special concession for every new customer.

Progress is being made in other Latin American countries, also. In Argentina, the government has awarded licenses to offer PCS in the greater Buenos Aires area by mid-1996.

Brazil is still trying to define its PCS frequency plans. The government there expects to begin licensing activity in 1997.

In Mexico, where analog cellular service is already available in many parts of the country, the government is considering U.S.-style auctioning of PCS frequencies. Such an approach would help the country's economic problems and relieve pressure on the peso.

In Venezuela, where the government views PCS as an important telecommunications service, a network is expected late in 1996.

9.8 Mobitex and RAM Mobile Data Network Coverage Around the World

There are 16 public Mobitex networks in 16 countries on 5 continents, as shown in Fig. 9.2. Included in this list are Australia, Belgium, Canada,

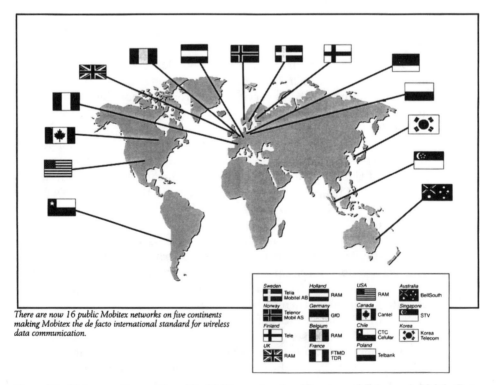

There are now 16 public Mobitex networks on five continents making Mobitex the de facto international standard for wireless data communication.

Figure 9.2 Sixteen nations with public Mobitex networks. (*Courtesy of Ericssson's* Mobile Data News.)

Chile, France, Holland, Korea, Finland, Germany, Norway, Sweden, the United Kingdom, the United States, Poland, and Singapore.

ARDIS has installations in the United Kingdom, Canada, Germany, Australia, Malaysia, Singapore, Sweden, and Thailand.

Summary

In this chapter, we have seen that there are different but vibrant wireless network scenes in Canada, Japan, Latin America, Europe, and Asia. Canada, Japan, and Latin America have opted for PCS, while the rest of the world has embraced GSM. Not only can we expect individual countries to be good markets for wireless networks, but we should also see GSM networks mushrooming everywhere except in North and South America and Japan. GSM/PCS integration issues will become increasingly more important as the two groups of countries proceed on their own distinct paths.

Remote Network Access Solutions for Mobile Computing

While mobile computing is moving so slowly, the remote network access industry based on switched wired networks has taken off. Whereas mobile computing is complex, RNA is simple and straightforward. RNA allows remote workers to connect to LAN applications and data resources transparently—in a few hours for a simple configuration to a few days for a complex one. RNA vendors will provide support for the wireless networks in their communications servers when the wireless network industry gets its act together by standardizing communications interfaces.

RNA INDUSTRY VIEWPOINT

About This Chapter

In Part 3 of this book we have so far looked at wireless networks and their ability to provide connections anywhere, anytime. However, wireless networks are not the only ways for mobile workers to make the connections they need. There are many more workers on the road who use standard PSTN networks to connect to their information servers—from their homes, their hotels, and their customers' offices—than there are who use wireless networks. And then there are the high-speed ISDN and ADSL connections provided by the remote network access (RNA) industry. It is to this topic that we shall now turn our attention.

10.1   RNA

RNA can be defined as the logical connectivity of either fixed or mobile remote client nodes to a central information server. From a networking point of view, RNA encompasses far more than does mobile computing. In fact, the dictionary definition of the term includes all forms of remote access, including mobile access and traditional non-fixed-link network access. It is not surprising, therefore, to see *Communications Week* magazine "demote" its Mobile Computing section to a subset of its Remote Access section. While such a broad definition and usage of the term is appropriate from a networking perspective, we feel that vendors and users alike have a more limited idea in mind when they speak of RNA. Moreover, mobile computing itself can be said to include more than just networking when applications and back-end systems are taken into consideration. Therefore, in the context of this book, we shall stay with the narrower definition of RNA as a subset rather than a superset of all forms of remote computing.

Thus in our context, RNA denotes the hardware and software technology used to connect remote users to central LAN servers using switched network connections. Additionally, RNA is distinct from corporate networks or LAN/WAN internetworking in the sense that it does not require a continuous connection to LAN resources. Unlike wireless network solutions, RNA predominantly uses PSTN links, including both regular dial-up and ISDN links. In fact, in the corporate world, the PSTN/ISDN-based RNA market is more developed than the wireless data market. While RNA does not provide the ultimate mobility that wireless networks provide, it does give mobile users access to mobile computing applications and corporate information wherever there is a telephone jack, be it in a customer's office, at home, or in a hotel.

Although the same business imperatives that are galvanizing mobile computing are spurring the growth of RNA (distributed operations, telecommuting business process reengineering, downsizing, etc.), there is one major difference between the two: RNA is cheaper and easier to use with existing applications.

10.2   Business Users of RNA

Anybody who is not permanently connected to a corporate network can be a remote network user. That means that in addition to internal mobile users, RNA services can also easily be extended to customers, suppliers, and business partners, as illustrated in Fig. 10.1.

10.2.1   RNA market size

RNA is one of IT's fastest growing markets. There are many reasons for this. The single most important factor is user demand for connectivity to home offices and corporate information resources—a demand that the vendor community has met with relatively inexpensive and simple entry-point solutions. Thus, with only minimal budgets, corporations can easily give remote users dial-up access hardware and software. Once a few users in any particular setting experience the convenience and efficiency of remote access, others are quick to join the band-

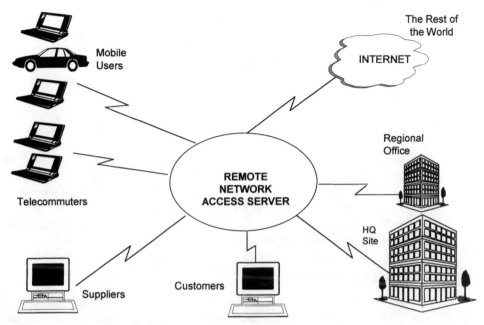

Figure 10.1 Remote network access—business user community.

wagon—in the process demanding access not just to e-mail, but to all the corporate applications and data. And fueling this demand is the availability of inexpensive, fast ISDN links that give remote connections LAN-like speeds. After a slow start in the 1980s and no growth at all for years, the number of ISDN links in the United States is estimated to grow by 24 percent in 1996. According to an IDC study reported by Shiva, 66,000 RNA systems and 740,000 ports valued at a combined $400 million were sold in the United States in the first half of 1995. According to INFONETICS of San Jose, California, the 62-percent penetration represented by the 300,000 RNA interconnected branch offices reported in 1995 will increase to a 91 percent penetration by the end of 1997. We do not believe that everybody who wants remote access from branch offices will have it by the end of 1997, but the majority of high-priority users will probably have the capability. The worldwide RNA market (hardware, software, and modems) was estimated at $1.7 billion for 1995, according to Electronic Trends Publications, as published in *Communications Week* in September 1995.

10.2.2 Rapid but haphazard growth

It is our observation that remote access is growing rapidly but in a haphazard fashion, and that network managers are largely unaware of what is going on in this area. Users are being exploited by vendors pushing products that may or may not be appropriate. Often, their low cost and ease of use are leading to wholesale adoptions of remote access solutions without proper research or planning.

This is not to suggest that there is anything wrong with users experimenting and trying new applications during the formative stages of a new technology. Far from it. It is just such bold experimentation that leads to breakthroughs and the emergence of new applications. However, in this particular case, we believe we have now reached a critical mass for this technology in medium- to large-size organizations. To progress beyond this point in an orderly, productive manner, we need to thoroughly grasp the business, technical-design, and management issues involved in providing remote access to the users. It is time for IT and network management staff to take charge and start advising those who are implementing remote network access solutions.

10.3   Remote Network Access Options

We briefly talked about these options in Chap. 5. It is time now to examine them in greater detail. The following are four generic approaches to providing remote network access to mobile users:

- Bulletin Board System (BBS)-style terminal emulation
- Remote control
- Remote node
- General purpose remote access servers, including mail gateways

We shall first describe each of these approaches briefly in order to clarify the differences.

10.3.1   Remote terminal emulation mode

In this implementation, also called *Network Bulletin Board System,* a terminal emulation software in a remote notebook is connected through a public switched network to a LAN as if it were a locally attached device. There are a number of terminal emulation software choices available, depending on the communications protocol of the host information server. These range from specialized terminals such as TN3270 for IBM mainframes and DEC LAT for DECNET-based servers, to the ASCII terminals implemented on CompuServe and similar information services. On the host end, there is a terminal emulation server software or a network bulletin board software such as Wildkat BBS from Mustang Software. With this approach, a LAN manager can provide access to LAN files, network e-mail, and some LAN applications. Most bulletin board or terminal servers operate in archaic character mode, though the interface can be upgraded to graphical user interface with the use of terminal emulation software that supports RIPscrip. Mustang Software's Qmodem is one such example. Access to e-mail is provided through a mail gateway.

The major advantage of this approach is the minimal demand on remote client nodes. All that is needed are a modem and terminal emulation software. Windows 3.1, Windows 95, NT, and OS/2 all provide terminal emulation software as an integral component of their communications driver suites.

The disadvantages are the limitations inherent in an old terminal interface and participation on the LAN as something decidedly less than a peer—a fact that will doubtless not be appreciated by users. Thus, this approach does not have much of a future. It is a short-term arrangement.

10.3.2 Remote control mode*

This scheme involves two PCs: the mobile user's remote client notebook and a surrogate PC at the LAN (Fig. 10.2). The remote user dials into the LAN and takes control of the host PC. With both running matching software, every piece of information accessed by the surrogate PC on the LAN appears on the client notebook PC. Carbon Copy

* "Using Remote Node with Remote Control," white paper by Shiva Corporation, an RNA vendor.

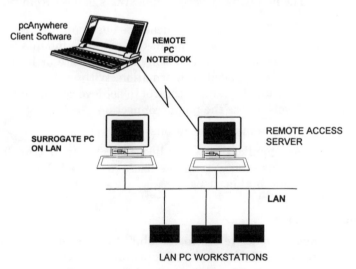

Figure 10.2   Remote-control concept in RNA architecture.

and pcAnywhere are two well-known examples of such an implementation. Remote control software is, in fact, a client/server implementation in so far as the host/surrogate PC runs the application and the client notebook displays the information.

This was originally devised as a solution to provide remote support to software technicians working from home or diagnosing remote systems from their offices. It did not take long for other mobile users to discover its benefits, however, and being very easy to implement, it grew rapidly in popularity.

An advantage of the remote control mode is its ability to access data at LAN speeds because of the fact that the remotely controlled or surrogate PC actually does reside on the LAN. Only displayable screen data is sent across the modem dial-up link. On the downside, as the remote node camp has been wont to point out, more often than not, as many surrogate PCs have been needed as the number of remote concurrent users. To counter this criticism, remote control vendors are now offering solutions, such as Symantec's (pcAnywhere) remote control server that runs on a single host PC and supports up to eight independent remote control sessions.

Because remote control applications are implemented using proprietary technologies between a remote control client and a server, security is weak. Theoretically, anyone can load remote communications software and dial into a host PC without the system administrator's knowledge. To counter the possibilities of this happening, remote control packages offer some security through features such as user passwords and dial back.

The following types of applications benefit from remote control:

1. Flat file database applications based on FoxPro, dBase, or Paradox that require the transmission of large amounts of data from the file server to the workstation (the surrogate or host PC) before a user can access needed information. With such applications, it is necessary for the remote control to achieve certain minimal performance levels across the phone connection.

2. Older e-mail applications.

3. Any application whose license prohibits its transfer to a remote PC. In this scenario, remote users are saved the expense of purchasing additional licenses.

10.3.3   Remote node mode

Remote node refers to the ability of a remote PC user to dial into a LAN-attached server and function as a full node (or a peer) on the network, thus permitting remote users the same privileges and level of

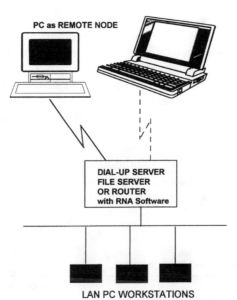

PC as REMOTE NODE

DIAL-UP SERVER
FILE SERVER
OR ROUTER
with RNA Software

LAN PC WORKSTATIONS

Figure 10.3 Remote node terminal emulation mode.

access to LAN-based resources as their office-based colleagues (Fig.
10.3). In this mode, packets are transmitted to and from the remote PC
over the phone lines. Network managers like the remote node option
because it is easy to manage and extend. An increase in the number of
ports is achieved by adding modems and multiport serial boards.

Remote node has the great advantage over remote control of full
security: several advanced products integrate with Novell's Netware
bindery software, and some also integrate third-party hardware security products that employ handheld devices or *tokens*.

While remote node is the method most preferred by network managers, its speed is a major drawback. Typical dial-up speeds of 28.8
Kbps are slow in comparison to the 10 Mbps of a LAN. Even with compression factors included, modem connections are three to four times
slower than LANs. This results in noticeable delays for remote users
while accessing LAN file servers. Even with ISDN connections increasing link speeds to 64 Kbps, it is unrealistic, we believe, for remote
mobile users to expect the same performance levels from remote connections as from LAN-based locally attached workstations.

Accordingly, remote node mobile applications should be designed
differently from local LAN applications, with as many functions as
possible being performed on the remote notebooks. Host information
servers should be used to access database information, and only the
results themselves of the queries should be transferred across dial-up
links.

Newer client/server applications that involve file transfers or access to SQL databases are more suitable for remote node operations. Many applications are now using the client/server design. These include Lotus NOTES, cc:Mail Mobile, Microsoft Mail Remote, and the like, as well as popular relational databases such as Informix, Oracle, Ingress, and Sybase. These applications operate well over remote node and it is expected that more such applications will be developed.

A remote node connection can take a number of different forms, depending on whether it is incorporated into a file server, or whether it is connected to a router or a stand-alone PC.

We summarize in Table 10.1 some of the advantages and disadvantages of the two key approaches in RNA: remote control and remote node.

TABLE 10.1   Comparison of Remote Control and Remote Node

Remote node		Remote control	
Advantages	Disadvantages	Advantages	Disadvantages
Availability: All LAN resources are available to the user.	*Performance:* Does not work well with legacy PC applications and databases that are executed from the host PC	*Performance:* Works best with older, non-client/server applications.	*Cost:* Generally more expensive for the RNA portion. May need one-for-one host PCs with nonserver implementations
Cost: Hardware is less expensive than remote control. Software license costs are higher.	*PC application software management:* More difficult to maintain the remote application software	*Client PC application management:* Easier to manage centrally maintained applications.	*Ease of use:* Interface is slightly different
Scalability: Can support more users.		*Hardware life:* Do not need powerful client nodes, because applications are run on the host PC.	*Security:* Does not require centralized security for authentication. Does not work with tokenized security packages
Ease of use: Same interface as on LAN.			
Performance: Good solution for client server, graphical applications that do a lot of processing in the remote client workstation.		*Performance:* Faster for PC database-intensive applications.	
Security: Centralized management of remote access servers and client nodes.		*Cost:* Application software costs are lower, since remote licenses are not required.	
Tunneling: Can combine remote control within a remote node session in one dial-up connection.			

SOURCE: Shiva Corporation White Paper modified by author.

10.3.4 General purpose remote access communications server

This category of remote servers involves sophisticated implementations of all of the three previous concepts in one box. Some communications servers, such as those from Shiva Corporation and Cubix, are hybrid in the sense that they support both remote control and remote node. Shiva currently ships a special version of pcAnywhere called Norton pcAnywhere, Shiva Edition. Using this product, users can switch between a network application, a DOS or Windows remote control session, and the file manager to access a mapped drive on the LAN—all in the same dial-up session. Novell's Communications Server and IBM's LAN Distance solution also support both modes.

If the communications server functionality of RNA is combined with the software sophistication of agent-based technology (to be discussed in Chap. 12), the result is a powerful communications solution. This is precisely what a partnership between Shiva (makers of hardware and software for communications servers) and Xcellnet (suppliers of a sophisticated agent-based remote application solutions) is expected to achieve: a complete solution for remote access customers (Fig. 10.4).

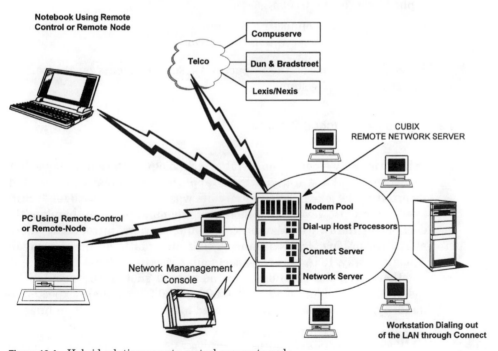

Figure 10.4 Hybrid solution: remote control or remote node.

10.4   Components of RNA

A remote network access solution consists of the following components:

- A remote client PC, typically a PC notebook or a MAC Powerbook
- A modem or PCMCIA ISDN adapter in the remote client notebook
- Communications software residing in the remote client notebook (provided by the RNA vendor, probably a terminal emulation)
- A dial-up or ISDN switched connection
- A modem pool or equivalent digital ISDN hardware at the server end
- Host PC or communications-server hardware at the location where information resides (supplied by the RNA vendor)
- Communications-server software (typically provided by the RNA vendor)
- Routing software (supplied by the RNA vendor)
- Security software (supplied by the RNA vendor or by a specialized security software vendor)
- Protocol-conversion or mainframe-gateway software (typically supplied by a third party)
- Backbone WAN connecting communications server to mainframe or other application servers in the network

Section 10.6 discusses the impact of these components on the performance of an RNA system.

10.5   ISDN as a High-Speed Switched Digital
Link for RNA*

ISDN originally emerged as a viable technology in the early 1980s, but its expense and a scarcity of ISDN applications dramatically stunted its growth. The joke in seminar circuits was that ISDN stood for "I Still Don't Know." However, with increasing demand for more bandwidth, decreasing hardware adapter costs, and carriers introducing more attractive pricing, ISDN is undergoing an equally dramatic revival. Along the way, RNA and the spectacular rise in Internet traffic have together given the technology an additional shot in the arm. ISDN installations doubled in 1995 (250,000 to 450,000) and are expected to quadruple by the year 1999 (estimated to increase to 2 million lines).

---

* "Understanding ISDN Solutions," white paper from Xyplex Corporation is a good source of introductory information on ISDN. Xyplex is an RNA system vendor. There are several books on ISDN available in the market for those who want more details.

The following is a brief description of ISDN to enable its evaluation as a viable option for RNA solutions:

10.5.1 What is ISDN?

ISDN is essentially a digital phone call—a high-speed digital service that carries simultaneous voice and data communications over existing twisted-pair phone cabling systems. Most companies offer ISDN service for home or office at attractive rates. As with regular long-distance phone service, ISDN calls are charged to the caller and billing is based on time connected rather than on data transmitted (as is the case with packet-switching services).

10.5.2 How does ISDN work?

ISDN is basically a WAN service available in two forms: Basic Rate Interface (BRI) and Primary Rate Interface (PRI) with speeds up to T1/E1 levels (1.5 or 2.0 Mbps, available in increments of 64 kbps). There are two main differences between ISDN and T1/E1, however. ISDN is a switched service with pay-as-you-go rates, unlike a dedicated T1/E1 line. Also, signaling in ISDN is done on a separate channel, meaning that more bandwidth is available for data transfer.

Channel types. ISDN uses *time division multiplexing* (TDM) to divide the available bandwidth into a number of fixed-size time slots, called *channels*. With ISDN, the local loop from the customer's premises comprises two basic channels, B and D, and several logical channels.

The B channel. The B (for *bearer*) channel carries user data as well as digitized voice and video information. It operates at a data rate of 64 kbps and can be used for both circuit-switched and packet-switched applications.

The D channel. The primary purpose of the D channel is to provide signaling and control for each ISDN line installed. However, since the exchange of signals on the D channel rarely uses all of its bandwidth, excess capacity can also be used for carrying data—though signaling always takes priority.

Access interfaces: BRI and PRI. The number of logical channels is a function of the local loop's bandwidth. As mentioned, there are two standard access interface types: Basic Rate Interface (BRI) and Primary Rate Interface (PRI).

BRI is commonly used for RNA and Internet connections and is available in most major metropolitan areas of the world today. BRI consists of two B channels (64 kbps each) and one D channel (16 kbps) for a maximum uncompressed data rate of 128 kbps (or 144 kbps if the D channel is used for data as well). The BRI is usually provided through two wires with an RJ-45 jack.

PRI is a higher bandwidth interface used at central communications server sites to aggregate BRI connections, or in applications where a native bandwidth higher than 128 kbps is required. Generally, PRI is less widely available because of cost and bandwidth. In the United States and Canada, PRI comprises 23 B channels and a single D channel. In Europe and most of the rest of the world, it comprises 30 B channels and a single D channel. The PRI D channel operates at 64 kbps, as do all B channels. This adds up to 1.5 Mbps in North America and Japan and 2.0 Mbps in the rest of the world. PRI is presented as a four-wire trunk circuit connecting an ISDN port on a customer's computer communications equipment (such as a bridge/router) to the local exchange. The physical interface is DSX-1 in the United States and Canada and 75-ohm G.703 (BNC) or 120-ohm G.703 (RJ-45) in the rest of the world.

ISDN functional devices. ISDN standards define several different devices that can be connected to a network, as well as interfaces between devices. Each interface, called a *reference point,* requires a communications protocol. Since ISDN was developed for telephony, common networking equipment like bridge/routers do not readily conform to ISDN standards.

Terminal equipment types. End-user devices such as digital or analog telephones, X.25 Des, or bridge/routers are called *terminal equipment.* There are two types of terminal equipment. Devices that use ISDN directly and support ISDN services are known as *terminal equipment type 1* (TE1). Non-ISDN devices are known as *terminal equipment type 2* (TE2) and require a terminal adapter (TA) and software that enables the TA to communicate with an ISDN adapter. For example, a bridge/router with a standard synchronous serial interface is a TE2, while a bridge/router with a native BRI interface is a TE1.

Network termination types. There are two types of network terminations. *Network termination type 1* (NT1) is a network terminal device, typically the local carrier's network termination unit at the customer site. This is the device to which all networking equipment is connected. The NT1 terminates the physical connection between the customer site and the local exchange and connects the four-wire customer wiring to two-wire local loop. The NT1's functions include line performance monitoring, timing, physical signaling protocol conversion, power transfer, and multiplexing of the B and D channels.

Network termination type 2 (NT2) equipment provides customer site switching, multiplexing, and concentration of multiple ISDN lines. The NT2 is a more intelligent piece of equipment and can include voice and data switching devices such as PBXs. The functionality of a bridge/router with a Primary Rate Interface would fall into this category. The NT1 and NT2 devices may be combined into a single physical device called NT12. This device handles the physical, data-link, and network-layer functions.

10.5.3   ISDN reference points

ISDN reference points define the communication between different functional devices. Four protocol reference points, called R, S, T, and U, are commonly defined for ISDN with different protocols used at each reference point.

The R reference point is between non-ISDN terminal equipment (TE2) and a terminal adapter (TA). There are no specific standards for the R reference point; the TA manufacturer will specify how the TE2 and TA communicate with each other. Typically, this protocol is either RS-232 or V.35.

The S reference point is between the ISDN user equipment (TE1 or TA) and the network termination equipment (NT1 or NT2).

The T reference point is between the customer site switching equipment (NT2) and the local loop termination (NT1). (In the absence of the NT2, the user-network interface is usually called the *S/T reference point.*) The protocols used at the S and T reference points are specified by the ITU (CCITT).

The U reference point is between the NT1 and the local exchange. The protocol used at this reference point is defined by the ISDN provider.

When you place your ISDN order, find out whether your carrier supplies NT1 equipment. (Carriers will sometimes ask if you have a U or a T interface at your site. A *U interface* means you have an NT1; a *T interface* means you don't.) Some carriers will lease NT1s for a small monthly fee, but others do not offer them at all. In the United States, NT1s can also be purchased from many DSU/CSU vendors.

10.5.4   ISDN bandwidth management options*

ISDN has the potential to be cost-effective, but has no inherent management features. Rather, it is the responsibility of the networking equipment vendor to provide these capabilities. Communications servers use sophisticated router software to manage available ISDN

---

* Information based on Xyplex's white paper on ISDN. These features are present in Xyplex's Internetworking software.

bandwidth automatically. Some or all of the following management features are available in currently available RNA products:

Dial-on-demand: Since ISDN charges are paid on a per-minute basis, this feature brings up a link only when there is data to be sent. When a router receives a packet for another network over an ISDN link, it establishes the call and transfers the data.

Bandwidth-on-demand: This feature in some routers (such as those from Xyplex) provides two capabilities: the ability to add bandwidth to a link and the ability to provide an alternate path when a primary link fails. Routers can be configured to add bandwidth whenever user-defined thresholds are reached.

Inverse multiplexing: This is the ability to combine multiple low-speed channels across leased lines or switched services to form a single, higher-speed data path. Bridges/routers can dynamically group two or more ISDN circuits to the same partner and dynamically balance the load over them.

Time-of-day circuit control: Network administrators can configure bridge/routers to automatically establish circuits for fixed durations at specific times during the day.

Compression: Various kinds of compression are offered over ISDN links. Stacker LZS (TM), V.42bis, Van Jacobson header compression for TCP/IP headers, and Compressed IPX (CIPX) for compression of IPX headers are examples. Stacker LZS compression has been known to achieve as high as four times compression ratio. Note, however, that compression ratios achieved depend on many factors, including the type of data items and files beings transmitted.

Multilink PPP: This provides load balancing over a variety of multiple circuits (such as ISDN, X.25, and leased lines). It provides more bandwidth by using WAN services more efficiently and quicker response times by using parallel transmissions.

Call priority: Call priority is critical for certain client/server applications. Network administrators can set priorities for circuits or circuit groups and assign WAN services to critical data calls during the day.

Telecommuting support: Internetworking software accommodates telecommuters by allowing PCs, workstations, and notebooks to dial-in and gain access to network resources. With PPP support across ISDN or any WAN service, telecommuters with password clearance can dial into their network through a router-based communications server, using the most economical and/or available network service.

WAN security: With the rapidly increasing use of remote networks for financial transactions, different forms of security must be available on switched links. Three common methods are dial back, PPP authentication protocol (PAP), and WAN data scrambling. PAP requires a password and a PAP ID from the remote PPP devices before a connection can be established.

WAN independence: As new switched services are offered, RNA server software should be able to mix any combination of WAN protocols: dial-up, ISDN, frame-relay, switched-56, and leased lines.

10.5.5 How much does ISDN cost?

ISDN prices vary from region to region and country to country. Table 10.2 represents 1995 prices (readers should obtain current prices from the local telephone company).

10.6 RNA Technology Issues

There are several technology issues that RNA planners need to keep in mind. However, because they are generic issues that apply to all mobile computing networks, not just to RNA PSTN implementations, they will be discussed in detail in Chaps. 15 and 18. We shall satisfy ourselves here with a brief point-form list:

- Synchronization of data on the remote client PC with that on the server
- Platforms and protocols supported by particular vendors
- Network management
- Performance optimization

10.6.1 Performance optimization in RNA configurations

Too often, users blame the technology, on the one hand, when they experience poor performance, without making any efforts at optimiza-

TABLE 10.2   ISDN Prices in 1995

Region/country	Installation	Monthly charge	Per minute
Nynex USA	$124	$38	$0.085
Pacific Bell, USA	$125	$20	$0.010
US West	Free	$60	Free
Bell Canada	$209	$91	Free, now
France	$119	$35	$0.19
Germany	$79	$42	$0.14
United Kingdom	$600	$30	$0.19

SOURCE: Xyplex white paper.

tion themselves, on the other. The reality is that presently very few technologies—RNA solutions included—are self-optimizing. A brief list follows of some of the optimization considerations to be kept in mind for RNA configurations.

End-to-end optimization.* This is the fundamental approach to mobile computing solutions recommended throughout this book. Thus, all hardware and software components should be evaluated in terms of their contribution to the overall optimization of the solution from the end user's perspective.

Remote client computer. This must provide as much local functionality as possible while minimizing the need to transfer huge amounts of data over switched links. In this context, a remote-control approach to all applications is not desirable. Rather, it should be considered for selected applications only.

PC serial port performance. This has an effect on the overall performance of a remote system. Unfortunately, the serial port seldom is given the attention it deserves. Older serial chips such as 8250 or 16450 were adequate for slow serial speeds. But ISDN and faster speeds need the more advanced 16550A serial chips for optimum performance. It is recommended, therefore, that both remote client PCs and serial boards in the communication server be equipped with 16550A serial chips.

Application software packages. These should be designed in such a way that they transfer only minimum amounts of data. Application designers should not fall into the trap of assuming that the users are necessarily going to be connected to high-speed LANs. When application designers cut corners in the interests of productivity, their users ultimately pay the price in terms of poor performance.

Network operating system (NOS). This can also play an important part in determining the overall performance of a remote-access system. The NOS protocol implementation is extremely important in the efficient utilization of link bandwidths. For example, protocols that acknowledge only a single packet at a time can be very inefficient (Novell's IPX/SPX is one such protocol). By contrast TCP/IP implements a sliding-window protocol.

Remote-access client software. This is the end user's point of entry into the remote-access system. Its primary function is to start a session, manage traffic flow, and then terminate the session with the host RNA server. Therefore, overall remote-session performance and

---

* "End-to-End Optimization in Remote Network Access," white paper from Shiva Corporation.

ease of use are highly dependent on this software. The error recovery built into it and its ability to interface with business applications are of the utmost importance.

Remote-access client software can be thought of as consisting of three parts: a dialer that manages the link, a driver that interfaces with the network protocol stacks, and a driver that interfaces with the serial port to send and receive network data.

The dialer is critical to the end user's ease of use and also determines the data transmission equipment that the remote-access system can use. It should support a variety of different modem types, phone connections, and security systems usable with a remote-access system and should be flexible enough to accommodate new modem types or other data communications equipment such as ISDN terminal adapters. A dialer must also be able to deal with complex connection sequences. Entering credit card numbers, passing through packet networks such as X.25, and navigating sophisticated security systems are all commonly required in establishing connections. A scripting capability for this type of navigation is extremely useful.

The *remote-access client's driver* is important in terms of the protocols it supports. Often, the broad variety of applications that mobile users expect to be available in the field requires multiple protocol support. Novell's IPX and TCP/IP are invariably mandatory. NetBEUI, used in LAN Manager and Workgroup for Windows, LLC/802.2 for IBM's LAN Server, and host connectivity are also required in many cases.

Modems.   These should support as high a speed as is sustainable on the link being used. A V.34 speed of 28.8 Kbps is common right now. Higher speeds will follow. Modems should also support the highest available compression levels. Users should be aware of the fact that vendor performance claims are based on performances measured under highly optimal conditions. A 4:1 compression ratio touted in sales literature is not always—or even often—possible. Rather than breaking connections at the slightest hiccup, modems should be able to cope gracefully with deteriorations in link conditions. This requires that they be able to fluctuate between faster and slower speeds in the same way that fax machine modems do.

Dial-up connection or ISDN.   Phone connection can have all sorts of problems—noise, distortion, and echoes. Substituting dial-up links with ISDN links can improve performance considerably. Users may find that the costs are comparable.

Remote-access server hardware.   Ideally, where large numbers of users are involved, hardware should be designed for, tested for, and optimized for remote access.

Remote-access server software.   Perhaps the most important component in determining performance is the remote-access server software. It should support shared use efficiently, and it should also support flexible TCP/IP addressing options.

LAN efficiency.   Since information resources reside either on a LAN or through a LAN gateway on a mainframe, LAN use and efficiency can also affect overall performance.

10.7   Understanding the Benefits and Costs of RNA Solutions

10.7.1   Benefits

A business case for an RNA solution can be justified as follows:

- Mail and distribution cost savings
- Form design and production cost savings
- Software distribution savings (as compared to post-office-based distribution)
- Telephone savings (in terms of long-distance voice communications)
- Hardware savings as a result of the consolidation of individual remote-access systems by different departments
- Productivity increases and increases in sales

10.7.2   Typical costs
of an RNA configuration

RNA planners need to examine the costs of the various options. An eight-port remote node system for 100 users may cost around $8,000 to $10,000. A corresponding remote-control configuration with individual surrogate PCs could cost up to $26,000. A remote-control configuration based on a shared server (such as pcAnywhere) will be much less expensive. The following ballpark estimates should be used as guidelines only:

- Cost of hardware and modem will run upward of $1000 per port for remote node. For remote control, add another $1000 for the cost of PC and Ethernet cards.
- Assume support for 10 to 12 users per port.
- Remote client software varies from $0 to $100 per user.
- Remote-access server software may vary from $0 (if it is bundled with hardware) to several thousand dollars for a 10-port system.

After a comprehensive survey, INFONETICS Research was able to show that initial acquisition costs are only 15 to 20 percent of the overall costs involved in implementing an RNA. Hidden user costs account for another 35 percent and recurring operational costs account for the remaining 45 to 50 percent.

10.8 Wireless Network Support in RNA Servers

During the 1994–1995 period, RNA vendors were busy trying to meet demands from customers who wanted to put in place solutions based on public switched telephone networks. While true mobile computing business requirements based on wireless networks are growing gradually but slowly, dial-in requirements are mushrooming. Accordingly, RNA vendors have been slow in incorporating support for wireless networks other than cellular networks. Cellular network connections do not require any additional hardware or software, with the possible exception of specialized modems in some instances. There is very little support for ARDIS, Mobitex-based RAM Mobile Data, and CDPD networks in communications servers offered by RNA vendors.

We expect this product strategy to change and that in 1996–1997, more and more RNA vendors will offer software drivers for the wireless networks mentioned above. As the wireless network industry develops standard communications interfaces, it will be easier to extend RNA solutions to wireless networks by simply adding software drivers. In future products, we also expect the RNA communication-server functionality to be integrated with the MCSS functionality (described in Chap. 8) because it makes sense from customers' business-requirements perspective, as well as from their network-management perspective.

Summary

In this chapter, we discussed RNA, a switched networking technology designed to connect remote mobile workers as well as customers and suppliers. We started with the premise that RNA is a de facto technology for network managers right now. Remote users are demanding it, business cases are strong, and the technology is ready, available, and easy to use.

While experimentation is desirable, we also need to approach the problem and evaluate our options from an end-to-end perspective. Network managers must provide technology leadership rather than stand by while users are sold solutions that may or may not be appropriate by self-interested RNA-vendor salespeople.

We described various options such as terminal emulation, remote control, and remote node. Each of these options has its advantages and disadvantages. Some vendors provide hybrid solutions that combine the best of both worlds. Application interfaces are extremely important. Agent-based communications servers have distinct advantages for large installations. Both dial-up links and faster ISDN links should be evaluated. ISDN links are becoming increasingly popular and cost-effective. The best solution for the organization should be selected.

Finally, we must consider compatibility/interoperability issues that exist between wireless network solutions that provide true mobility, but are more expensive, and RNA network solutions. We should optimize the solution for the organization by combining RNA and wireless network solutions for different sets of users. In due course, we expect the two solutions to merge into one technology base.

Mobile Computing
Communications Server/Switch

The communications server/switch is the traffic cop in a mobile computing system. No data goes anywhere without passing through this traffic gateway. It connects wireless network to the wired backbone of an organization. It is the link between applications on front-end client devices and client/server and legacy applications at the back-end systems. It is the electronic heart of the system.   CHANDER DHAWAN

About This Chapter

In Chap. 8 we surveyed wireless networks. In the process we reviewed underlying technologies and the ways in which wireless networks provide untethered connections between end-user devices and information resources. In Chap. 9 we reviewed the European and international scene with respect to wireless data technologies. In Chap. 10 we discussed the RNA industry as a superset of remote networking.

In the course of moving from end-user devices on through networks to information servers, we come to the mobile communication server/ switch, an important component—vital to the overall architecture of a mobile computing solution—that acts as an intermediary between end-user devices and the platform(s) on which information resides. In this chapter we shall discuss the role of the server/switch and the functionality it provides. We shall also review a few vendor offerings.

11.1 The Logical Architecture of a Mobile Communications Server/Switch

In Chap. 5, we introduced the relatively new architectural concept of the communications server and/or switch. Since it is a new concept, no universal agreement exists on its precise shape and form. There is no common terminology for it either, and in fact, it is known by a variety of names. In the remote network access industry, the term *communications server* is common. In some vendor implementations, such as those from Motorola, it has been called a *communications controller.* In financial and retail industries, where queries can be destined for any number of database servers, *communications switch* is frequently used. In yet another context, *communications gateway* may perhaps be appropriate. In this book, we shall use *MCSS,* short for *mobile communications server/switch.*

End-user devices in the field can be connected to information servers in a variety of ways. There may be PSTN links via dial-up lines or ISDN connections, or wireless networks may be used to provide ubiquitous connectivity without the need of locating a telephone jack. While the former wired links support common hardware and software interfaces, the latter (wireless networks) support only specific interfaces, such as X.25, asynchronous serial, or proprietary interfaces like Motorola's RD-LAP. It is the MCSS and its accompanying software that ultimately provide the cohesive link across this variety of connections (Fig. 11.1).

In fulfilling this, its most important role, an MCSS also fulfills other roles. If a mobile network and its information servers do not speak the same language, the MCSS bridges the gap between the protocol of the mobile network and the protocol of the LAN, minicomputer, or

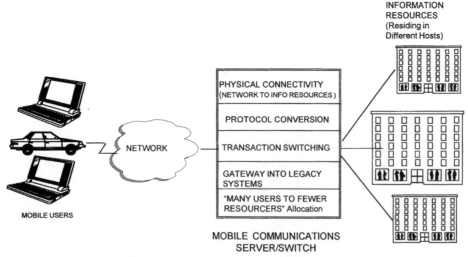

Figure 11.1 Logical role of mobile communications server/switch.

mainframe applications with which it is trying to communicate. Even though network providers are standardizing on transport protocols such as TCP/IP, there are literally thousands of large legacy databases in government organizations, financial institutions, and manufacturing companies that use their own unique protocols and will continue to reside on these platforms until these applications are redeveloped in the future. One obvious example is the set of database applications running on IBM mainframes under MVS/IMS/VS using older communications interfaces. While the new open operating system environment of OS/390, announced by IBM in late 1995, may support TCP/IP transport, older IMS/VS applications may still require significant changes before they can run under the new operating system. As an example, many message-switching-based command and control public safety applications reside on the DEC platform under DECNET. Therefore, you need a gateway function to handle this incompatibility between the network transport protocol and the software communications interfaces hard-coded either in the application logic or defined in configuration tables on mainframes and minicomputers.

New application and communications middleware will eventually help to minimize such incompatibilities. Right now, however, their main focus is to provide transparent connectivity between existing suites of applications and wired networks. Therefore, mobile computing solutions, for now, must recognize the dependencies of the existing applications while they hope for new application interfaces to help out in the future.

In situations where information needed by mobile workers resides on multiple machines, an MCSS is needed to switch transactions to appropriate information servers as needed. In more sophisticated implementations, a software agent residing in an MCSS obtains responses from different databases on multiple computers and returns the information to the user.

Finally, the MCSS addresses the "many-users-of-few-dedicated-resources" problem in a communications server. The total number of users is always higher than the available number of hardware ports and software resources (such as the number of active sessions, control blocks, or TCP/IP sockets supported on a given configuration). Moreover, at any given time, only a fraction of all users are engaged in an active communication session. Passage of a message through these resources may take only a few milliseconds. It is simply not economical to use relatively scarce communications resources on a one-to-one basis. Whether it is the number of modems in the modem pool or the number of logical sessions (e.g., 255 SNA sessions per physical unit), the MCSS ensures that these resources are efficiently allocated and de-allocated.

11.2   Functional Description
of a Generalized MCSS

This section describes a generalized MCSS suitable for a variety of customers. It should be pointed out, however, that the author has yet to see an implementation that includes every one of the specifications listed. There are many reasons for this. First, the industry is currently focusing more on other components of mobile computing solutions. Second, in the absence of leading-edge designs, individual vendors are working on proprietary approaches that are, in essence, upgrades of current products. Third, remote-network access vendors are addressing the needs of switched network connectivity for smaller organizations with LAN-based information resources before they address general purpose needs of wireless network connectivity.

An MCSS comprises two logical sets of functions, defined here as a *front-end component* that interfaces with the network and a *back-end component* that interfaces with application servers or databases.

These two functional components can be housed together in a single hardware box or in two separate boxes, as indicated in Chap. 5.

Based on our discussions with some of the industry leaders, we have come to the conclusion that many of them still do not have a specific product strategy with regard to MCSSs. Perhaps as more and more mobile computing solution implementers look to the development of an open mobile computing solution, they will ask vendors to develop an MCSS that includes the following functional specifications:

- Communications services (front-end functionality)

- Protocol conversion (front-end functionality)

- Gateway into legacy applications (back-end functionality)

- Transaction switching (back-end functionality)

- Miscellaneous (see Sec. 11.2.5)

11.2.1   Communications services

One of the more important functions of an MCSS is to connect mobile devices to networks used to transfer information between clients and servers.

Thus, the front-end component should support the chosen network or networks (in some situations, the solution may be based on more than one network) involved in the overall mobile computing configuration. For example, as many as 60 to 70 percent of mobile workers access remote networks through public switched network services (dial-up and ISDN) from facilities where telephone jacks are available. The following is a selection of such networks:

- PSTN (dial-up), including frame relay
- ISDN
- Private SMR network (wireless)
- Circuit-switched cellular
- CDPD (wireless)
- Packet networks such as ARDIS and RAM/Mobitex
- PCS (narrowband now and broadband in the future)
- Specialized networks such as paging, ESMR, and Metricom Richochet

In order to provide communications services, an MCCS must have a hardware and software configuration that supports the selected PSTN/radio/wireless networks. In terms of the hardware, this consists of wired-line modems connecting the MCSS to the network provider's premises. Note that the wireless or radio modems which interface with the mobile devices are typically installed on a network provider's premises. Multiple telecommunications circuits from different base stations are multiplexed on to a thick pipe from the network provider's premises to the customer's premises. It is good systems-engineering design and practice to divide the traffic over at least two distinct circuits for redundancy and fault tolerance.

Support for wireless networks is provided through software drivers implemented on the hardware platform of the switch. ARDIS and Mobitex packet drivers, for example, are available on several different platforms. When drivers are not available for a specific platform, they can be ported easily to the chosen platform, so long as they were written in a common language such as C. In other cases, network vendors provide a *system development kit* (SDK) to enable software developers to do this work themselves.

The MCSS can be connected directly to a local area network. Token-Ring LANs, however, are rarely supported by MCSSs, except by remote network access vendors who specialize in connectivity to LANs. Ethernet connections between an MCSS and a LAN is more commonly supported.

For large configurations (over 1000 users), an MCSS should support the required bandwidth on the incoming side (from the mobile users) as well as on the backbone network side connecting the MCSS to the information servers. Depending on the system requirements, this could be in the range of multiple T1 lines. The back-end connection should be either at a sub-T1 speed for low-end configurations or at a T1 speed for large configurations.

11.2.2 Protocol conversion

In an ideal network environment, an open transport protocol such as TCP/IP should be the protocol of choice. However, it is not realistic to have a pure TCP/IP implementation all the way to IBM mainframes for the reasons cited earlier. The MCSS should provide protocol conversion across the following protocols, from the front end to the back end:

- X.25
- TCP/IP
- LAN protocols such as IPX (Novell) and NETBIOS
- Mainframe protocols such as SNA LU 2 and SNA LU 6.2
- Minicomputer protocols such as DECNET

The systems architect should evaluate whether it is easier to change the application interface from minicomputer/application server–supported protocols such as DECNET or to make the conversion in the MCSS itself. The decision will depend on the number of application servers and the precise changes involved. If only table definitions are involved, it is easier to change the application interface. If, on the other hand, the protocol-dependent communications logic is embedded in many different applications, it may be easier to provide this capability in the MCSS. It is also implied that functions such as code conversion from ASCII to EBCDIC are inherent in the MCSS.

The MCSS should support asynchronous (unblocked two-way) logical communication between the mobile devices and host computer application programs written in supported protocols.

11.2.3 Gateway into legacy systems

Legacy applications are more difficult to change for a variety of reasons. Rather, they are candidates for future redevelopment, using the new client/server design paradigm. In order to provide a migration path for these applications in the meantime, the MCSS needs to include a gateway function that makes access to these legacy systems possible for mobile devices.

Reusable or parallel sessions. The number of sessions available through an MCSS is often limited. To provide support for larger numbers of users, the concepts of *parallel sessions* (also called *reusable sessions*) or *TCP/IP sockets* are used in the design of MCSSs. In parallel sessions, a session or socket is temporarily allocated to an incoming and outgoing message through the switch.

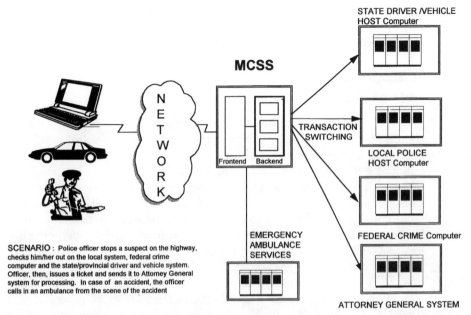

STATE DRIVER /VEHICLE
HOST Computer

MCSS

Frontend   Backend

TRANSACTION
SWITCHING

LOCAL POLICE
HOST Computer

FEDERAL CRIME Computer

EMERGENCY
AMBULANCE
SERVICES

SCENARIO : Police officer stops a suspect on the highway,
checks him/her out on the local system, federal crime
computer and the state/provincial driver and vehicle system.
Officer, then, issues a ticket and sends it to Attorney General
system for processing. In case of an accident, the officer
calls in an ambulance from the scene of the accident

ATTORNEY GENERAL SYSTEM

Figure 11.2   Transaction switching in MCSS: public safety application.

11.2.4 Transaction switching

Transaction switching, also called *message routing,* is the functionality through which transactions can be switched to different host computers depending on where the application and/or the data reside (Fig. 11.2).

There are at least two good reasons to have the MCSS server perform transaction switching, rather than have the information server or host computer do it. First, using the MCSS server results in improved performance and lower response times for mobile device transactions. Remember that the fewer the number of processors in the path of a transaction, the better the overall response time will be.

Second, some MCSS computers used in mobile computing solutions have transaction-switching capabilities already available. Nonstop VLX computers from Tandem have several implementations of switching software. Similarly, IBM's MQI series middleware has transaction switching built into the basic software.

Message queuing and recovery functionality. In order to support asynchronous messaging and to guarantee message delivery, an MCSS should have message queuing and recovery facilities. Middleware software products such as IBM's MQI series and Momentum's Xipc provide this capability, but do not provide generic wireless network communications services. Should this be a highly desirable functionality for a mobile application, a custom solution can be proposed with a front-end MCSS for the wireless network connection and a back end based on MQI software on one of the supported hardware platforms.

11.2.5 Miscellaneous MCSS specifications

An MCSS should have the following additional functions:

1. *Message processing capacity.* An MCSS's hardware architecture should support the mobile computing requirements of small, medium, and large organizations and should allow for both centralized and distributed network designs. Thus it should be scalable in terms of the number of mobile users a single box can support (see Sec. 11.2.6 also). This number ranges from a few hundred users in relatively small or medium-size organizations to several thousand in large organizations. For example, the Ontario government is studying a 4000-vehicle public safety application and Sears is deploying a service representative application across 14,000 vehicles.

In order to meet capacity requirements for such networks, MCSSs should be built on processor architectures that are more powerful and robust than Intel 486s or Pentiums running under Windows/NT or OS/2. Of course, low-end capacities can be met with such platforms but high-end capacities need more powerful platforms such as Tandem Guardian, VME, AIX RS/6000, and NT/Alpha. Apart from the processor, an MCSS should be capable of supporting a wide variety of medium- and high-speed communications ports (ISDN, sub-T1, and full T1s), especially for back-end server connections.

Different vendors have different ways of expressing the processing capacities of MCSSs. Some MCSS vendors express capacity in terms of data packets handled per second. Be cautioned against too readily accepting vendor claims of very high packet-processing rates. Quite often, router manufacturers quote router capacities in terms of 10,000, or even 15,000, packets per second. Remember that routers do very little other than route packets. What is being discussed here is the transfer of user-information packets through all hardware/software layers—with all other MCSS functions happening simultaneously. A better measure is user messages or system-level transactions processed per second.

2. *Fault-tolerant design.* Since the MCSS is in essence the heart of the system, its uninterrupted functioning is critical to the uninterrupted operation of the mobile computing solution. Therefore, it should be fault-tolerant-designed for 7-day, 24-hour operation, with redundant components that can be hot-swapped. This redundancy should extend to communications interface boards, processor boards, disk storage, and power supply. Tandem or Stratus fault-tolerant hardware architecture is appropriate for larger MCSSs.

3. *Security.*   Although security requirements will be covered fully in Chap. 15, we would like to emphasize here the role the MCSS plays in providing overall security. Unless there is an Internet-style firewall protecting the information servers, the MCSS should provide a direct level of user authentication or have an interface with a security server through a software API.

4. *Encryption.*   Without proper security measures in place, unauthorized personnel can easily hack into sensitive financial or law-enforcement data traveling on wireless and switched networks. While user-identification/authentication security can be provided by the MCSS, actual encryption and decryption of data is a processor-intensive task that is generally best handled by specialized onboard, sideboard, or off-board encryption engines.

5. *Compression.*   Bandwidth on wireless networks is scarce and expensive. It behooves us, therefore, to exploit efficient software compression technology over and above the compression provided by modem manufacturers. The MCSS is well suited to provide compression on outgoing messages and decompression on incoming messages. In practice, however, very few MCSS vendors provide this capability today.

6. *Open standards–based design.*   Too often, mobile computing vendors have been building proprietary solutions, with the result that there exists today a general lack of standards in most areas—except in end-user-device hardware. Some efforts are being made to establish de facto or de jure standards. The IEEE 802.11 standard for wireless LANs is one such effort. We need more of these standards in all components of mobile computing solutions, including MCSS.

As implementers of mobile computing solutions, readers should insist upon an open standards–based MCSS solution. At a minimum, the hardware and the OS platform should be based on common industry standards, thereby ensuring the ability to take advantage of future price reductions.

7. *MCSS network management.*   In view of the importance of the MCSS, care should be taken to ensure that it has standards-based network-management capabilities. The software should include problem diagnosis, continuous monitoring and statistics collection, and reporting. SNMP-based support and the availability of a suitable MIB for mobile users is highly desirable.

11.2.6   Capacity and distribution issues

The MCSS should support a technology architecture based on either a centralized MCSS or multiple distributed MCSSs.

The choice between a centralized versus a distributed architecture should be based on an evaluation of the following factors:

- The geography of the area in which the mobile users operate
- The location of the information servers
- The capacity of the MCSS
- The nature of the wireless network provider's wired-line backbone network
- Technical support skills available at remote offices where distributed MCSSs may reside

Figure 11.3 shows a distributed MCSS design where MCSSs are distributed in the east, south, north, and west.

11.2.7   Physical versus logical design: one/multiple boxes

Discussion thus far of the functionality of MCSSs has shown that existing products do, in fact, have many of the most desired features. Therefore the question arises as to whether or not we should have a single box that combines front-end and back-end functions.

Consideration should be given to the existing infrastructure, but ideally there should be only one box—albeit, in many cases, a very full and complex box. Having a single box makes for easier management of the network and reduced MCSS transit times through the switch, as well as for (hopefully) reduced costs.

11.3   Low- and High-End MCSSs

This section features schematics for two different mobile computing solutions: a low-end scenario and a high-end scenario (Figs. 11.4 and

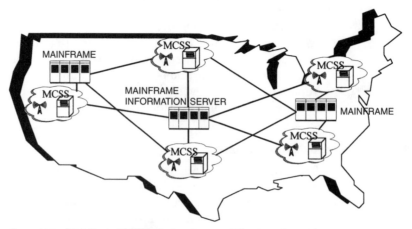

Figure 11.3   Distributed MCSS design for a multihost configuration.

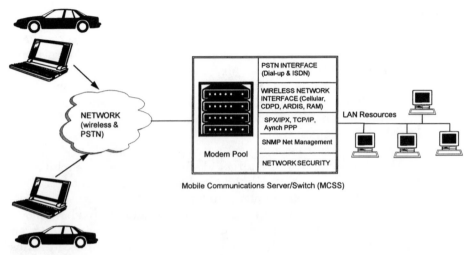

Figure 11.4   Low-end mobile communications server/switch.

11.5). The low-end configuration provides typical remote network access to LAN-based information resources from mobile devices in the field. It is suited to situations where small numbers of mobile users want to use their notebooks to connect to their home-office LANs using dial-up modems or wireless connections. The high-end configuration is more comprehensive in functionality, capacity, and internetworking complexity. It matches the generalized requirements described in Sec. 11.2.

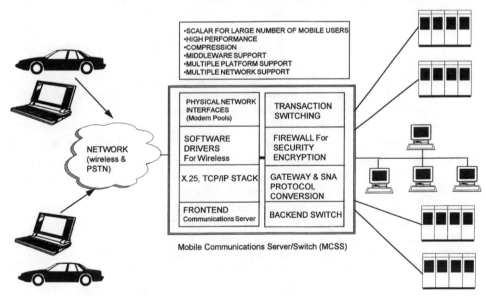

Figure 11.5   High-end mobile communications server/switch.

11.4   Prominent Industry Implementations of MCSS

Sections 11.1 and 11.2 of this chapter have described a conceptual and generalized MCSS which, as far as the author knows, does not actually exist. Perhaps this conceptual MCSS is too rich in functionality to be feasible. Perhaps it is also too complex. And perhaps, too, some of the proposed functionality already exists in other network components. Nevertheless, an MCSS such as the theoretical one that has been described here could very well make a great deal of sense in a situation where one was starting out from scratch with a totally clean slate. Failing that, it can be used as a target ideal—a target switch architecture to aim for while progress is made in discrete steps as subsets of the overall functionality are implemented in successive interim solutions.

In the course of selecting such interim solutions, readers should review prominent vendor implementations, a sampling of which are described here. They can be divided into the following three categories:

1. Vendor implementations of RNA MCSSs

2. Specialized wireless switches

3. Custom-developed switch solutions

11.4.1   RNA vendors' communications servers as MCSSs

RNA vendors have developed solutions that enable remote users to access LAN resources through switched wired-line (PSTN or ISDN) connections. Now that salespeople, telecommuters, and other remote users are accustomed to the idea of being able to connect to their office LANs from remote locations via switched telephone connections, they are eyeing the next logical extension of the idea and are asking for wireless connections so that they can achieve the same connectivity from their cars or from nonwired field locations. In response to this demand, many vendors have added support for wireless network connections to their existing software.

While most vendors claim to be able to support cellular network connections as dial-up connections, only a few have actually provided drivers for packet radio networks and built specific error-recovery procedures into their software. Thus, although RNA solutions are relatively inexpensive and easy to implement for switched network connections (including ISDN), as of 1996, existing RNA vendor solutions tend to be acceptable only as temporary solutions suitable for occasional cellular network connections. They are simply not reliable enough for mission-critical and interactive applications. On top of that, they take almost 45 seconds to establish a dial-up connection.

Cellular network costs are high enough as it is, without the added expense engendered by 45-second dial-up sequences, especially in cases where frequent connections are necessary.

Refer to Chap. 10 for details on RNA solutions by vendors such as Shiva and Xyplex.

11.4.2 TEKnique's specialized wireless gateways

TEKnique is a consulting and systems integration company based in Chicago, Illinois, that specializes in wireless-communication-based integration work.

The company offers two wireless MCSSs: the TX5000 on a PC platform for low-end applications and the TX5000 on a VME bus for high-end applications. It supplies its MCSSs to several large telecommunications-network service providers, including Motorola in the United States and Cantel in Canada. A description of the TX5000 configuration follows.

Hardware description. The TX5000 is based on a VME chassis with a fully redundant hardware design. It includes the following features:

- Hardware for mounting in a standard 19-inch (48-centimeter) rack with a 400-watt power supply.

- A 12-slot monolithic backplane that includes P1/J1 and P1/J2 connections, again on a standard 19-inch (48-centimeter) rack with a 400-watt power supply.

- Up to eight communications processors (68020 with 8 MB RAM, 128 KB static RAM, 256 bytes battery-backed SRAM). Each communications processor has three 68302 intelligent communications processors to support three lines.

- Each processor supports nine communications lines for a total capacity of 72 lines.

- Each communications processor has a maximum throughput capacity of 5 Mbps with all nine lines simultaneously active. Actual throughput is a function more of the software than the hardware.

- The TX5000 has a single line-interface board for each communications processor board. Supported line types include RS232 and V.35.

- An administrative processor for network management with remote connection and diagnostics.

Software and protocols supported. The TX5000 is controlled by proprietary software that supports TCP/IP, X.25, and SNA.

Functions provided.   The TX5000 provides the following functions:

- Protocol translation and routing
- Session management, including multiple sessions for session-oriented applications
- Session disconnect and recovery
- Accounting and statistical data
- Authentication services/multiple sessions
- Menu services, including greetings and user messages
- Direct connection (directly routing all packets from specific IP addresses or sets of addresses to specific gateway IPs)

11.4.3 Research in Motion's RIMGate protocol converter

RIMGate, supplied by Research in Motion, based in Waterloo, Ontario, is a specialized gateway for RAM/Mobitex packet networks (Fig. 11.6). It is not really an MCSS as described in Sec. 11.2, but it can be used in tandem with other RNA servers to provide the functionality of an MCSS.

It is unique in that it provides a simple way to add wireless access to existing applications without changing them either at the client or the server end. With its software installed in mobile computers and remote terminals using MTP/1 (the Mobitex transport protocol) and compatible radio modems such as the Ericsson Mobidem, it provides all the

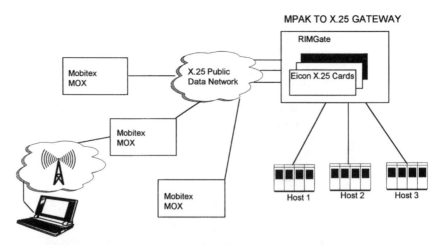

Figure 11.6   RIMGate Mobitex to X.25 network gateway. (*Note:* MOX is a Mobitex network's hierarchical exchange. MPAK is a Mobitex packet.)

protocol conversion needed to make a seamless connection to the host by supporting the MTP/1 and MCP/1 (the Mobitex compression protocol) used in the radio modem.

RIMGate supports X.25/MPAK communication with a Mobitex area exchange (MOX) and either native X.25 or PSTN communication with the host. RIMGate is used by Mobitex network operators such as Bell-South.

The following are selected features of RIMGate:

- Supports up to 64 users per RIMGate.

- Compatible with Intel and Ericsson AT-compatible modems.

- Uses X.25 network adapters from Eicon technology. Support for multiple X.25 channels (up to six in a single box) allows physical connection to multiple hosts or multiple Mobitex MOX ports.

- Three different compression algorithms (V.42 bis, run-length encoding, and table lookup).

- A pass-through mode is available for the *Mobitex Asynchronous* (MASC) protocol. MASC is a machine-level protocol for transferring Mobitex packets.

- Host-initiated calls can be made to a mobile terminal.

- A RIMGate management tool provides full-screen monitoring of each session, including MAN number, name of host connection, and packet transfer information.

- Remote monitoring of RIMGate is supported.

- A *time-efficient reporting scripting* (TERS) facility improves network performance by filtering data transmitted by the host and by transmitting information detected in the data stream back to the host. TERS automates log-on procedures for remote Mobidem-AT users connecting to RIMGate.

General comments. RIMGate is a specialized gateway for low-end applications. It should be considered in situations where it is difficult to change mobile client or server applications, or where there are a limited number of users accessing a single RIMGate. It is also useful for pilots and prototypes to validate business-case assumptions for a mobile application.

11.4.4   IBM's ARTour

IBM's networking software division has developed a product called ARTour that makes possible the immediate use via a wireless telephone network of client/server applications and proven mainframe

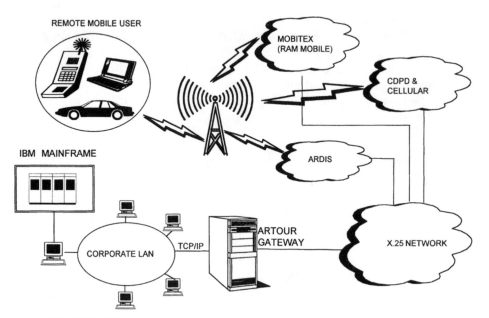

Figure 11.7 IBM's ARTour gateway for wireless network integration.

applications like IBM 3270/5250 terminal emulations. It is based on client-agent-server architecture (Fig. 11.7).

Developed in Europe and designed to run on the OS/2 engine, ARTour has been ported to run on the AIX/RS 6000 platform for the North American market. The distinctive feature of ARTour is that it integrates several different wireless networks into a single MCSS architecture. Currently supported are many of the European networks, such as Mobitex, Modacom, GSM, and Inmarsat. Since the architecture of ARTour is open, support is now available for U.S. networks like ARDIS, RAM, and CDPD. RS/6000 implementation makes ARTour highly scalable to handle thousands of active users.

Additionally, it supports data compression features, as well as client/server-based mobile applications using middleware software. A strong capability exists for interfacing with IBM mainframe applications such as those running under IMS/VS and CICS/VS without any changes in the application code. This is achieved through an agent software application running on ARTour. ARTour also addresses the basic problem with wireless networks, i.e., low and expensive bandwidth. It uses storage-catching techniques to ensure that only changes are transmitted over the airlink. It uses an optimized version of TCP/IP. It also has SNMP-based network management capabilities that can be linked to IBM's NETVIEW/6000.

Since IBM is aiming this product specifically at the banking and insurance industries, special considerations are given to security and reliability issues. Based on our initial assessment, ARTour is a strong candidate for a scalable MCSS for mobile computing solutions where IBM mainframes are involved as information servers.

11.4.5   Motorola radio/wireless controllers

Motorola has several proprietary MCSS solutions. The RNC 6000 will be discussed here as a component of an evolutionary growth path for the private DataTAC radio networks used by hundreds of public safety agencies (Fig. 11.8).

The RNC 6000 is the hub of a DataTAC mobile communications system. It controls the routing of data messages from one or more host computers through a private radio network to mobile devices. The RNC 6000 performs this function by keeping track of the location of the terminals within the radio network via an exchange of information with DSS2 base stations. It provides the following critical functionality:

- Standard host-link connectivity

- Logical message segmentation and recompilation

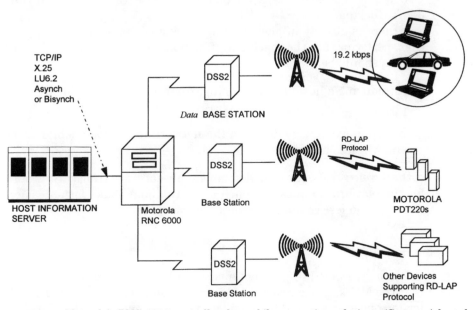

Figure 11.8   Motorola's RNC 6000 controller for mobile computing solutions. (*Source: Adapted from Motorola schematics.*)

- Packet transmission (and retransmission in the case of errors)
- Duplicate packet elimination
- Data system station-transmitter control
- Limited network-management facilities

The RNC 6000 also requires terminal registration to prevent uncontrolled access. Only those terminals properly registered to the RNC 6000 are given access to the network.

DataTAC RF link supports 19.2-kbps raw data rates from a mobile device to the controller through a private network. It has an optimized protocol for maximizing the efficiency of a radio link—the most precious systems resource in a mobile computing solution. The high speed and the efficient data-link protocol support a maximum number of users per radio channel. One countervailing factor to be considered is that modern Windows applications transmit more data than do finely tuned older applications.

11.4.6 Custom-developed MCSSs

Several systems integration companies have built custom MCSSs to cater to the specific needs of individual customers. These solutions are based on a variety of software platforms, including OS/2, Windows NT, UNIX, DEC VMS, IBM's OS/400 on the AS/400, Stratus VOS, and Tandem's Guardian.

If there is already in place in a network a custom-developed MCSS that performs all the back-end functions required, the following options should be evaluated:

1. Introducing a new full-functioning (front-end/back-end) MCSS

2. Introducing a front-end-only MCSS

3. Enabling existing switches in the network to support mobile users and wireless networks

If the wireless network provider can supply software transport drivers for the platform in question, such a solution may make for an easier, cheaper, and less complex network environment.

11.5 Selecting the Best MCSS

Choosing the best MCSS is not a straightforward exercise. Remember that the MCSS is the heart of any mobile computing solution. The following questions should be considered in selecting an optimal MCSS:

1. Does it support selected switched and wireless networks?
2. Does it support preferred transport protocol(s) for the networking environment in question?
3. Does it provide a back-end gateway function that enables access to legacy information servers?
4. Does it meet performance standards in terms of numbers of users and numbers of logical business transactions per second during peak loads—while still meeting response time criteria for interactive transactions?
5. Is it scalable?
6. Does it support multiple platforms?
7. Is it an open solution based on industry standards?
8. Does it provide security that meets the corporate standard?
9. Does it have a suitable network management capability that can be integrated with the overall systems management umbrella software?
10. How easy is it to install and integrate into the overall mobile computing solution and the overall networking environment?
11. What is the five-year cost of acquisition, maintenance, and support?

While these are important questions in considering what is, after all, a most vital component of a mobile solution, it should be born in mind, too, that an MCSS is but one component in the broader scheme of things. See Chap. 18 for overall evaluation criteria.

Summary

In this chapter, we have discussed the mobile communications server/ switch, or MCSS, at the heart of a mobile computing solution. The MCSS acts as mediator between end-user devices and information servers logically connected to wireless or switched networks.

We reviewed the technology architecture first discussed in Chap. 5 and described the role of the MCSS and its logical architecture and functionality, including the functionality pertaining to both the front end and the back end. We looked at the similarities between traditional remote network industry communications servers and a front-end MCSS. We discussed the importance of transaction switching in enabling users to retrieve information from multiple information servers. We noted that the capacity, location, and distribution of

MCSSs are important design issues that systems designers need to confront.

Industry implementations of MCSSs were described. Since integration of the MCSS into communications and applications software interfaces is one of the most challenging aspects of systems integration, care should be taken in selecting the most suitable MCSS.

Mobile Computing Software

If hardware is the engine, software is the fuel. Whereas hardware constitutes the body, software supplies the intelligence and brain power. Without software, hardware is brain dead.

CHANDER DHAWAN

About This Chapter

In earlier chapters we discussed client devices and network infrastructures, including mobile communications server switches. In so doing, we have covered most of mobile computing's hardware components—with the notable exception of information servers (which, because they are seldom implemented exclusively for mobile computing projects, are not considered to be within the scope of this book).

Now we shall turn our attention to software. In this chapter we will cover system services software supplied by the vendor community. Discussion of application development tools and design strategies will follow in Chap. 13.

For the purposes of our discussion we are assuming that the basic mobile end-user software environment is Windows 3.1 running on PC notebooks. In reality, there is in fact a significant variety of both operating systems and mobile devices in use. OS/2, MacOS in Powerbooks, GEOS in PDAs, and DOS in handheld terminals are common. Yet, Windows 3.1 (with Windows 95 gaining in popularity) remains the most popular operating system—and therefore we are using it as the de facto operating interface standard in our discussions.

12.1 Mobile Computing Software Architecture Models

A discussion of any mobile computing component starts with an examination of its architecture. We should therefore review at this point the software architectures described in Chap. 5.

We can discuss four software models, each one more sophisticated than the last in terms of connectivity between the client device and the server platform:

1. Stand-alone workstation with occasional file transfer

2. Traditional terminal emulation

3. Client/server model

4. Intelligent agent-based client/server or client-agent-server

These models also represent a gradually increasing maturity in the application development paradigm. Let us briefly review them one by one.

1. *Stand-alone workstations with occasional file transfer.* In this model, the end-user devices are fully functional and self-sufficient intelligent devices without any real-time connection to information servers. Application and data are loaded directly onto the PC notebooks, Macintosh Powerbooks, or handheld terminals themselves. A database stored on such a device is a subset of a larger database stored at a central site. Thus, users of these devices are highly mobile.

Stored data is updated on a regular basis by way of asynchronous modem connections, or, in some cases, by way of diskettes. Similarly, data entered on a mobile device is uploaded to a PC host or to a LAN server.

This is by far the most common mode of mobile computing. Sales force automation software typically falls into this category. The software tools used are popular application development tools like dBase, Clipper, Microsoft Access with Visual Basic, Delphi, and Lotus NOTES.

2. *Terminal emulation model.* Terminal emulation software enables mobile devices to use older terminal-based applications. IBM 3270, IBM 5250, and VT 220/240 are among the most common terminal emulations used in the industry for accessing legacy databases on mainframes and minicomputers. Many of the handheld terminals employed in automatic data collection in the manufacturing and distribution industries use this model.

There is a lot of application development software available for AS/400, DEC VAX/VMS, and UNIX RISC servers, but a discussion of these tools is beyond the scope of this book.

3. *Client/Server model.* This is the model of choice for most newer applications. Our discussion here and in Chap. 13 assumes this model as the application-design paradigm.

4. *Intelligent agent–based client/server.* As discussed in Chap. 5, an intelligent agent–based client/server implementation, also called client-agent-server, constitutes one of the best approaches to the devel-

opment of mobile computing applications. However, despite the fact that many vendors have begun to use the terminology as a marketing and advertising ploy, this model is only now emerging as a true design philosophy. There are a few genuine implementations available through the likes of General Magic, Oracle, and Sybase—and in due course we shall discuss these vendors' implementations.

12.2   The Overall Software Architecture of Mobile Computing

Let us recap our discussion from Chap. 5, in which we examined the overall technology architecture of mobile computing, including software architecture. There are three areas where software integration poses its greatest challenge to the practitioner of mobile computing. They are the *mobile client device,* the *MCSS,* and the *information server.*

MCSS software functionality was discussed in Chap. 11. MCSS software is either supplied by a vendor with a degree of customization or table definition capability or it is developed by a systems integrator. Development tools for MCSS software are not the focus of this book.

Rather, we want to concentrate here on those aspects of software development that a mobile computing systems integrator has to wrestle with.

Figure 12.1 shows the software components that are important from a software-development and -integration perspective.

The following software issues are important in a mobile computing project:

- Selection of an operating environment and user interface for the client application
- Availability of software device drivers on the client device
- Selection of client workstation and server APIs
- The role of middleware in application development and integration
- The choice of a transport layer and the need for protocol conversion
- The value of software gateways to legacy applications
- The use of emerging application engines such as Oracle Agents
- Emerging two-way paging software for developing frugal mobile communications applications
- The use of intelligent agent–based client/server designs
- The use of specialized messaging protocols (as in paging and e-mail, for example)

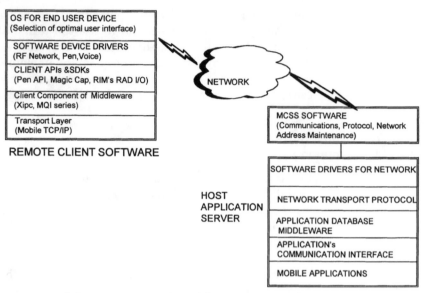

Figure 12.1   Software components for mobile computing integration.

12.3   Client Operating Systems and User Interfaces

The following questions should be asked when developing software architecture for a client workstation:

1. Which operating system (Windows 3.1, Windows 95, Windows/NT, OS/2, Solaris, UNIX, or specialties like GEOS or Magic Cap for PDAs) will be used for *new* mobile computing applications?

2. Which user interface shell will be used for these applications? (Remember, the user interface can differ from the operating system for the device. For example, Windows 3.1 could be selected as the user interface, with UNIX being implemented as the OS for the server.)

3. Are old DOS applications on the existing platforms going to be converted to the new operating system or will they be run in Virtual DOS Machine (VDM) for the time being?

The answers to these questions will depend on a number of factors, some of which we shall discuss.

Unless an organization has decided to standardize on a specific desktop platform as a result of an IT strategic planning exercise, mobile computing applications should follow the standards in the industry. There is general agreement among practitioners that for new PC applications, Windows 3.1 and Windows 95 are the default or de facto standards because of their wide use in the industry, the large number of applications available for them, and an abundance of trained programming staff.

Other considerations are support for a particular shrink-wrapped application by the chosen OS platform and the availability of specialized device drivers for pen, voice, multimedia, etc. Of course, if the chosen hardware platform in the organization is a Macintosh, then MacOS might be the preferred operating system environment. Similarly, if the organization has standardized on OS/2 Warp for custom applications and if the mobile application is going to be developed in-house, OS/2 Warp would obviously be a serious contender.

The maturity of any software and the number of versions it has gone through before its release are good indicators of the potential number of bugs it might contain, especially for untested configurations. Mobile computing projects involve many emerging technologies. A good project manager should try to strike a balance between the functionality of new OS development features (e.g., 32-bit development under Windows 95) and the risk of encountering bugs in developing applications for the uncharted expanses of wireless network interfaces and pen computing.

Mobile computing applications are generally network-intensive and depend on newer network and device interfaces. We shall see in Sec. 12.4 how Windows 95 enhancements will provide transparency to network types and transport protocols for mobile computing applications.

12.3.1   Do we need a new OS for mobile devices?

There is an effort in some research and development circles to design a highly optimized OS for mobile applications. While there are unique demands for low-end devices such as PDAs, communicators, and wearable computers, the vast majority of business applications are well served by existing operating systems. Hardware miniaturization is constantly setting new expanded standards for memory capacities, hard disks, and other peripherals. Specialized applications on hardware such as PDAs may yet warrant a new optimized OS, but these devices will have communications interfaces and applications built into shrink-wrapped packages. Any distinctly unique OS will be transparent to users. When, for example, has any user ever wondered about the operating system used by a Casio organizer? The point we are making is that as long as a device has a vendor-provided shrink-wrapped application that users do not have to modify, it does not matter what the OS is.

12.4   Client Device Software Drivers
and Network Connectivity

It is imperative that there be software drivers for a variety of peripheral devices on mobile client workstations, including drivers for the following:

- Pen

- Voice

- Special-purpose devices for telemetry, manufacturing applications, such as OCR readers

- Wireless networks—ARDIS, Mobitex, CDPD, and, soon, PCS (circuit-switched cellular networks use the same software drivers as do switched networks)

12.4.1   Early mobile connectivity software approach

In the early phases of developing connectivity solutions for packet radio networks such as ARDIS and Mobitex, objectives were relatively modest. One important goal was to provide a means of allowing a radio modem (such as Mobitex Mobidem) to emulate a conventional Hayes AT command set. This emulation enabled users to connect radio modems to the serial ports of PCs and thus to use any standard communications software to exchange information on wireless networks.

12.4.2   Emerging mobile connectivity
software requirements

The fact that network connectivity requirements are very sophisticated has thus far tended to hold mobile computing back. If we observe the way people actually work, we discover that office workers tend to have local connections to office LANs, WAN connections to corporate networks, and dial-up connections to external services such as CompuServe *in their offices*. For true mobility, all these connections should be maintainable outside the office. Thus, because a single wireless network does not have ubiquitous coverage, mobile workers often need to connect to multiple wireless networks—cellular in one area, Mobitex/ARDIS in another. These requirements will change as more modem digital wireless networks such as PCS are implemented.

12.4.3   New client workstation connectivity
software architecture

Connectivity software in mobile workstations should cater to the above requirements. It should present a consistent interface that makes connection to different physical networks transparent while providing a uniform set of services similar to those provided by wired-line products. When required, it should bind appropriate network device drivers to the application.

The client workstation software model for wireless data applications shown in Fig. 12.2a has the following attributes:

1. It supports mobile-aware application designs.

2. Interfaces between applications and networks are transport-independent.

3. The applications themselves are network-independent.

Since *mobile-aware* is a new concept that is relevant not only in vendor-provided system services software but also in application designs, we shall discuss mobile-aware design requirements in Chap. 13.

Transport independence and network independence are issues related to communications architectures. Basically, applications should not assume that data will be formatted as Netware IPX or Mobitex MPAK packets. Furthermore, no assumption should be made about the physical network. If it is a wireless network, it may be Mobitex or ARDIS, CDPD, or circuit-switched cellular. Therefore, new system services from the OS are required. We shall see how Windows 95 initiatives support this requirement.

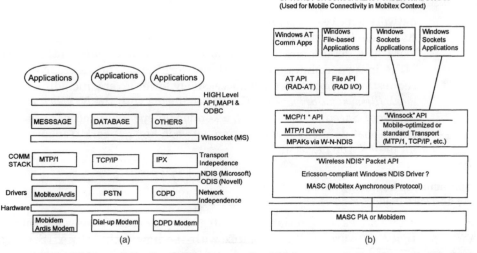

Figure 12.2 (a) Wireless application model; (b) upgrading Mobitex connectivity software for Windows 95. (Note: MTP/1.MCP/1, MASC, MPAKS are Mobitex terms. (*Source: Ericsson* Mobile Data News.)

12.4.4   New client workstation software
support of mobile application model*

Several Windows 3.1 and 95 systems services enable the implementation of these requirements in future generations of mobile-aware software. In all fairness, it should be pointed out that most of these capabilities are also available in OS/2 and MacOS as well. In this book, we are describing Windows facilities because of Microsoft's dominance of the market.

Windows 95 contains specific enhancements for mobile computing, including Remote Access Wizards that walk users through the process of remote dial-ups; software that recognizes security passwords; a system intelligent enough to know when it is connected to a network, or when it needs to dial in; true plug-and-play support that enables hot docking; software that automates the process of updating desktop and mobile files (Briefcase); and simplified faxing and remote mail.

One Windows architectural facility that is particularly useful in meeting mobile application requirements is the *Windows Open Services Architecture* (WOSA). The WOSA concept, introduced by Microsoft and accepted by several vendors, is designed around a *service provider interface* (SPI) that permits multiple software components with similar functionalities to coexist. The WOSA concept has been applied in a number of application programming interfaces (APIs) for messaging (MAPI), telephony (TAPI), TCP/IP sockets interface (WINSOCK), and open database connectivity (ODBC). Each of these interfaces allows software from different vendors to be accessed by the operating system in a uniform manner.

With the introduction of Windows 95, WOSA has been extended to network services, providing support for multiple network connections. At the top level, the Windows WNet functions, which comprise the network API, communicate with a *multiple provider router* (MPR). The MPR, in turn, communicates with network service providers. Network requests are routed through successively lower layers, including a *network file-system driver*. This driver conforms with the *Network Driver Interface Specification* (NDIS), a vendor-independent software specification that defines the interaction between any network transport and an underlying device driver.

12.4.5   Approaches to improving mobile
connectivity software using Windows 95

There are several initiatives to integrate packet radio network connectivity software such as Mobitex with the new architecture. (See Fig.

---

* *Mobile Data News* from Ericsson.

12.2*b*). One approach builds on existing file-transfer and AT-modem emulation software that supports traditional applications oriented toward file transfer and conventional modem communications. Another approach takes as its starting point the WINSOCK TCP/IP socket interface and is more oriented toward distribution applications and client/ server systems. Both approaches, however, point to a need for standardized low-level components.

Several companies, including Ericsson, RAM Mobile Data in the United States, and RIM in Canada, are evaluating proposals for a network device interface specification (NIDS)-compliant network device driver that will interface directly with a modem and provide level 1 (physical layer) and level 2 (data-link layer) services according to the SO network model. NDIS, originally defined by Microsoft, is now widely accepted in the industry.

The Portable Computer Communications Association (PCCA) is providing valuable input to this process by defining a specification for a wireless application model that is NDIS-compliant and supports device drivers for both Mobitex, ARDIS, and CDPD networks. A forthcoming proposal builds on previous work that defined an extended AT command set for use by all wireless modems. The new proposal will define a number of extensions to NDIS for wireless communications.

Currently, network device drivers for Mobitex are usually implemented as a DOS memory-tested TSR. These device drivers are being modified to provide NDIS compliance and support for the forthcoming wireless NDIS extensions from the PCCA.

This design has several drawbacks, however, one of which is difficulty in sharing a serial port with other software. A more general solution being considered by Ericsson is to implement a wireless NDIS driver as a so-called virtual device (VxD) able to run under Windows 95.

12.4.6 Packet radio networks look to Windows 95

Both Mobitex and ARDIS network service providers are moving toward the above goals in different ways. Several wireless integration companies are working with Ericsson on developing a new generation of Mobitex software connectivity tools in the next year or so. ARDIS has already announced that it will incorporate Windows 95 drivers in Microsoft Exchange. As more network providers get into the public shared network services business, the need for standardization and interoperability becomes all the more important. During this interim period, network-implementation decisions should be kept to an absolute minimum. Contact the sources mentioned in this chapter and in App. B to obtain up-to-the-minute details regarding current product features.

12.5   Client Workstation APIs

We will look at four different application interfaces—spanning input APIs (pen and speech), communications API (extensions to the AT command set for wireless data), and PDA APIs for the electronic marketplace—that will give application designers and programmers four different perspectives:

1. Pen API for Windows

2. Voice input drivers

3. Research in Motion's RAD I/O

4. Magic Cap's application interface

12.5.1   Pen computing API for Windows:
A step up from the keyboard

Since the keyboard and/or mouse are often unwieldy in mobile environments, pen-based devices have been introduced as replacements for these input peripherals. Several software companies, including Microsoft, have extended the basic Windows operating software to include pen-based workstations. The pen interface can be used for handwriting recognition or to replace the mouse and the keyboard in many applications. In fact, the pen interface has led to the creation of a whole series of interesting mobile applications in many different industries. Trade magazines such as *Pen Computing* review these applications.

A pen-based operating environment enables the creation of applications that extend current graphical user interfaces (GUIs) to a pen-and-paper level—a familiar (and therefore comfortable) approach that oftentimes can be a lot less intimidating than a keyboard and/or a mouse.

A compatible pen-based operating environment interacts with existing Windows applications without any modifications. Microsoft Windows for Pen Computing (and other compatible pen extensions) thus offer increased effectiveness to the installed base of Windows applications and hardware platforms.

The Pen Extensions are a series of modular extensions to the Windows 3.1 operating environment. They include a set of dynamic-link libraries (DLLs) and drivers that enable pen-based input and handwriting recognition. PENWIN.DLL is the manager of all pen-specific components in the Windows 3.1 system.

Pen services are available through a new set of APIs, referred to in the following pages as *Pen APIs*. If a pen is present, the Pen API informs applications so they can activate advanced pen-specific features. Pen extensions can be broken down into four general classes. The following paragraphs summarize them briefly:

Pen device drivers. To use Windows for pen computing, a pen driver must be installed along with a modified display driver.

Pen drivers. A *pen driver* is an installable software device driver. Its primary purpose is to import data from a digitizing device into the Windows system. The data from a digitizing device consists of *x*-axis and *y*-axis coordinates that indicate the pen's position. All the information reported by the pen driver is available to all compatible applications.

Pen drivers are different from mouse drivers. Handwriting recognition, for example, requires that they report data at higher sampling rates and higher resolutions than mouse drivers. They also may contain angle and pressure information.

Modified display drivers. A pen interface has the ability to *ink,* or draw, lines on the screen. To support the inking process, a communication is established between a Windows display driver and the pen interface in such a way that lines can be drawn at interrupt times.

The recognizer. The software that turns streams of *x*-axis and *y*-axis coordinates into recognized characters is a Windows DLL referred to as the *recognizer.*

Together, these pen computing drivers enable application programmers to use the pen interface as a preferred method of input for mobile computers.

12.5.2 Voice input software: Providing mobile applications with a natural interface

In a mobile environment for permanently mounted laptops, perhaps the most convenient method of input is voice recognition. This area is still in the early stages of development and is expensive. IBM has done considerable pioneering work in voice recognition technology on the Power PC platform with its Sensory software suite. A similar capability is available with VoiceType on the PC. VoiceType makes possible the creation of text and the replacement of pull-down graphical mouse-driven commands and menus by dictating into a hands-free microphone. Kolvox, Kurzweil, and PureSpeech are other companies active in this field. Kolvox has OfficeTalk and LawTalk voice recognition products.

So far, this technology has basically been used in offices for dictation and administrative applications only. However, law enforcement agencies are now experimenting with voice-activated inquiries into motor vehicle databases. They are also interested in the possibility of being able to dictate daily notes. As it is, many police services already dictate reports through telephone networks into transcription centers. This software (and hardware) technology will eliminate the transcription process—and the errors inherent in this process. Other professionals trying this technology are physicians and journalists.

12.5.3 RIM's RAD-I/O: Easing mobile application development and conversion

RAD-I/O is a set of system utilities and application development tools developed by Research In Motion (RIM) of Waterloo, Ontario, and available from RAM Mobile Data and other resellers. This tool set enables new or existing applications to run on Mobitex wireless networks. The availability of RAD-I/O for Windows, Macintosh, and UNIX allows developers to create cross-platform solutions (Fig. 12.3).

Rapid application development, insulation of Mobitex low-level network protocols, sharing of radio channels by multiple RAD-I/O-developed applications, and interoperability with existing applications are major features of this software tool.

Essentially, RAD-I/O is a set of three different interface routines: RAD-I/O for simple file-based protocol, RAD-Link for device-level interfaces to Mobitex radios, and RAD-Sock for WINSOCK-based sockets programming.

RAD-I/O routines are system services available to application programmers. Application-related services (the RAD-AT command set and methods of sending files to mobile devices) are described in Chap. 13.

The following system services provided by RAD-I/O are useful for a variety of development purposes:

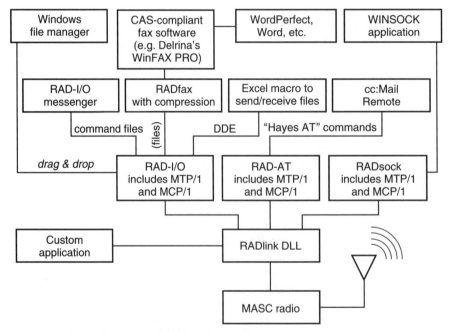

Figure 12.3 Research in Motion's RAD tool architecture.

Multisession functionality.   The main window in a RAD-I/O displays files being sent or received by different MAN numbers. Jobs can be canceled, held, or restarted using commands in the job menu.

Guaranteed transport.   RAD-I/O sends information from one machine to another quickly and reliably, using RIM's Mobitex-certified and standard MTP/1 transport protocols. Three modes of transport are available:

- *Express:* RAD-I/O does not provide confirmation.
- *Standard:* RAD-I/O sends the file and returns a confirmation that the right number of bytes has been received.
- *Confirm:* RAD-I/O sends the file with a 32-bit CRC and confirms that the file was received correctly.

Real-time information status.   RAD-I/O and RAD-AT both feature real-time information about radio connections in a radio status bar in the main window. Mobitex coverage can be monitored, along with battery level, signal strength, and MPAKS sent and received.

Data compression.   RAD-I/O includes Mobitex-certified MCP/1 compression and provides a choice between V.42 bis, RLE, or lookup compression protocols.

Configuration options.   Both RAD-I/O and RAD-AT include configurable parameters (such as baud rate, parity, transport mode, and compression type) that can be accessed and selected through dialogue boxes. Different users (MAN numbers) can be grouped together and assigned a group of configuration options.

Security.   RAD-I/O includes several security options. The MTP/1 (Mobitex Transport Protocol) provides security features for data at the local level. Read/write permissions on directories can be imposed. Mobile computing security issues will be covered at greater length in Chap. 15.

12.5.4   General Magic's application interface: A revolutionary new approach

We have so far discussed three relatively conventional application interfaces for mobile applications. Now we will look at a very interesting application interface from a company that has introduced software for futuristic consumer products that enable people to communicate in new and interesting ways. General Magic is a Sunnyvale, California, company that received venture capital funding through share participation or licensing rights from several international companies, including Sony, Motorola, AT&T, Apple, Phillips, and Nippon. Two of the company's best known products are Magic Cap and Telescript (Figs. 12.4 and 12.5). Sev-

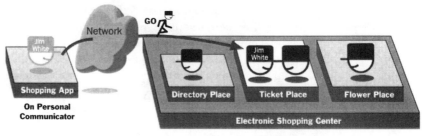

Telescript technology seeks to integrate the electronic world of computers and the networks that commect them into an electronic marketplace.

Telescript *places* lend structure to the electronic marketplace. Each place represents an individual or organization in the electronic world.

Places are occupied by *agents*, independent processes that transact the business in the electronic marketplace.

Figure 12.4   Magic Cap technology.

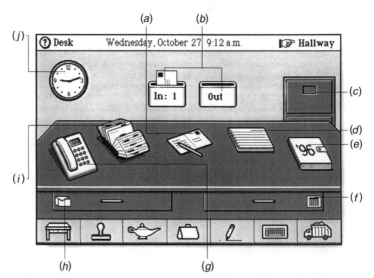

Figure 12.5   Telescript technology: (*a*) electronic messaging is automatically sent to multiple e-mail services; (*b*) in and out boxes are used to automatically file messages; (*c*) the file cabinet is used for message filing; (*d*) the notebook is a place to jot down ideas, maps, shopping lists, etc.; (*e*) the datebook is used to track the user's daily schedule; (*f*) the multifeature calculator includes a paper tape capability; (*g*) the telephone is equipped with many features of a smart phone, including speed dial and phone logs; (*h*) messaging forms are for business and personal uses such as faxes, e-mail, or paging; (*i*) the card file is integrated with modem and other applications, is customizable, provides auto entry and group addresses; and (*j*) in Magic Cap, the clock is integrated with all applications.

eral interesting implementations of these and other General Magic products have appeared in the marketplace in the form of PDA-based personal communicators. Sony's Magic Link and Motorola's Envoy communicators were among the first.

For several reasons, the acceptance of these new consumer devices has been slow. Perhaps most noticeable among the reasons are their high cost, their lack-luster marketing, their nonavailability on common client and server platforms, natural resistance to new concepts, and a lack of shrink-wrapped applications. Nonetheless, the technology inherent in Magic Cap and Telescript is promising and in one form or another will doubtless show up in future products. The software architecture may well form the eventual basis for a new generation of specialized mobile devices that are more portable than notebooks. (Already General Magic has already announced availability of Magic Cap for Windows in 1996.) The Internet revolution will help in this trend. Perhaps, JAVA will borrow from these concepts and achieve what General Magic hoped initially. Nonetheless, because of this potential and promise, the concepts and products are described here.

Fundamental concepts behind the Magic Cap and Telescript architectures. It is important to understand the following concepts inherent in Magic Cap and Telescript products.

Places. Telescript exists in a world that it sees as an electronic marketplace full of Telescript *places*. One place, representing home, might exist in a user's communicator. Another place, representing an electronic shopping center, might exist in a mainframe operated by a network service provider such as AT&T. Places can be *nested* (e.g., individual stores in a shopping center). Each place has a Telescript name.

Agent. A place is occupied by Telescript *agents*. Where places give the electronic marketplace its static nature, agents are responsible for its dynamic activity. Agents transact the business for a place over which they have authority (e.g., an electronic agent for a movie theater place would sell tickets). Telescript agents represent providers of consumer goods and services. *Agenting* is an independent process programmed in Telescript language.

Travel. The electronic marketplace is full of buyers' and sellers' agents. The marketplace is in tension because the agents are separated by networks. Telescript language enables the two parties to travel and meet in places on networks via a GO instruction that makes this travel possible.

Meetings. Once in the same Telescript place, two Telescript agents can meet and interact. One agent initiates the meeting using MEET, a Telescript instruction. The second agent, if present, accepts or declines the meeting (using another Telescript instruction). The agent initiating the meeting presents a *petition.* The petition identifies the other agent, for example, by *telename,* and specifies the terms of the meeting.

Connections.   While in the same place, two agents interact by meeting. While in different places, they interact by *communicating.* Two agents communicate via a *connection* between them. Using Telescript instructions, one agent requests a connection, the other accepts or declines it.

Smart agents, smart places, smart objects.   Everything in a Magic Cap and Telescript place is intelligent. Whether it is a message, a mailbox for electronic mail, an agent, or a place, they are all *smart* (i.e., they carry the user's preferences with them). For example, a user can send an invitation message containing a smart button which the recipient can tap to indicate acceptance. The user can specify that if the message is not read by the recipient, it should be forwarded to a manager. Users can customize their mailboxes to send messages from certain people to the users' pagers.

Magic Cap.   Magic Cap provides the software foundation for personal intelligent communicators. Because these communicators need to be small and inexpensive, they have limited computing resources. To meet the challenges of these limited resources, Magic Cap is compact and efficient, but also rich and flexible enough to support a wide variety of sophisticated software packages. Magic Cap can be customized by users to reflect individual personalities and tastes.

Because communicators are used by people who are not necessarily familiar with computers, Magic Cap presents an interface that is far easier to use than even the friendliest PC GUI.

Magic Cap provides several built-in packages, including a date book, name-card file, notebook, and electronic mail. Users can add more packages on PC Cards. They can also receive them via electronic mail or by infrared beam, or they can copy them into memory via a link to a personal computer.

The built-in packages provide core features for communication and personal information management. Magic Cap provides programmers access to some of the information that users enter into the built-in packages. For example, entries in the date book, people and companies in the name-card file, and the name of the communicator's user are all available.

Parts of Magic Cap.   Magic Cap provides a powerful, flexible platform for communication and personal information management software packages. The software in Magic Cap is made up of sets of classes, the operations the classes perform, and related items such as constants and other symbols. Magic Cap includes classes that provide a broad range of features for handling user interaction, objects that are drawn on the screen and manipulated by users, and abstractions from hardware details.

All objects displayed on the screen by Magic Cap are *viewables*. Users perform actions in Magic Cap by touching and manipulating viewables. Magic Cap provides extensive features for organizing viewables on the screen. Viewables can contain other viewables, thus creating a hierarchy of viewables. The current *scene* fills most of the screen and contains mostly viewables. The title that appears in the upper-left corner of the screen is the name of the current scene.

In addition to classes that support viewable objects, Magic Cap provides classes for high-level communication concepts, user-interface tools, text, graphics, sound, and much more.

The underlying foundation of Magic Cap includes an environment for persistent objects, a scheme for maintaining multiple active packages at once, hardware support, an operating system kernel that provides multitasking, basic support for communication, and peripherals.

Magic Cap provides an object-oriented *run-time environment* that defines the format and behavior of objects. This object run time provides the basic features for all objects in Magic Cap. The object run time is at the lowest level of Magic Cap and can accommodate different programming languages. However, for various practical reasons, C is most often used when creating Magic Cap packages.

Hardware platforms for Magic Cap. Magic Cap is designed to work on a variety of hardware platforms. The first platforms for Magic Cap have been families of handheld communicators created by various manufacturers under license from General Magic. Each communicator contains a set of essential features that support the software, including the following:

- 480 by 320 pixel display with four levels of gray per pixel (black, white, light gray, dark gray)
- touch-sensitive screen
- 1 Mb RAM and 2 Mb ROM
- built-in modem, 2400 bps or faster
- built-in microphone and speaker
- PC Card expansion interface
- connector for MagicBus peripherals
- connector for external telephone line
- one button, the option key

This is a list of essential minimal features. Different communicators vary in actual features provided. For example, most communicators include an infrared transmitter/receiver, and some communicators include a two-way data radio for advanced wireless communication. The ROM includes a system area filled by Magic Cap and a vendor area that contains software chosen by the manufacturer.

Magic Cap communicators use a Motorola Dragon I 68349 central processing unit. In addition, as was mentioned earlier, Magic Cap will soon be available as an application program that runs under Microsoft Windows on PCs. Eventually, Magic Cap will also be available for communicators based on other microprocessor platforms.

Telescript.   Telescript is a set of technologies that provides the software foundation for electronic messaging, distributed processing, and remote programming with communicators, computers, telephones, and the networks that link them together.

Telescript includes a programming language that enables software developers and users to implement and customize powerful messaging systems. Telescript lets users send messages that include not only static message information but also intelligence in the form of Telescript programs or agents that give additional instruction to the message, such as telling it how to move through the network or what to do when it arrives or is removed.

Telescript provides an environment in which messages are active programs, not just passive data. In many ways, Telescript is to messaging what PostScript is to printing.

Programs written in Telescript are executed by a Telescript engine running on a computing device in the network. Telescript engines are available in several forms and exist in many places. Every communicator equipped with Magic Cap includes a Telescript engine, as do some communicators without Magic Cap. Personal computer manufacturers will soon provide Telescript engines as extensions to their system software. Workstations and larger computers will run Telescript engines to provide an environment for Telescript agents. Creating Telescript services requires an additional product, the Telescript Developer Kit, which is available through a special license.

The first service to use Telescript was AT&T PersonaLink services. NTT of Japan has announced a joint venture with AT&T and Sony for an NTT Future Agent Network, which will offer similar services in Japan. AT&T PersonaLink offers advanced electronic mail, news, electronic shopping, and other information services. It also provides a Telescript-based platform for software developers and information providers. Telescript developers can create software packages that work with Magic Cap and use Telescript to send intelligent agents to and from the AT&T service.

12.5.5   Other PDA operating systems:
GEOS and Newton Intelligence

GEOS from Geoworks and Newton Intelligence from Apple are two other operating systems designed especially for PDAs. Casio Z-7000 and Sharp PT9000 run under GEOS. Apple's Newton and Motorola's Marco run under Newton Intelligence.

GEOS 3.0 (from Geoworks) is a communications-centric graphical operating system for PDAs and palmtops. It has a systems-level in-box and out-box called a universal mailbox library. This library enables users to page in different ways (e-mail, two-way pages, voice mail, faxes, etc.). GEOS 3.0 supports wired and wireless networks (ARDIS, RAM, GSM, and PCS) and includes support for TCP/IP, thus enabling it to interface with PSTN networks over PPP or SLIP initially, and with CDPD eventually.

12.6   Middleware's Role in Mobile Computing*

Mobile computing applications are based on the concept of a single mobile user accessing information residing on multiple information servers on different communication platforms—thus presenting interesting application programming challenges. This is where middleware can play an important role.

---

* *Data Communications,* July 1994, has a good introductory article on middleware.

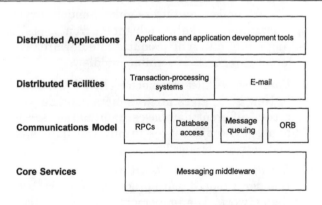

Figure 12.6 Middleware in mobile computing. (*Source:* Data Communications *magazine, July 1994.*)

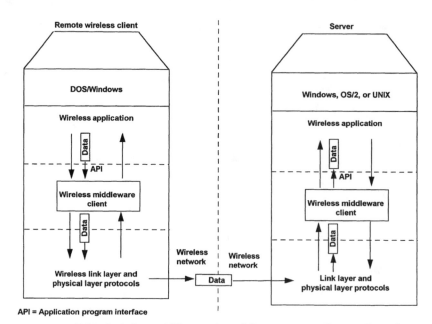

API = Application program interface

Figure 12.7   Role of wireless middleware in mobile computing. (*Source:* Data Communications *magazine, March 1995.*)

We shall give a brief overview of what is meant by middleware, the different types of middleware products that exist, and the pragmatic advantage of middleware technology in mobile applications (Figs. 12.6 and 12.7).

12.6.1   What is middleware?

Middleware is another industry buzzword that is being overused (misused?) by vendors and users alike. Vendors, trying their best to leverage maximum marketing advantage from users' lack of knowledge, are calling many different products middleware. There are probably five different categories of genuine middleware:

1. *Database middleware,* such as EDA/SQL from Information Builders in New York, gives programmers a single API that they can use to access different databases (information servers). This middleware handles the differences in physical implementations of databases and network protocols.

2. *Remote procedure calls* (RPCs) from vendors, such as Netwise in Boulder, Colorado, are programs to reach procedures anywhere on the network. The implementation of a distributed computing environment (DCE) by the Open Software Foundation (OSF) is based on RPCs.

3. *Object request brokers* (ORBs), such as the ORB from Digital Equipment Corporation, are a way for objects (applications and resources) on any computer to request and receive services offered by any other object on a network.

4. *Message-oriented middleware,* such as Xipc from Momentum in New Jersey, Pipes Platform from Peerlogic, and MQI series from IBM, uses messages to communicate across programs on different computers.

5. *Specialized mobile application engines,* such as Oracle's Agents and Sybase's Enterprise Messaging System (EMS), are unique software engines built to deal with the requirements of mobile computing.

12.6.2   What does middleware do?

A common thread uniting different categories of middleware is a single API across platforms and protocols. By coding to the messaging API rather than to the communications protocol, developers are assured that programs that communicate now will continue to do so as the network grows and changes—as long the network continues to be of a supported type.

We shall discuss messaging middleware and specialized middleware engines for mobile computing.

12.6.3   Basic principles of messaging middleware

All messaging middleware products conform to one of two models: *process-to-process messaging* (on the same or different platforms) or *message queuing.* Only the first model assumes that both processes are available. Because queuing does not depend on the availability of both processes, no active network session is required, thereby reducing network traffic—an important requirement in mobile computing.

Message queuing can be nonpersistent, or persistent and transactional. Nonpersistent message queuing relies on memory-based queues, which means that messages will not survive or persist across failures. Persistent queues are based on disk storage rather than memory, which means that messages will survive across failures. Persistent queuing uses transaction semantics to indicate what happens to a message—that is, whether or not it makes it to the queue on the disk. Each step is a separate unit of work that middleware logic deals with by the successful return of GET and PUT commands.

Messaging middleware supports *asynchronous communication*—another important requirement of mobile computing. Multiple requests can be processed simultaneously and can thus use the network more efficiently. In contrast, RPCs are inherently synchronous.

12.6.4 Role of middleware in mobile computing; To use it or not to use it?

It is our observation that, currently, middleware products do offer certain benefits in specific situations—though at a tradeoff. A good systems designer and architect needs to weigh the price and make an appropriate recommendation or decision. The following are a few general observations regarding the use of middleware in mobile computing applications:

1. Middleware decisions should be taken in the context of the broader IT strategic-planning discipline in an organization. We should consider the impact on all applications, not just on mobile computing applications.

2. Middleware should be used only when faced with multiple client/server application/database platforms. Every extra layer of software adds complexity, increases cost, and inevitably degrades performance.

3. When developing a business application on the unique API of a middleware vendor, a developer needs to select a product that is going to endure and be able to support future networks. There is no point in solving a short-term problem at the cost of a long-term crisis by choosing a vendor who cannot or will not support new wireless network protocols based on CDPD or PCS.

4. Middleware always adds more network traffic than a native application written to a specific network protocol. This can be detrimental to the point that it can make an application almost unusable. The author worked on a project where Tandem's RSC middleware added such a significant overhead to messages on a wireless network that it increased the log-on time to 40 seconds on Mobitex. Ultimately, it was thrown out and replaced by a simple and effective native Sockets protocol.

5. If mobile computing applications on wireless networks constitute a significant portion of the interactive application suite, and if these are mission-critical operational applications (e.g., public safety, financial services, airline reservations), it is absolutely imperative that the following options be evaluated:

   ■ Specialized mobile computing middleware engines mentioned here and described further in Chap. 11. However, homegrown middleware should be avoided, except as a short-term isolated development with a quick payback.

   ■ Writing to a native application interface such as Sockets in the case of TCP/IP-WINSOCK on the PC.

6. Large mobile computing projects are extremely complex undertakings in and of themselves. Do not add complexity to such projects by introducing immature middleware technologies, unless the organi-

zation is committed to such a course. It should make this commitment only after the planning group fully understands all the many requirements of the middleware-based application and the cost of implementation.

12.7 Protocol Conversion and Gateways to Legacy Applications

While middleware is an option that should at least be considered for new application development, there are many situations where customers need to provide connectivity to existing applications without rewriting the application's communications interfaces. In such cases, protocol conversions and database gateway functions can be provided by MCSS or database middleware such as Enterprise Data Access/SQL, or EDA/SQL by Information Builders (Fig. 12.8).

EDA/SQL provides SQL and stored procedure-based access to more than 60 relational and nonrelational data structures across all major operating platforms and networks. This middleware technology allows customers to build departmental and enterprisewide client/server systems that integrate heterogeneous data, application tools, and operating environments transparently through a common application interface.

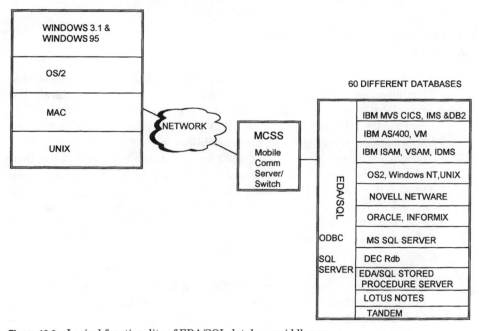

Figure 12.8 Logical functionality of EDA/SQL database middleware.

One other important consideration in using middleware is the degree to which it degrades performance relative to any savings in application development and integration efforts.

With the comprehensive support of all the major APIs (ODBC, DB-Library, DDCS/2, OCI, X/Open, etc.), EDA/SQL simplifies the challenge of operating in environments comprising multiple standards. EDA/SQL's leadership in the marketplace has been highlighted by its adoption by IBM, Microsoft, Hewlett-Packard, Digital, Informix, and Oracle into their information warehouses and DBMS products.

12.8   Emerging Mobile Application Engines

As we have already mentioned, mobile computing requirements are unique when compared to conventional wired-line network applications. Accordingly, several vendors are now bringing out mobile application engines. IBM has had an AS/400 mobile application engine for some time (since 1993). Oracle introduced Oracle in Motion in 1994 and enhanced it under the new name of Oracle Mobile Agents. Sybase has also introduced extensions to its EMS to support mobile computing applications.

These software engines have both system services as well as application services. Since their main objective, however, is to simplify the application development of mobile applications, we will wait until Chap. 13 to describe them more fully.

12.9   Two-Way Messaging Software Protocol for Paging Networks

Pagers and paging networks have constituted a very specialized shrink-wrapped market so far. Paging network service providers sell a complete service—a pager and network time—for a bundled monthly service fee. During the period 1993 to 1995, many e-mail and other applications began to include a capability to send brief alphanumeric messages to outside paging networks, and there are several applications available to activate pagers from PCs or Macintoshes. These applications are all based on a simplex or one-way messaging protocol called *Telocator Alphanumeric Protocol* (TAP). TAP is the industry-standard protocol for submitting alphanumeric pages over dial-up phone lines. The protocol consists of check-summed blocks of data containing the message and the destination PIN.

Paging networks are currently highly optimized to use available channels most efficiently and, as a result, can send only limited-length messages of from 15 to 80 characters (with a few exceptions that extend to 225 alpha characters). Many PDAs, palmtops (e.g., HP 200LX), and notebooks can accept PCMCIA pager cards that effectively turn these devices into pagers.

That is about as far as paging has come to date. However, with the introduction of new two-way messaging protocols such as TDP, which brings non-ASCII data to the pager, paging is now poised to become a mainstream mobile computing application.

Two-way paging allows the pager to verify reception to the transmitter, and to respond to it in one of four ways:

1. Automatic message confirmation

2. Multiple choice responses

3. Predefined responses

4. Free-form responses

In full implementation of the two-way messaging protocol, paging could eventually replace traditional e-mail. Many interesting simple dispatching applications will be developed using very small devices, including PDAs. The reader should also review SkyTel's two-way message service discussed in Chap. 14.

Several software development kits (SDKs) for developing horizontal and vertical industry applications are now available from network providers such as SkyTel, Motorola, Ex-Machina, and Socket to qualifying third parties.

12.10   Agent-Based Client/Server Software Solutions

In Chap. 5 and earlier in this chapter, we introduced the concept of an agent-based client/server paradigm of application development. We also introduced General Magic's intelligent agent concept. In Sec. 12.8 of this chapter, Oracle's Mobile Agent middleware was also mentioned. All these concepts and products are extensions of the client/server paradigm, with unique variations. Certainly there will be many more implementations of this concept in vendor products in the coming months. We shall, therefore, describe in detail an agent-based client/server implementation, using as an example a product called RemoteWare, developed and marketed by Xcellnet, an Atlanta-based company.

RemoteWare is essentially an application engine for mobile communications with a few interesting enhancements over a straightforward application gateway. In terms of the discussion of software architectures and MCSSs in Chaps. 5 and 11, RemoteWare addresses the client application portion of the client/server paradigm, the application interface portion of the switch, and e-mail server functionality. It assumes the selection of supported communications-server hardware.

Simply put, RemoteWare is a 32-bit OS/2 platform (NT version available in 1996) that lets mobile users send messages, transfer files, and query remote databases. The software comprises e-mail and forms applications, OS/2-based server management, and client software that runs on Windows, OS/2, Macintosh, and other environments.

12.10.1   Agent-based software technology

Agent-based design is a combination of real-time interactions with batched store-and-forward transmissions. The store-and-forward implementation is appropriate because mobile workers are more often than not interrupted in their work sessions by other activities and can seldom sit down for long interactive sessions. We shall look at several fundamental concepts in an agent-managed mobile work session (Fig. 12.9). These concepts are:

- Communication session
- Staging of session work
- Session scheduling
- Agent processes

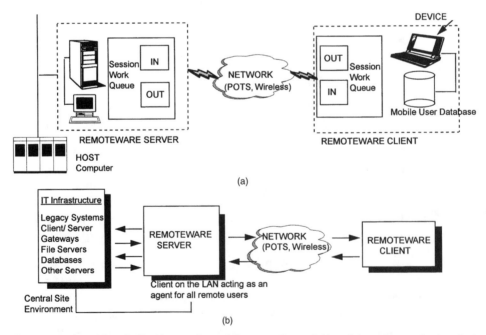

Figure 12.9 Agent-based client/server for mobile computing solution: (*a*) session work staged at server or client; (*b*) agent process on server and client. (*Source: Xcellnet RemoteWare Technology Overview White Paper.*)

Communication session. The *communication session,* in this context, may consist of the following:

- Security validation at the start of each session
- Communication tasks, such as file transfers initiated by the client application, file synchronization, operating system commands, or processes invoked as a result of central server commands
- Application tasks, such as servicing of specific remote client application-directed activities (e.g., posting of a sales order into a sales order database)
- Electronic software distribution under the direction of central server
- Resource monitoring and diagnostic tasks (disk space and memory consumption)

With RemoteWare, these tasks are completed automatically using integral run-time data compression and checkpoint restart techniques where appropriate—as if there was an agent supervising the completion of these tasks as one set.

Staging and session work. *Staging* is a process by which activities to be accomplished (*session work*) during subsequent communications sessions are queued up at the central server as well as on the remote notebooks. In this context, staging can be considered as batching of work to be handled at a later point. For example, when a remote salesperson submits an inquiry, the central server needs to access information from several distributed databases. In agent-based implementation, the salesperson can book an order off-line and whenever the next communications session takes place, the agent will both update the data and provide answers to any previous inquiries.

Session scheduling. *Session scheduling* is the process by which session work is queued and initiated at a central RemoteWare server. In this context, the RemoteWare server can intelligently stage session work for thousands of remote notebooks. For example, if information is to be retrieved from 100 notebooks and there are only 10 modems in the modem pool, RemoteWare schedules 100 sessions—launching the first 10 and holding the remaining in a queue. The scheduling of sessions and their execution can be controlled by time of day, type of connection (e.g., use only PSTN and not wireless network for file transfers), or the availability of resources (disk space in the remote notebook).

Agent processes. *Agent process* is simply a process that does work on behalf of a user (Fig. 12.10). Whereas users interact with an applica-

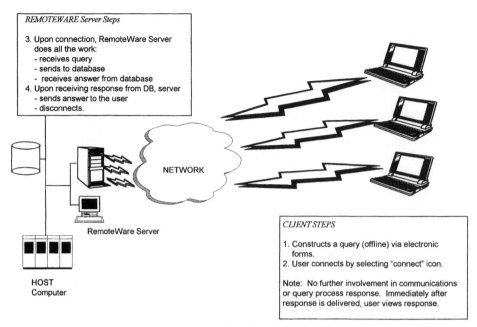

REMOTEWARE Server Steps

3. Upon connection, RemoteWare Server
   does all the work:
   - receives query
   - sends to database
   - receives answer from database
4. Upon receiving response from DB, server
   - sends answer to the user
   - disconnects.

NETWORK

RemoteWare Server

CLIENT STEPS

1. Constructs a query (offline) via electronic
   forms.
2. User connects by selecting "connect" icon.

Note: No further involvement in communications
or query process response. Immediately after
response is delivered, user views response.

HOST
Computer

Figure 12.10   Agent-based client server example. (*Source: Xcellnet RemoteWare Technology Overview White Paper.*)

tion program with keystrokes, mouse clicks, a pen, or voice-activated commands, agent processes do not require user interactions. With RemoteWare, agent processes themselves interact with other program processes on behalf of the remote user in order to accomplish specific session activities. For example, an agent process may post an order into a central DB2 database on behalf of the remote user if certain conditions (e.g., credit verification) are met.

In RemoteWare, agent processes can handle both communications and application tasks. RemoteWare applications have agent processes automatically built into the system (e.g., querying from an ODBC-compliant database). Agent processes can reside at the central server or the remote client.

The above concepts of sessions, staging, session work, session scheduling, and agent processes provide a context for implementing extensions to the traditional client/server model. This new extended model is known as the agent/server or agent-based client/server model. The RemoteWare server acts as a central agent capable of performing complex functions on behalf of a large number of remote clients. With continuous connection to a central IT infrastructure, the RemoteWare server buffers the remote clients from the complexities of connection to a number of distributed computers. It improves the overall system integrity and performance by acting as a central information wall a la a firewall in the Internet context.

The agent/server model does not require continuous connection—which fits in nicely with an important characteristic of the mobile workers' lifestyle. All work takes place off-line and is staged for the next communications session. When a connection takes place with a specific RemoteWare client, the RemoteWare server takes over control and performs all the session work that needs to be done. It automatically synchronizes all session work and drops the connection after completing the tasks. This allows for all staged tasks for each user to occur in a single, automated communications session.

12.11 The Transport Layer: Mobile TCP/IP

The mobile computing industry has been focusing on building component technologies: laptops, wireless LAN adapters, wireless modems, the networks, pen-based computing, and voice recognition. What is also required is an overall network architecture with standard specifications that address the networking needs of mobile workers. In the absence of a user push in this direction, however, vendors have only a limited interest in this area, preferring instead to build proprietary solutions. The only organization addressing this problem is the mobile working group of the Internet Engineering Task Force (IETF). Several vendors, including DEC and IBM, are active participants in this effort.

TCP/IP appears to be the most common transport layer supported by the major software vendors and by wired and wireless network providers. Those vendors who do not support it now will have to do so eventually, if they intend to remain in business.

It should be noted, however, that TCP/IP was not designed for wireless networks. Accordingly, some researchers, especially the IETF organization mentioned earlier, are looking at enhancing TCP/IP for mobile environments. Note that this work is still at a research stage, and, as far as we know, there are no products yet conforming to the proposed enhancements. Nonetheless, this is a promising development that mobile computing planners should keep in mind. Therefore, a very brief overview of this work follows.

Mobile TCP/IP should address the following characteristics of mobile computing:

Temporary connection. In mobile computing, connections are temporary and intermittent. Quite often, a mobile user will switch between a stand-alone mode and a connected mode during a short time span in order to minimize network charges.

Different physical networks used by the same or different users.   On any one occasion, a user may be connected via PSTN, on another, via a wireless network.

Security.   Security poses greater problems in mobile environments than it does in fixed networks.

Roaming.   Since users really are mobile, they move from base station to base station, cell to cell.

User support.   In the absence of quick and easy access to support personnel, users must be able to solve many problems themselves.

Error recovery unique to wireless networks.   Since most wireless networks are inherently less stable than wired-line networks, communications software must have a higher tolerance for temporary errors. Time-out parameters must be configurable, based on the particular network.

The task force is still evaluating these and other requirements and standard approaches to meeting requirements. The author suggests that those readers who are interested in this area should contact IETF members through the Internet for further information in this area.

Summary

In this chapter we have reviewed various mobile computing software components. Starting with a discussion of application models, we discussed overall software architectures first introduced in Chap. 5. Major software issues were outlined. Starting with a discussion of operating systems for client workstations, we pointed out that current software offerings do not support network transparency and concurrency. New operating software features in Windows 95 will help in building transparency in future versions. We looked at pen-based APIs and voice as natural forms of input for future applications.

Mobile-enabling current applications is an important requirement. In this context, we introduced RIM's RAD-I/O capabilities and promised a more complete discussion of the subject in the next chapter.

Moving from conventional mobile computing solutions to a new concept that shows great promise, we briefly introduced General Magic's Magic Cap and Telescript technology.

Middleware's role in mobile computing was reviewed in the second half of this chapter. We also gave an example of a database middleware. After a brief discussion of two-way paging, a more advanced agent-based client/server design was reviewed.

While the primary focus in this chapter was on a full-function note-book and the Windows software environment, Magic Cap software for PDAs, palmtops, and handheld devices was also discussed.

References

1. White paper on Mobile Computing available from Xcellnet, Atlanta, Georgia.
2. *Mobile Data News,* no. 2/95, available from Ericsson.
3. Magic Cap Telescript Technology white paper, 1994.
4. *Data Communications,* July 1994.

Mobile Computing Application Development Tools and Strategies

Develop and market a shrink-wrapped application and you will have all the customers in one vertical industry. Develop and market an application development tool, and you will have thousands of third-party software developers creating shrink-wrap applications for all the industries.   CHANDER DHAWAN

About This Chapter

In the previous chapter we discussed system services software—software that controls hardware and communications links but does not address the business application needs of mobile computing users. Having examined the system services and how those services glue components together, we can now turn our attention to application development tools and strategies. This is where a substantial portion of a project implementor's time will be spent. We shall describe typical application development tools available in the market today. There will be a detailed description of special mobile computing application development engines, such as Oracle Agents. After the discussion of tools, we shall turn to application development strategies. We will also describe the concepts behind a mobile-aware design. Finally we will recommend strategies for mobile computing application development, focusing attention on custom applications. Prepackaged applications are described in App. A.

There are two, possibly three tiers where business application or data logic may reside for mobile computing applications. In a conventional client/server design, there are only two tiers: the client and the server. In an agent-based client/server design, there is an intermediate tier: the agent application that handles requests from the user. In this book we are concentrating on development tools for client and agent applications.

When it comes to server applications, modifications for mobile computing may be required, but the changes do not warrant any special tools other than those used for nonmobile computing applications—and there are virtually hundreds of application development tools for server-based applications. These server-based tools are described in a host of other books and will not be dealt with here.

Since it currently represents the largest market share, we shall concentrate on Windows 3.1 applications development, with a brief mention of PDA application development tools as well.

13.1   Pen Application Development Tools

In Chap. 12, we briefly discussed various application development platforms, including a brief description of OS extensions to support pen features. Here we shall turn our attention to a closer study of application development for pen in mobile environments.

There are several pen-oriented development tools for Windows applications. We can discuss these tools under three headings:

1. Pen for Windows with Visual Basic (VB)

2. Pen-based Application Software Development Kits (SDKs)

3. Proprietary tools for handheld computers/terminals

13.1.1   Pen for Windows with VB

Windows 3.1 and Windows 95 have certain software capabilities that are readily exploitable for pen-based applications. VB has become a very popular GUI development environment and there are now several VB controls for pen. As a result, VB is now by far the most common development environment for pen. Since the market stability and longevity of Windows and VB is not in question, it can be seen as one of the less risky development options, compared to some offered by lesser known SDK vendors. But then, it may also not be as productive as some of the higher-level development environments available from specialized tool vendors.

13.1.2   Pen-based application SDKs

There are many, many SDKs available in the marketplace and the following list is more illustrative than exhaustive. Please refer to App. A for additional information on vendor products:

1. PenDOS SDK, from CIC at (415) 802-7888, provides direct access to CIC's handwriting-recognition engine.

2. Layout For DOS Pen applications, by Objects at (508) 777-2800, uses objects to model the business flow created by the PenDOS SDK.

3. Menuet/CPP, by Autumn Hill Software at (303) 494-8865, is a forms-based application development environment for PenDOS, Windows, and OS/2 for Pen and has an interface designer, icon and font editor, and C++ libraries.

4. Pen Developer, from Clarion Software at (305) 785-4585, is a full-featured development system for database applications (dBase, Fox-Pro, Paradox, Btrieve, Clipper Files) in the Clarion environment.

5. Power Pen Pal, from Pen Pal Associates at (415) 462-4888, is based on PenDOS and PenRight, a powerful, integrated pen development environment.

6. padBase, from R2Z at (510) 792-1477, is a library of Clipper routines for pen.

7. PenRight Pro, from PenRight at (510) 249-6900, is a powerful cross-platform development tool that will be described further in this section.

8. PenOp, from Peripheral Vision at (212) 262-1588, has a C++ interface and is available in PenDOS and Pen for Windows. It secures, captures, and manages hand-written signatures.

Most SDKs have a similar set of base functions, with special enhancements for each of them. We shall illustrate pen-based application development by using the features and capabilities of PenRight.

13.1.3 PenRight for Windows

PenRight for Windows is one of the most popular SDKs for pen applications. It supports DOS-, DOS DPMI-, and Windows-based pen-enabled platforms.

With newer PDA hardware improving in speed and memory capacity, the slow-running pen code associated with an older generation of PDAs is becoming something of a problem in today's IT world. PenRight helps to address the need for a development engine that can produce faster execution, especially in situations where pen applications are to be deployed on palmpads and notebooks with fast Pentium processors. PenRight also enables the migration of Pen DOS applications to the Windows environment. Other features of PenRight are as follows:

1. *Integrated graphical environment* (IGE). This allows forms to be drawn rather than coded (similar to VB). The IGE runs as a *multiple document interface* (MDI) application, allowing simultaneous editing of multiple projects. A project window supports a graphical tree of databases, forms, and code. Figure 13.1 is a schematic view of the application where every form and database has an associated icon. For example, a HELP button can be connected to the help text.

2. *Hierarchical property sheet* (HPS). This is associated with every form, button, text field, or other visual object. For example, a text line has properties that include ID, height, width, color, font, size, font style, and pattern.

3. *Create-database toolbar.* When the user clicks on this, a database icon is added to the project. Double clicking on this icon brings up a database schema manager window that controls the database table layout in dBase format.

4. *Code editor window.* This is invoked by double clicking on icons representing specific events or functions. The code can be written by a programmer or provided by PenRight. A *function* can be anything that a programmer wishes to create, such as a function to look up customer names in a database.

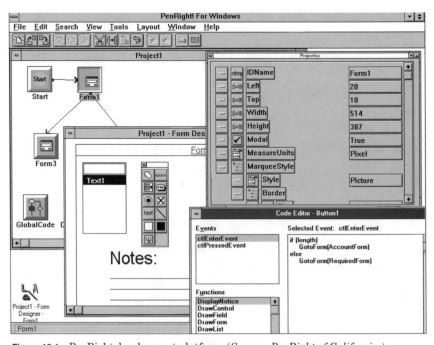

Figure 13.1 PenRight development platform. (*Source: PenRight of California.*)

5. *Macros.* These allow abbreviations to be used for long words or phrases (e.g., STL for *shipping to location address*).

6. *Support for keyboards.* There are times when keyboards are preferable in terms of data-entry speeds.

7. *On-line help facility.* This provides the entire API reference manual, as well as in-context searches.

According to the trade press, over 1000 vertical applications had been written for PenRight by the end of 1995.

Environments and languages supported

- PDAs, Palm Pads, and Notebooks
- DOS and Windows, currently

Familiarity with C language code development is necessary to use Pen-Right.

13.1.4 Pen for OS/2 SDK

This is an SDK from IBM for developing pen-based applications under OS/2. IBM also offers its Pen Developer Assistance Program (PenAssist) to the developers of pen software. Under this program, the following products and services are offered:

- Pen for OS/2 software at a special developer price.
- Access to IBM test centers located in Palo Alto, California, and Atlanta, Georgia.
- IBM computers with digitizing pads. These computers have PenDOS SDK and Pen for OS/2 installed.
- Electronic mail. Enrolled developers (ISVs and corporate developers alike) can send and receive messages directly from the software developers' support organization.
- Pen for OS/2 application catalog.

13.1.5 SDKs for PDA application development

The following SDKs are available for PDAs:

- *Newton SDK.* A Macintosh desktop-based development kit that uses Newton script, an object-oriented interpreted language
- *NSBasic.* From NS Basic Corporation (416) 264-5999, features Basic implementation on Newton with a toolbox for Window objects and graphics

- *Geos.* For Casio, Sharp, and other PDAs
- *Magic Cap SDK.* Discussed in Chap. 12 since it is both an operating system and an application development platform.

13.1.6   PDA enterprise SDK

Wayfarer Communications Corporation, based in Mountain View, California, has released a messaging middleware/development tool kit for integrating PDAs into enterprise networks and legacy systems (Fig. 13.2). PDA Enterprise Server provides access to databases for inquiry as well as for the real-time update of sales orders and price lists frequently used by mobile users.

Most hosts for such applications are low-end servers. In case of Wayfarer, up to 100 PDAs can simultaneously access the enterprise server.

Independence Technologies of Fremont, California, has also developed a middleware module that enables Newton-based PDAs to access and communicate with database applications running on a variety of servers with transaction monitors like Tuxedo, TopEnd, and Encina. These applications typically run on minicomputers and UNIX servers like the RS/6000.

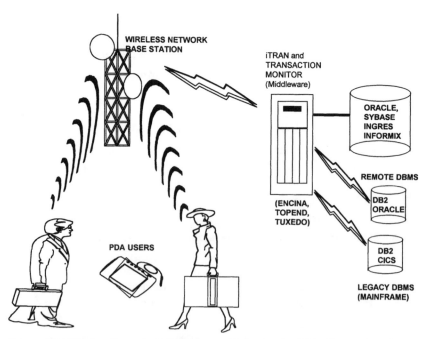

Figure 13.2 Wayfarer's and Independence Technology's PDA Enterprise Server. (*Source:* PC Week.)

13.1.7   Development tools for handheld computers

Many handheld computer suppliers provide proprietary tools for developing mobile applications. These tools generally support the C language development environment. Other vendors offer higher-level GUI-based application-generation tools, such as KeyWare from Racotek in Minneapolis, Minnesota. KeyWare is actually a suite of tools that includes the following:

- A messaging services module that facilitates multiple-information-type requests/responses, message queuing and storing, forwarding of messages. The messaging service uses agent-like technology.

- A naming services module that enables the assignment of logical names rather than serial numbers or network-specific designators for users, devices, and applications on the network.

- Time or scheduling services (task scheduling).

- An optimal delivery services module that bundles messages, assigns priorities, and optimizes packet sizes.

Racotek's KeyBuilder application development tool allows developers to quickly integrate existing and new applications to KeyWare-provided application services (Fig. 13.3). With KeyBuilder, applications can be developed in conjunction with Visual Basic or Sybase's PowerBuilder. Other features of KeyBuilder are:

- A GUI code generator that results in KeyWare-compliant source-code generation

- Sample applications available as models for wireless application development

- Application test capabilities that exercise an application under both normal and abnormal conditions

- On-line help

KeyBuilder creates mobile applications conforming to Racotek's KeyWare architecture, which is essentially an agent-based client/server model.

13.2   Network-Specific SDKs

Wireless and radio network providers have been doing a considerable amount of missionary work in promoting mobile applications. Many of them provide SDKs that enable applications to be developed or ported to their networks, including the following.

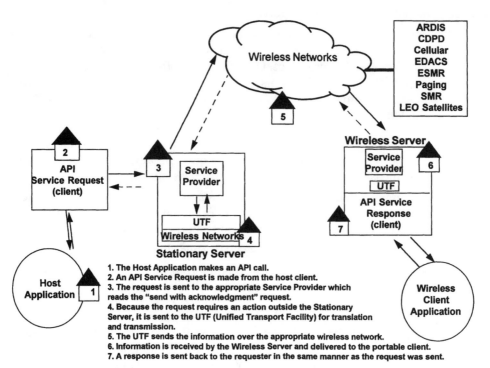

1. The Host Application makes an API call.
2. An API Service Request is made from the host client.
3. The request is sent to the appropriate Service Provider which reads the "send with acknowledgment" request.
4. Because the request requires an action outside the Stationary Server, it is sent to the UTF (Unified Transport Facility) for translation and transmission.
5. The UTF sends the information over the appropriate wireless network.
6. Information is received by the Wireless Server and delivered to the portable client.
7. A response is sent back to the requester in the same manner as the request was sent.

Figure 13.3   Racotek's KeyWare information flow. (*Source: Racotek.*)

13.2.1   RAD-I/O for RAM Mobitex networks

RAD-I/O (supplied by Research in Motion, based in Waterloo, Ontario) is a hybrid tool that provides system-level connectivity services for applications that interface with Mobitex networks. We originally mentioned RAD-I/O's system services in Chap. 12. We will describe here certain application development-specific services that RAD-I/O provides.

RAD-AT.   RAD-I/O is accompanied by RAD-AT, which provides an AT command-set emulation for telecommunications packages.

Four file-based methods of sending data to wireless devices.   The first method involves building a command file that specifies the file to be sent, the destination MAN, and other parameters, and then placing the command file in RAD-I/O's command directory. RAD-I/O reads the command file and completes the transmission as specified.

The second method involves the definition of subdirectories according to MAN numbers to be sent and their subsequent placement in RAD-I/O's out-box. For example, a subdirectory called *1234567* could be created and then sent to MAN number 1234567.

The third method involves dragging a file from Windows File Manager, dropping it into RAD-I/O, and then filling in the user name in the dialogue box that pops up. RAD-I/O looks up the MAN number and sends the file to that address.

The fourth method is based on DDE interface. Two example Macros for Microsoft WORD and EXCEL are supplied to demonstrate DDE.

Rapid application development.   Interfaces between RAD-I/O and existing applications (such as e-mail or fax) can be created quickly and easily to add real-time wireless data transfer capabilities without modifying original application software. This can be done in days, rather than months, according to the vendor.

13.2.2   ARDIS's wireless SDK

ARDIS offers a software development program similar to IBM's that includes not only software, but test facilities as well, including a test port, discounted ARDIS network charges, and technical and online support for ISVs and large corporate users.

ARDIS supports several development-software platforms, including InstantRF from NetTech (repackaged by Motorola as AirMobile), Enterprise Management System (EMS) from Complex Architecture (now a Sybase company), and FieldNet from Aironet (recently acquired by Novell). The following is a description of AirMobile: AirMobile SDK runs under the Windows and OS/2 operating environments and can be used to develop both client and server software for wireless access to LANs and host-based applications, as shown in Figure 13.4.

AirMobile was specifically designed for wireless environments and offers optimized transport—insulating the software developer from the complexities of wireless data communication in the process. Programmers work with intuitive, well-documented APIs to build reliable, network-independent wireless applications quickly and easily.

AirMobile's major features.   The APIs provide for a comparatively small number of function calls compared to middleware. Communication is based on OPEN, SEND, and RECEIVE commands. Sample source codes for DOS, Windows, OS/2, and PenRight is provided for a quick start up. The APIs provide other features as well, including file transfers, blocking or nonblocked data transfer, multiple sends and receives, optional negotiated permissions between end-points (for permission to transfer), wireless device status reports, and extensive error codes.

Programmers can use the SDK to develop peer-to-peer or client/server applications, or it can be used to develop pencentric applications with PenRight and Windows for Pen.

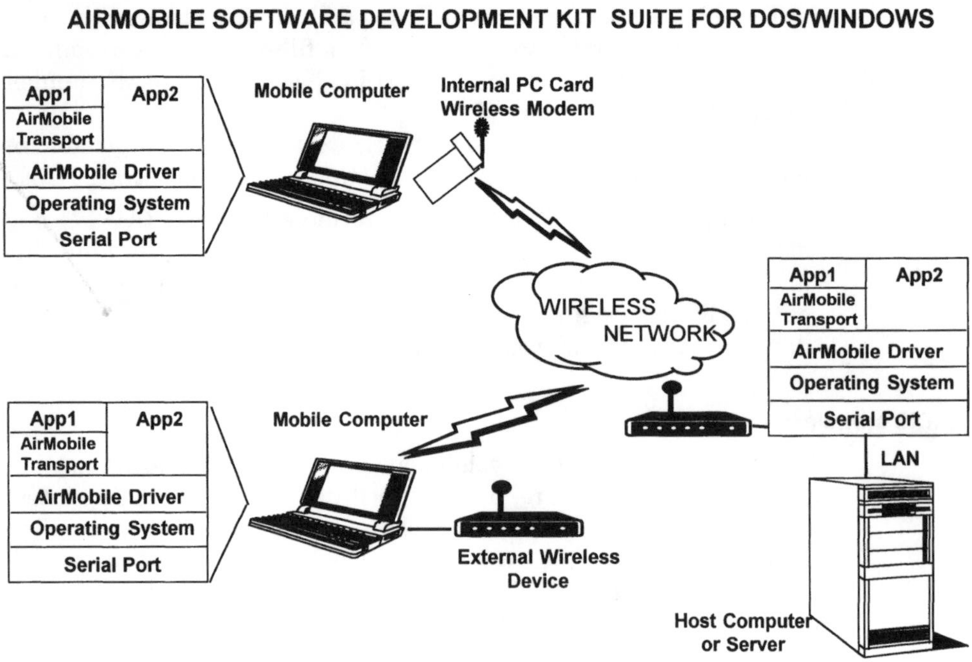

AIRMOBILE SOFTWARE DEVELOPMENT KIT SUITE FOR DOS/WINDOWS

Figure 13.4   Motorola's AirMobile SDK. (*Source: Motorola.*)

The SDK suite provides a transport-layer API that transmits data reliably and efficiently, and a driver-layer API that provides network independence as well as device independence. The suite also provides a communications interface between personal computers and various wireless devices, including packet radios such as Motorola and Ericsson's Mobidem.

The SDK also supports all major wireless packet networks, including Motorola's DataTac, ARDIS, Mobitex, and CDPD.

Unlike TCP/IP, which was designed for reliable high-speed WANs and LANs, the AirMobile SDK minimizes overheads and adapts automatically to fluctuating wireless conditions. It provides guaranteed delivery, checkpoint/restart capability, unlimited message lengths, sequencing, retry, and pacing algorithms fine-tuned for wireless networks. It is a connectionless protocol.

The SDK comes with C run-time libraries for MS-DOS development, and dynamic-link libraries (DLLs) are provided for development under Windows using C, VB, or any other language that is capable of using C function calls for DLLs. Device drivers are included to manage interfaces to selected wireless devices.

13.2.3 CDPD SDK

CDPD network service is now available in many large urban centers in the United States and in a limited number of locations in other countries. A CDPD consortium composed of major CDPD network service providers offers support to ISVs and large users developing CDPD-enabled applications.

As well, Wireless Connect, based in Santa Clara, California, has developed Desktop CDPD, an interesting tool kit that allows third-party software vendors and corporate developers to monitor, test, and debug native CDPD applications, and applications that have been modified to run over CDPD networks.

Essentially, desktop CDPD emulates cellular-network peculiarities (packet drops, packet delays, duplicate packets, out-of-order packets, loss of signal as a result of poor coverage, etc.), thereby providing a network emulation environment in which an application can be developed and tested before it is actually launched. This is particularly important in view of the fact that CDPD is not available everywhere.

CDPD is based on the IP protocol of the TCP/IP transport stack. Theoretically, therefore, applications should work if programmed to the TCP/IP standard. Even so, such programs do still require extensive testing in a CDPD environment, and this is where this particular tool comes into its own.

13.2.4 Motorola RadioWare program

Motorola's RadioWare is not a single software package. Rather, it is a portfolio of mobile applications, software tools, software integration and project management utilities, and maintenance services from one organization. The software packages and tools themselves come from both Motorola development staff and third parties. The portfolio's initial applications include WaveSoft-Fire (fire safety applications) and applications from UCS (based in Florida) and OCS (based in Vancouver, British Columbia) in the public safety area. Among the tools are FlashPort, used for loading software/firmware into portable and mobile devices, and Magic Pipe, a kind of middleware.

13.3 Agent-Based Client/Server Mobile Application Development Engines

We shall describe three application development engines based on client-agent-server model: Oracle Agents, Sybase's EMS, and IBM's AS/400 Mobile Network Access engine.

13.3.1 Oracle Mobile Agents

Oracle Mobile Agents, previously known as Oracle-In-Motion, essentially is an application development engine. It enables users to develop new applications to run on wireless networks or to modify existing applications for the same purpose. Version 1 was introduced in 1994 and was followed by version 1.1 in 1995 and version 2 in 1996. At the time of its original announcement, several business partners, notably BellSouth (RAM Mobile), Air Touch, Compaq, Dauphin, Dell, Ericsson, Fujitsu, McCaw, Motorola, and NetManage announced their support for the new engine.

Oracle Mobile Agents runs on RAM's Mobitex network, CDPD and spread-spectrum service from Metricom. It also works over asynchronous dial-up lines, using PPP (Point-to-Point Protocol), and on TCP/IP networks. Since most mobile applications require support on both RNA switched networks as well as wireless networks, this is a good combination.

How the Oracle Mobile Agents engine works. The Oracle Mobile Agents engine is built on the agent-based client/server design model introduced in Chap. 5 and further examined in Chap. 12. Given that Oracle's package deals with custom applications more than systems functions, it will be discussed further here. The engine comprises three pieces of software inserted between the client and the server portions of an application. As illustrated in Fig. 13.5, these three pieces are:

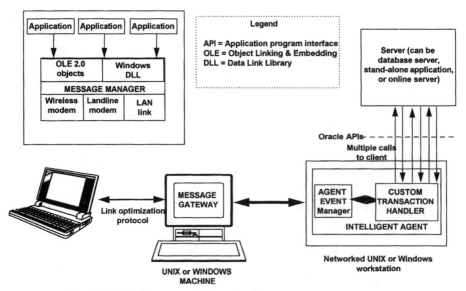

Figure 13.5 Oracle's Mobile Agents application development engine.

- A message manager
- A message gateway
- A message agent

The Oracle message manager. The *message manager* resides on the client workstation. Its main function is to receive messages from various user applications and then to send the messages to the server over a designated communications link.

The Oracle message gateway. The *message gateway* runs on a UNIX machine or on a Windows NT server. Typically, it is local to the information server. It comprises an event manager supplied by Oracle and a transaction manager written by the user or a third-party software-integration company. The message gateway provides the following functions:

1. It passes messages between mobile users and intelligent agents by way of an optimized wireless protocol.
2. It provides a queuing function for users who are not within range of the wireless network, or who have turned off their notebooks.
3. It provides system management and administration functions (it looks after security—Oracle Mobile Agents version 1.1 supports user password, encryption, and message authentication—and keeps track of client configurations and agent configurations).

Wireless link optimizer. As we have seen, there are two major problems with wireless networks: unreliable links compared to wired-line networks and low bandwidth capacity. Oracle Mobile Agents addresses both these limitations.

To deal with the low bandwidth issue, it tries to keep traffic to a minimum. For example, if a mobile user forwards multiple requests to access a mainframe server and a minicomputer for two information sets, the client application passes the requests to the message manager (in the mobile workstation), which then picks a communications link and forwards the requests to the gateway. From the gateway, these requests are forwarded to the agent, which acknowledges receipt. The client can then disconnect and avoid further cellular connect charges while responses to the requests are being prepared. If the agent were not present, the user would have to stay online and wait for the responses. This process can be useful also when multiple requests are forwarded to the same server. Now multiple responses can be batched together and sent to the user in one transmission, thus avoiding a back-and-forth chitchat of control messages (which are essentially overheard in data communications traffic).

To deal with the unreliability issue, Oracle has implemented message integrity by using an enhanced higher-level protocol that acknowledges packets even in the event of a broken connection. The protocol uses a dual-windowing scheme that allows the sender to continue sending data, whether or not a previous packet has been acknowledged. This is similar to IBM's synchronous data link control (SDLC) or X.25 HDLC protocol at the link level.

The size of both windows can be set by the system administrator to create the notion of a *soft acknowledgment* (indicating only a likelihood that the packet has been successfully transmitted) and a *hard acknowledgment* (confirmed transmission). For example, if the window sizes were set at 4 and 8, the user would send an initial four packets and wait for a hard acknowledgment, and then another four on the strength of the hard acknowledgment of the first four.

When the recipient acknowledges the first four, the sender resets the packet count and starts sending more. Under this scheme, packets are sent only when necessary, even when a connection is dropped. By keeping track of packet sequence numbers, the protocol ensures that duplicate packets are not sent. Only if soft acknowledgment alone is required can packet duplication take place.

Intelligent agents. Making agents intelligent opens opportunities to monitor servers and take appropriate action under specific sets of circumstances. For example, if a price database changes, an application in the server could automatically send updates to its client notebooks and a corresponding agent in the client notebooks could then incorporate the updates. Urgent e-mail could also be sent to remote users from e-mail servers such as cc:Mail from Lotus Corporation and MS Mail from Microsoft. Agents in the e-mail server would look for mail messages at regular intervals and forward them to users when they connect.

Features in Oracle Mobile Agents version 2.0 (1996)

- *Data compression.* Automatic data compression reduces message size and increases throughput.

- *Systems management support.* An SNMP agent allows Oracle Mobile Agents to be managed by existing management applications. Additional logging and statistics information provide enhanced audit and trend capabilities.

- *Developer kit enhancements.* A new prototyping capability lets developers build and test applications on Windows with no UNIX-based gateways or agents required. Additional tools and programming controls simplify and accelerate development.

- *Multiple network support.* Simultaneous support for multiple network services allows third-party information services to function in tandem with corporate information. For example, a user can receive updates from an online travel service application accessing its specific gateway while also receiving sales information and e-mail from headquarters.

- *Integrated dial-up PPP support.* Leading vendors, Persoft and Distinct included, have worked with Oracle to integrate their point-to-point protocol (PPP) stacks with Oracle Mobile Agents so that users can easily change dialing locations when traveling.

- *Other usability improvements.* An autoconfiguration capability has been added to sense and address new mobile devices automatically, along with a redesigned, more informative user panel.

- *Database paging system.* This is a Version 2.0 feature that will link narrowband PCS paging networks. The software features triggers that can be configured within a database to create, queue, and send pages to end users when a specific event or transaction occurs.

- *Automatic synchronization of databases.* This is a Version 2.0 feature in the mobile client device that works by setting up database triggers.

- *Large enterprise deployment.* This feature offers deployment of Oracle database technology in concert with Oracle Mobile Agents to support up to 10,000 users.

Flexibility in Oracle Agents' APIs. Oracle's APIs are able to connect to the client portion by either passing an object linking and embedding (OLE 2.0) object or through a Windows dynamic-link library (DLL). Thus, any development tool that can create OLE 2.0 objects or DLLs can be used.

13.3.2 Sybase Enterprise Messaging System (EMS) for mobile computing

EMS is a product originally developed by Complex Architectures, which was acquired by Sybase in February 1995. EMS provides reliable access to corporate data via wired and wireless networks, enabling businesses to support a truly distributed enterprise, including a growing number of mobile users. EMS has been shipping since 1992 and is one of the industry's leading messaging products available today.

The currently available version 3 of EMS (introduced in 1995), is a message-based communications management environment that provides a comprehensive solution for reliable, intelligent, and cost-effective connectivity to a range of platforms over wired and wireless networks. EMS insulates commercial programmers from the complexities of network communications, allowing them to focus on business applications by simply writing to the EMS API. EMS handles underlying communications, providing transparent bidirectional, asynchronous transfer of messages (data, transactions, etc.) across a network. Key components of EMS include:

- An EMS API interface providing applications access to EMS communications services

- An event manager that manages routing and message queuing

- A communications manager that formats queued messages into packets for a communications agent chosen by the event manager

- Communications agents that handle the data-link and physical-layer requirements of each network, protocol, device, or communications facility

Sybase EMS features. The comprehensive scope of the EMS message-based computing environment provides customers with cost-effective, reliable communications across a network even when a system fails. It also provides single interface support for both wired and wireless communications.

Intelligent cost-effective communications. A key feature unique to the EMS 3.1 communication environment is a quality of service (QOS) capability that allows applications to request a class and quality of service. EMS accepts messages and dynamically bases its delivery of them on available network bandwidth and resources, thus providing customers with a highly flexible and cost-effective means of matching computing processes to business operations. For example, the delivery of time-sensitive shipping data can be prioritized on the one hand, and bulk data transfers from expensive satellite communications can be deferred on the other.

Reliable messaging for different networks. EMS provides reliable messaging for all networks it supports. Its store-and-forward capabilities, adaptive routing, and other recovery mechanisms are designed to ensure that messages are not lost in the event of a network or system failure, or when a target platform is disconnected from the network. Additionally, EMS provides guaranteed delivery through multinetwork routing and delivery, and the ability to reroute around network connection failures.

Optimized for mobile computing. EMS is optimized for mobile devices, requiring a mere 120 to 190 Kb of hard disk space in support of the smallest Windows, Mac, and DOS portables, as well as DOS-based personal digital assistants (PDAs). Its store-and-forward capabilities ensure that previously queued messages will be transmitted to connected portable devices via either wired or wireless communications. Additionally, EMS provides support for directory and security services. However, as far as we can tell, the current versions of EMS are not truly *mobile-aware,* as defined in this book. It does not have features that minimize bandwidth requirements on wireless networks.

New EMS versions for mobile computing. In 1995, Sybase introduced Sybase EMS 3.2, an enhanced version of the EMS communications environment that features superior performance and additional communication agents.

Sybase EMS 4.0 (introduced in 1996) includes an embedded authentication feature and support for symmetric multiprocessing (SMP), additional public APIs, and support for more PDA operating systems and networks.

Sybase Open Client and Open Server to include messaging. In the immediate future, Sybase plans to build on the strengths of Sybase EMS by integrating the messaging technology into Sybase's Open Client and Open Server APIs. These new, extended APIs will automatically manage both the session-oriented communications common to corporate environments and message-based networking, which is rapidly gaining acceptance. The integration of Sybase EMS messaging into Open Server will also enable Sybase's Open Server-based product (Replication Server) and Sybase's Enterprise connect interoperability software to operate on a common message-based model.

The integration of messaging into Open Client and Open Server represents a significant extension to the Sybase architecture, since these APIs are the foundation of every Sybase product. Through the extended Sybase architecture, Sybase expects that users will be empowered to develop a new breed of client/server applications that reflect their business operations. These features are expected to be available in mid-1996.

13.3.3   IBM's AS/400 Mobile Network Access engine

Please note that the AS/400 Mobile Network Access is not an agent-based client/server design. It is described here as an example of a proprietary mobile application engine that may be appropriate in certain situations.

The AS/400 Mobile Network Access is IBM's application engine for mobile computing. It uses the computing and database architecture of IBM's AS/400 server product line to provide an interface with IBM-compatible PC-based workstations through both radio networks and public switched networks. It supports the ARDIS, Ram Mobile Data (or

Mobitex), and Motorola private radio networks. IBM RadioPac/400 and PagerPac/400 are two third-party programs developed by Business Partner Solutions, Inc. (BPSI). Using the AS/400 as a communications hub, PagerPac/400 supports one-way wireless private and public pager services, while IBM RadioPac/400 supports two-way wireless communication through public and private radio frequency networks. These programs share a common command set and user interface.

The AS/400 provides a communication server, a transaction manager, and router functionality. The transaction manager function is built into RadioPac/400. The remote devices can be palmtops or PC notebooks equipped with an appropriate network modem.

RadioPac/400 provides easy-to-use APIs which eliminate the need for complex communications programming and which also insulate business application functionality from communications software.

PagerPac/400 immediately delivers critical messages to individuals or service vendors locally, regionally, or nationally, using private or public paging networks. PagerPac/400 monitors AS/400 queues, identifying all important messages that can be sent to an alphanumeric pager. Event Manager is another AS/400 application that provides a tool for sending AS/400 system messages through a pager. Users define priority messages that must be sent to a pager. Other features include *group definition* (for broadcast purposes), *response groups* (if a user does not respond, send the message to a list of users until someone does respond), and *look-ahead* (in high volume situations, batch messages to the same service company in one call).

The AS/400 is a popular midrange computing platform that has thousands of application packages and is very easy to maintain, support, and manage. Nonetheless, AS/400 is a proprietary hardware and software platform. Therefore, this application engine should be used for mobile computing applications only when the organization has already invested in the AS/400 for other applications, or if there is already a significant IBM technology infrastructure in the organization, or if the application requirements are a close fit. Introducing the AS/400 as a communications hub for mobile computing applications with custom code is not recommended where an open solution with multiple mobile devices is the aim. AS/400 is not a preferred platform for many third-party mobile applications.

13.4   Strategies for Mobile Application Development Architectures

The whole discipline of application development is subject to heated debate in our industry, and a tremendous amount of literature exists on the subject. Many well-known experts in the field have spoken to countless packed seminars. What we would like to do here is focus only on those strategic issues that are relevant to mobile applications. We will highlight the issues and discuss the pros and cons of different approaches. No specific solutions will be offered. Readers must determine for themselves the strategies that best apply in their particular organizational contexts. It is the only way.

13.4.1 What is different about mobile applications?

In order to appreciate the mobile user's requirements accurately, we should try to understand the differences between the ways in which mobile workers actually perform their work and the ways in which they would like to perform their work, if technology were not a hindrance. Only with such an understanding is it possible to develop good solutions that make the most of existing technologies—even if technological restrictions do bar the way to a perfect, low-cost solution. The following stand out:

- *Urgency.* Mobile workers are always in a greater hurry than office workers. They want to be able to work faster and more efficiently than inside the office.

- *Lower tolerance for error.* Mobile workers cannot afford to make mistakes in the mobile environment. A slipup in presentation, an erroneous piece of information, an incorrect answer can all cost a sale.

- *Self-reliance.* Professionals on the road often have no access to administrative staff; nor in many cases can they take notes for later action.

- *Time is of the essence.* The value of a mobile worker's time outside the office (or the factory) is at a premium.

- *Low tolerance for technology faults.* Since mobile workers do not have access to external help, and technical support is far away, the technology configuration (software and hardware both) must work more reliably than inside the office.

- *Efficiency of use.* The application dialogue must be highly efficient and lead to the completion of the task at hand in a minimal number of steps.

- *Essential only.* Outside the office, mobile workers do not have the time to tend to nonurgent tasks.

- *Criticality.* Many mobile computing applications (fire control, ambulance dispatch, law enforcement, etc.) are truly mission-critical, far more so than accessing a bank account for a withdrawal or a travel agent for a ticket reservation.

- *Need for alternative inputting devices.* While the keyboard is generally accepted as a major way of interacting with computers, it is not as acceptable outside the comforts and convenience of home or office. Pen and voice, or a combination of both, are often more intuitively useful.

- *Sporadic work bytes.* Because of the stop-and-go, multitasking nature of mobile workers' lifestyles, it is inefficient (and unnecessary) to make them wait for the completion of communications tasks or inquiries. It is better to be able to perform many of these tasks in the background while users are doing other things that do not require computer input or attention.

- *The personal touch.* Mobile workers are the ones who know their clients best. Thus, correct information about clients or subjects under review should be procured at the source, without intermediate processes or persons being involved, unless value is added or operational efficiency is achieved thereby.

- *Optimal technological performance.* Hurried, harried mobile users deserve as good a technological performance as can be economically justified.

13.4.2 Current state of mobile computing application development

Such are the criteria of mobile computing (or at least, some of them). Unfortunately, current mobile computing applications do not measure up to these criteria very well. Following are some of the common problems:

- The enabling of current legacy applications for mobile computing without any change

- The generation of highly inefficient code from high-level application development tools

- A lack of in-house development of custom communications code for wireless applications

- A lack of LAN-type GUI applications in the current generation of wireless data networks

- A preponderance of closed proprietary solutions for critical parts of customers' businesses

13.4.3 Strategic application development issues

The following application development issues must be dealt with by systems development managers:

A different application development tool set for mobile computing applications. Should a mobile computing application development tool set be different from the corporate standard for desktop applications? This is the key question for application development managers. In the early stages of an emerging technology, there is a tendency to pick a specialized development platform because it is the only tool available, or because it happens to fit the need. While the features in a tool set and the efficiency of the code it produces are extremely important considerations, for any given organization, consideration should also be given to prevalent standards, existing skill sets, and the availability of trained staff in the marketplace. Ultimately, it comes down to a trade-off. If an application is to be extensively rolled out across an organization, specialized training, and the acquisition of expensive experimental skills can be worth the price.

Multiplatform tool sets. A universal end-user device for mobile computing has not yet appeared in the marketplace. The IBM-compatible PC is the de facto standard, but PDAs, palmtops, and other devices are grouping in an uncertain marketplace to find their true personality. Perhaps there will always be more than one standard for client devices. Certainly, information that mobile workers seek will always reside on multiple platforms. It is extremely important, then, to select a tool set that supports multiple client devices and allows logical communication to take place with multiple information-server platforms.

Developing applications for target platforms other than the desktop platform. It does not always make sense to select a specialized client device that differs from the desktop device. If the user is going to switch between the two modes frequently, there should be as much similarity as possible. In fact, the concept of a notebook-based universal workstation has a certain merit. However, mobile workers do generally accept simple-to-operate, specialized devices in cases where they are custom-designed for a specific purpose and are supplied completely prepackaged, as in the case, for example, of vertical manufacturing and distribution applications.

Modular development for network and protocol independence. It is extremely important in mobile computing to design applications that are modular, and where business rules and logic are not embedded with communications logic. We need to protect our investment in business applications and allow for the advent of newer and more cost-effective networks and transport protocols.

4-GL or Case-based application development tools for mobile computing. Except in very rare instances, 4-GL and Case tools are not recommended for mobile computing applications, unless volume is low, wireless network traffic is not an issue, and the number of mobile users is small. 4-GL tools (and Case tools) do not currently recognize the unique needs of mobile computing applications. They tend to create highly inefficient code that, though easily processed on the new generation of notebooks, generates very large amounts of data on wireless networks. In fact, most 4-GL programmers do not even understand what network traffic the tool is going to generate because their focus is purely on the flow of business data. A very large amount of system control packets may be generated by generalized communications software used in these packages. 4-GL tools may make the job of the application programmer easy, but they also make response times very high, and the cost of operating an application prohibitive.

Remember that the application coding task is generally 20 to 30 percent of total application development costs, and application development costs may be a small percentage of total mobile wireless infrastructure costs. Thus, network operating costs could easily outweigh any savings achieved with the use of 4-GL tools.

Middleware's role in application development.   Please review Chap. 12 for comments on this issue.

An open mobile computing solution for mobile computing.   On occasion, rare tools from small vendors may be discovered that have all the desired features. Use such tools only as tactical solutions for quick short-term paybacks. They just might prove the experience and tactical advantage needed until such time that a full strategic solution is found. For full strategic solutions, pick an open standards–based solution from a vendor who has a sound track record and will be around in the future.

13.5   Strategies: Making Applications Mobile-Aware

Finally, we come to the important topic of designing mobile-aware applications. This is a term proposed by the author as a descriptor for application designs and software products that meet the unique criteria previously described for mobile computing applications. Application designs and software products must meet the following criteria to be worthy of a mobile-aware label.

We are recommending the following systems software, infrastructure, and application-design considerations in order to make your applications mobile-aware.

13.5.1   Systems or infrastructure design considerations

System Design Criteria Number 1: Compression.   Wireless networks will continue to be slower in speed and more costly to operate. Mobile-aware systems software should have as good a compression algorithm as possible. We are not only talking about modem-based compression of blanks and repeated characters, but phrases and common data elements being replaced by codes of shorter length.

Systems Design Criteria Number 2: Security.   Wireless networks pose greater security risks. The possibility of temporary disconnections should be minimized, and holes in error-recovery software that enable reconnection without full user validation in the event of a disconnection should be avoided. Mobile-aware system designs must provide encryption and authentication of users.

Systems Design Criteria Number 3: Judicious use of general-purpose software such as middleware.   While middleware has a role in application development where network bandwidth is not an issue, mobile computing will continue to demand efficient design and special-purpose middleware, such as Oracle Agents, which optimizes the communications layer.

Systems Design Criteria Number 4: Shortest and fastest logical path to the information server.   In the past, many message-switching applications were built with back-end applications acting as message switches. Since transaction response time is a critical performance requirement in many mobile applications (e.g., in public safety applications) there should be a minimum number of intermediary software components involved in communications switching. Ideally messages should be switched to the destination information server by the MCSS (described in Chap. 11) directly.

Systems Design Criteria Number 5: Minimum hops in a wireless network.   Many packet networks are hierarchical in topology. Intermediate switches are connected by wired line at 56 Kbps or T1. Nonetheless, they add delays. A mobile-aware infrastructure design should have minimum hops in the path.

Systems Design Criteria Number 6: Agent-based client/server.   An agent-based client/server design is more in-line with the ways in which mobile workers work. In real life, they call to pick up their messages, check the status of customer orders, and ask administrative staff to do errands till they return. Intelligent agent software (as in the Magic Cap Telescript paradigm) should know users' working preferences and act on tasks in their absence so that they are ready with results when they call back. An advanced mobile-aware design should incorporate an agent-based client/server design.

13.5.2   Application design considerations

Application Design Criteria 1: Fast-track dialogue for the user interface.   An application's user interface should be similar to the one on the desktop, but it should also provide a fast track to the intended operation. Mimicry of CompuServe's GO concept or hot-button concepts in other applications could be incorporated.

Application Design Criteria 2: User interface screen as a subset of the desktop menu screen.   Mobile users do not want to be burdened with large numbers of control buttons, multiple tool bars, too many menu items, too many entries in scroll bars, and large numbers of options in menu items. Instead, mobile workers want a simple, customizable, uncluttered screen interface.

Application Design Criteria 3: Only minimal amounts of information presented, with options for more.   Mobile users should be presented with a minimum amount of information required to do the job. Summary information only should be presented initially. Any further details should follow only at the request of the user.

Application Design Criteria 4: Data input interface should be compatible with the task.   Too many desktop keyboard-dependent applications are outfitted with a wireless communications interface and given to mobile users, even though they may be extremely inconvenient to use outside the office. Pen and voice will become increasingly important requirements of mobile-aware applications.

Application Design Criteria 5: Abundant user help.   A mobile user does not have access to help desks. There must be as much local application help available as is required to make users self-reliant.

Application Design Criteria 6: High error recovery at network levels.   The application interface to the communication layer must be extremely sturdy and must recognize the roaming, coverage, and handoff issues of wireless networks. Mobile workers should not have to worry about being in or out of range, application errors, or network disconnections.

Summary

In this chapter, we discussed mobile computing application development tools and strategies, starting with client application-design tools. We reviewed various tools, including Visual Basic, which has become very popular for Windows-based application development.

We reviewed the importance of pen as an input source for mobile applications. Using PenRight as an example, we reviewed the features of a specialized pen-based application development tool for Windows. We also briefly mentioned development kits for PDAs and handheld computers.

Next we described network-specific software development kits—one each from Mobitex, ARDIS, and CDPD.

After covering the client application development scene, we moved over to the agent-based client/server application development engines. Oracle Agents, Sybase's EMS, and IBM's AS/400 Network access software were described. The AS/400 Mobile Network Access Engine was also described in this section.

Finally, we discussed strategies for mobile application development architectures. We started by distinguishing mobile applications from desktop applications. We pointed out various strategic issues and then developed the concept of mobile-aware application design.

This concludes our discussion of the software components of mobile computing. In Chap. 14 we shall move on to vendor implementations and their strategies and products.

References

1. *Pen Computing,* June/July 1995 and Aug./Sept. 1995.
2. *Data Communication,* July 1994, Oracle Mobile Agents.
3. White paper on ORACLE Agents, available from Oracle Corporation.
4. See Internet home pages of Sybase and Oracle Corporation for description of ORACLE Mobile Agents.
5. Software Development Binder, available from ARDIS.

Understanding
the Vendor Offerings

Mobile Computing Vendor Product Strategies

Mobile Computing technology is still evolving. Unlike many other computing markets, no single vendor is dominant in the entire mobile computing arena. Some provide more pieces of the mobile computing puzzle than others. Some lead the market; others follow it. While it has not and will not reach the levels of hysteria of the client/server or the Internet market, most major vendors want to participate in this industry.   CHANDER DHAWAN

About This Chapter

In Part 1 of this book, we reviewed the potential of mobile computing, business applications, role of business process reengineering, and business case–development methodologies. In Part 2, we discussed the technology architecture of a mobile computing solution. In Part 3, we addressed the components of that architecture, attending in detail to wireless LANs, wireless WANs, and mobile communications server switches. In Chaps. 12 and 13 we also studied mobile computing software components and application development strategies.

Now we are ready to embark on another part of our journey: the investigation of what is available from the vendor community today (an assortment of vendor products is listed in App. A)—and what might be available tomorrow. Product availability today and product strategies for the future are important. While we can influence the product design through our requirements, ultimately it is what the vendors bring to the market that determines the makeup of mobile computing solutions.

In this chapter we shall focus on the strategies of vendors, which, unlike their products, tend to be relatively more stable. An examination of vendor product strategies, therefore, can be of considerable help in our own efforts to make better planning decisions.

Since it is not possible to cover all vendors in a book of this size, we have opted to cover only the relatively few vendors of products and services who have demonstrated, in our opinion, well-orchestrated mobile computing strategies. We have also included a few vendors who hold out the promise of innovative, futuristic new technologies. We apologize to any vendor we have omitted who has a legitimate claim to inclusion. We will be happy to review the list and look at recommendations in any future revisions of the book.

14.1   Mobile Computing Vendor Types

Although several companies offer products in a number of different areas, we have divided vendors into the following categories, based on core product lines and competencies:

- Major infrastructure product suppliers
- Major network services infrastructure suppliers
- End-user hardware suppliers
- Switch suppliers
- Software suppliers—systems and application software
- Application integration services suppliers
- Systems integrators

14.2   Major Infrastructure Products Vendors

We have included Motorola, Ericsson, Northern Telecom, and IBM in this category.

Motorola and Ericsson are two multinational giants with similar product and service strategies and core competencies in network infrastructure products. As such, the two companies compete vigorously and provide infrastructure products to different sets of customers. Motorola owns and supplies the ARDIS packet-switching wireless data network service. Ericsson is the major supplier and supporter of RAM Mobile Data and Mobitex networks in over 15 different countries. Between them, Motorola and Ericsson provide the infrastructure for the majority of cellular and other wireless data networks.

Northern Telecom differs from Motorola and Ericsson in that it is primarily an infrastructure and an end-user equipment supplier to the carrier industry.

IBM provides products for mobile computing's nonnetwork infrastructure segments such as end-user devices, notebook PCMCIA peripherals, software-based network gateways, application servers, and overall systems integration for end-to-end business solutions.

14.2.1 Motorola

Motorola is one of the largest American corporations in the world, with an international presence that is playing a leadership role in accelerating and driving the growth of the wireless communications market. Motorola delivers wireless solutions that include LAN and WAN wireless data networks, mobile and portable data terminals, personal wireless communicators, and data network services. The company's major businesses include cellular telephones, two-way radio, paging, data communications, and personal communications. It maintains sales, service, and manufacturing facilities across the globe, conducts business on six continents and employs 132,000 people worldwide. Motorola's sales in 1994 were $22.2 billion.

Half of Motorola's wireless communications revenue is derived from the cellular market. The other half includes land mobile radio, paging, and wireless networking components. Motorola is the second largest manufacturer of cellular infrastructures in North America and is acknowledged as a world leader in paging.

Business strategy. Motorola's mobile computing business strategy includes the following:

- Research and develop both hardware and software wireless data solutions.

- Develop, manufacture, and supply (on an OEM basis) building-block semiconductor components, such as microprocessors and chips used in building systems.

- Pioneer the introduction of emerging technologies in bringing two-way wireless communication to the user community (Motorola was among the first select group of companies to introduce PDAs).

- Provide a leadership role in developing, supplying, and supporting products used for building private and public shared wireless data infrastructures.

- Play a key role in supplying private radio network solutions to the public safety industry. (In the early 1980s, Motorola built for police forces specialized, rugged, intelligent mobile data terminals known as MDTs, and now upgraded to the 486-based Forte product line. Motorola also plays an active role in the APCO 25 project—an industry group of communications players dedicated to standardizing mobile computing and radio-network-based voice-communication solutions for police, fire, and emergency health services.) Motorola offers Astro product line to meet APCO 25 requirements.

- Help smaller companies exploit promising technologies in emerging consumer and business markets. (Motorola welcomes teaming proposals for such ventures. It was one of the first companies to provide development funding to General Magic for its Magic Cap and Telescript technology.)

- Team with third parties to provide complete solutions aimed at filling product and service gaps.

- Promote global wireless network infrastructure solutions by facilitating the formation of a worldwide consortium for the IRIDIUM project. (Motorola is playing a key role in designing and managing the implementation of IRIDIUM, even though many independent subsidiaries have been formed.)

- Provide a vertical product line (semiconductor components, chips, system components, complete systems, supporting mobile computing infrastructures).

- Where warranted from a business perspective, partner with other companies to implement infrastructure solutions for emerging markets. (ARDIS was set up exclusively for IBM's computer-aided dispatch application initially, but with an eye to the much larger long-term goal of a national packet data network.)

- Develop and supply custom solutions where standard solutions do not meet customers' needs. (Motorola designed and supplies UPS's custom DIAD terminal.)

Motorola's core strengths in wireless and mobile computing

- Motorola has a significant product-and-service offering that ranges from semiconductor components to complete infrastructure systems. In this respect, the company is unique in providing a majority of components for end-to-end mobile computing solutions. These are either Motorola-brand components, sold and supported by the company's own full-time organization, or they are brands supplied by third-party business partners. This product range includes business applications in certain instances, especially in the public safety industry, which it has earmarked for special attention. Motorola installs private networks or public shared networks operated by its customers through their internal organizations.

- Motorola has one of the most extensive ranges of products and services in the mobile computing industry. The company supplies ruggedized notebooks, PC Servers, PCMCIA cards, PDAs, pagers, cellular phones, wireless LAN components, wireless WAN modems, ISDN adapters, private radio network components (base stations, controllers, etc.), and infrastructures for public shared networks. Nonnetwork business-systems integration is done by third parties, including IBM.

- Motorola is well known for its 6-sigma initiatives in manufacturing quality products and maintaining business processes to support these products. The company won the first Malcolm Baldrige National Quality Award in 1988, in recognition of superior management of quality processes.

- Motorola emphasizes technology research and development (R&D) in all areas of mobile computer components, systems, telecommunications, wireless data products, and communications software. As a result, it invests significant amounts in R&D, both inside the corporation and externally, in young entrepreneurial companies.

- Motorola continues to be a major force—perhaps the single most important force—in influencing the future shape of mobile computing infrastructures. Therefore, the company is certainly one of the prime vendors that every mobile computing user organization should investigate in the course of technology planning and acquisition processes.

14.2.2 Ericsson

Ericsson is the largest European player in the telecommunications and mobile computing field—and, indeed, one of the largest players in the world. In many respects it parallels Motorola in worldwide geographic coverage and product breadth and depth. It employed over 75,000 people in 1995 in more than 100 countries. The company specializes in switching, radio, and mobile data networking solutions. Net sales in 1994 amounted to more than 82.5 billion Swedish krona (Skr), equivalent to US$12.7 billion (up 31 percent from the previous year).

Ericsson is active in almost all sectors of the telecommunications field. The company is divided into five business areas with a common core technology and strategy. The five areas cooperate closely and to a very large extent provide one another with products and services. These five business areas are:

- Public telecommunications
- Radio communications
- Business networks
- Components
- Microwave systems

Products. Ericsson's production resources are divided among approximately 50 facilities worldwide, with an emphasis on Sweden and Europe. The range of Ericsson products is extensive, but we shall mention only those that are relevant to mobile computing:

- Mobitex systems and equipment for mobile data communications
- Radio base stations that handle both analog and digital standards for mobile telephony
- EDACS digital systems for land-based mobile radio
- Various systems for local and nationwide paging
- ERIPAX data network products
- Microwave links

Services. Ericsson specializes in six major service areas:

Mobile telephony. Ericsson develops and supplies network and terminal equipment for all the major mobile telephone network standards. The company claims exceptional expertise in analog and digital mobile networks, including the emerging Personal Communications Services (PCS).

Fixed radio access. There is growing interest in both developed and developing countries in providing fixed radio-based telephony services, rather than traditional copper wire connections. Ericsson has two solutions for fixed radio access: one provides radio access for fixed-network local exchanges; the other is based on a cellular network infrastructure and offers the possibility of mixed mobile and fixed services on a single network.

Wireless data. As both portable computers and the need to access and send information remotely become more widespread, wireless data services—whether for mobile or fixed access—are growing in importance. Ericsson provides two main alternative solutions: dedicated (private) wireless data networks and cellular-based data systems.

Wide-area paging. Wide-area paging is enjoying increasing popularity as a convenient, low-cost and (in some markets) fashionable means of keeping in touch while on the move.

Private mobile radio. Private mobile radio (PMR) systems have traditionally been widely used for communications in public safety and security applications. Now PMR is finding a wider market in industry as an efficient and reliable way of coordinating mobile workgroups. The latest digital systems from Ericsson offer high-speed data communications and internetworking with other systems.

Network engineering and construction. Ericsson has a substantial business in turnkey public and private telecommunications networks. The company provides custom solutions, including wireless local loop, VPN, and CATV systems, as well as DRX-1, a rural telephone exchange system for remote locations.

Research and development.   Ericsson invests heavily in R&D to maintain its competitiveness in the marketplace. Research focuses on the key technological areas and is concentrated on systems that are used in Ericsson's core business. Investments in R&D amounted to SKr13 billion in 1994 (approximately US$2 billion), corresponding to 16 percent of net sales. In 1994, 22 percent of the entire workforce (22,000 employees) was engaged in developmental work. Although R&D is conducted worldwide in approximately 40 development centers in 20 countries, over 60 percent of the total is done in Sweden. The main areas of research from a mobile computing perspective are digital wireless telephony (voice and data) and broadband communication for rapid data transfer, plus conventional research into optical fiber cable transmissions, microwave technology, optical switching matrices, and very-high-capacity processors.

Ericsson's product strategy and core competencies.   When it comes to core mobile computing infrastructure products, Ericsson's strategy is similar to that of Motorola—but with distinct differences in emphasis. Ericsson concentrates more on infrastructure products than does Motorola (from a total mobile computing solution perspective, Motorola has greater overall product breadth). Ericsson is less active in consumer and end user–oriented products than Motorola.

Ericsson's strategy regarding business solutions is to let third parties and software developers compete for best-of-breed solutions. We consider this a superior strategy to one which sees infrastructure suppliers designing and developing business-solutions software themselves.

Ericsson is well established in the European market and has a strong presence in Asia and Africa as well. As a result of this strategy, 15 countries have installed Mobitex wireless packet-switching data networks. Ericsson was the first company to provide several GSM networks in Europe and a PCS network infrastructure in the United States.

In short, Ericsson covers the entire spectrum of cellular networks, packet-switching networks, PCS, and GSM networks.

14.2.3   Northern Telecom*

Northern Telecom (NorTel), based in Mississauga, Ontario, is one of the world's largest companies specializing in the development, design, and supply of telecommunications switching equipment, including emerging wireless data networks. The company has development and manufacturing facilities in Canada and the United States.

---

* Information obtained from NorTel press releases on its web pages on the Internet. For additional technical information, refer to *Telesis,* a magazine published by BNR and Northern Telecom of Mississauga, Ontario.

Northern Telecom is backed by Bell Northern Research (BNR), a major research and development organization that has centers in Ontario, Quebec, Georgia, North Carolina, Texas, England, Australia, Japan, and China.

The company's product strategies emphasize the following:

- Provide in-building mobility for mobile workers. NorTel has unveiled Meridian Companion wireless communications systems, which enable wireless telephones and personal communication devices to be connected to advanced in-house PBX switches in an arrangement analogous to wireless LANs (as compared to wide area wireless data networks). A recent assessment indicated 5500 companion wireless communications systems installed worldwide in a variety of applications. NorTel's Meridian Homelink Ethernet access package links teleworkers to corporate LANs or WANs via high-speed digital data compression connections over public switched networks.

- NorTel and BNR have developed a broad set of wireless network products that support all major digital cellular technologies, including GSM, CDMA, TDMA (PCS), and AMPS. The company's solutions comprise controllers, base stations, GSM terminals, digital wireless switches, microwave backhaul, and network intelligence (such as databases and location registers) in GSM 900, PCS 1900, and CDPD technologies. In February 1995, Nortel's N1901 end-user device became the first terminal in the telecommunications industry to be certified by the FCC for GSM-based PCS 1900 systems in the United States.

- BNR is partnering with QUALCOMM to develop CDMA-based technologies for the world market. In Canada, BC Tel is implementing NorTel's CDMA technology.

- NorTel has entered into a business relationship with U.S. Intelco Wireless Communications (USIW) to develop the PCS area. The companies plan to extend the use of existing infrastructures to provide PCS, and the procurement of CDMA-based PCS infrastructure for USIW partners (primarily rural telephone companies). USIW provides partnership formation services, build-out services, marketing, and auction planning services to its partners.

- The company is intent on developing advanced technologies (multimedia, speech recognition) in future devices and on mobile computing networks.

- BNR/NorTel have designed several innovative devices that have won awards for futuristic and ergonomically pleasing design.

Acceptance of NorTel's wireless technology in the marketplace. NorTel has succeeded in selling its wireless systems to several prominent customers. In 1995, BellSouth selected Northern Telecom to build a $100 million GSM PCS network for North Carolina, South Carolina, and East Tennessee. NorTel is supplying a $175+ million infrastructure for the Mercury One-2-One PCS network in England. Western Wireless will spend $200 million with NorTel for its PCS 1900 network. It should be noted that NorTel sells equipment primarily to network service providers, or to customers implementing private wireless networks that will be operated and managed by network providers.

Northern Telecom employs approximately 22,000 people. It's 1994 revenues were US$8.87 billion. Bell Northern Research employed 10,000 people in 1995.

14.2.4   IBM as a mobile computing vendor

IBM, the largest computer products and services company in the world, participates in almost all major emerging computer-networking-related businesses. IBM's network-computing strategies include active participation in the Internet and mobile computing. We believe that IBM is now formulating its overall mobile computing strategy. As a first step, the company has appointed several senior executives to head up a mobile computing market push. The following are observations of this as yet unchronicled issue gleaned from a number of sources by the author:

- There are two organizations within IBM focusing on mobile computing strategies: the IBM PC Company and the IBM Wireless Group within networking division in Raleigh, North Carolina. IBM is trying to bring these two groups together from an external marketing perspective.

- With the success of the company's ThinkPad technology (it has won several innovation awards), IBM is eager to participate actively in the end-user-device segment of the mobile computing market for two compelling reasons. First, in dollar terms, the money spent on end-user devices represents a significant percentage of the total dollars spent on any mobile computing solution by a customer (this is especially true when multimedia features are included). Second, it is a very visible component of any solution. The opportunity thus presented to bring the company's logo into prominence has important promotional and public relations connotations for IBM, particularly in light of the company's efforts to market semiruggedized ThinkPad 730T pen-based notebooks for field applications.

- IBM has also been promoting (through an association with Bell-South) a PDA called *Simon*. So far the effort appears to be somewhat half-hearted—it seems IBM is content to let BellSouth test-market the product until technology innovators and users agree on the logical and physical personality of the device. Since PDAs and PCAs are unlikely to gain widespread acceptance before a PCS network infrastructure is in place and network service costs come down, such an approach is a sound one.

- The company is taking advantage of its design and manufacturing prowess in component design to pursue other mobile computing market segments more vigorously. One such segment is the PC Card (PCMCIA) peripheral devices market, for which IBM has developed the following devices:
  - IBM ARDIS and Mobitex modems
  - IBM cellular/CDPD modems
  - IBM wireless LAN adapters
  - IBM V.34 dial-in modems for PSTN access

- With its highly adaptable desktop products and its significant presence in the retail industry through point-of-sale terminals and back-end systems, IBM is also targeting a complete suite of wireless LAN products at stores, hospitals, and factories. They have two offerings, each in both Token-Ring and Ethernet versions—for main wireless LAN and for entry wireless LAN. The first addresses the desktop market and the second (which comprises the new IBM 8227 Access Point and PC Card adapter that links the adapter to a wired Ethernet network) addresses the mobile notebook sector.

- Having addressed end-user requirements with its highly successful ThinkPad line and its PC cards, IBM is now poised to pursue the wireless switching and gateway sectors (see Chap. 11)—but more from a strategic than a revenue perspective. A share of the MCSS market would tie IBM's ThinkPads to applications on LANs and mainframes through network providers' wireless networks. In this regard, the company is fielding an OS/2 and AIX-based product called ARTour that was developed in Europe. Our early assessment indicates that the product has very attractive features. We expect that organizations with a significant IBM investment and mainframe applications will find it attractive. Please review ARTour in Chap. 11 for further details. It is the overall application and systems integration service opportunities that IBM is after through ARTour MCSS.

- The next logical product area that IBM has targetted is application servers bundled with specialty software for mobile applications. The company's AS/400-based mobile application server will be attractive to organizations with AS/400 technology already installed as part of a computer infrastructure. We feel that as a specialized technology (its ease of operation notwithstanding), the AS/400 should be used for mobile solutions only if it is compatible with existing applications. Unfortunately, at the time of writing (in late 1996) there was no IBM software offering similar to, say, Oracle Mobile Agents. IBM's MQI series, for example, contained no specialized features for the construction of mobile-aware applications as defined in this book, and there were apparently no plans to build them in MQI series. It will be interesting if IBM were to integrate ARTour with MQI series middleware in the future.

- Unlike the first three vendors described earlier in this chapter, IBM does not supply network infrastructure components. Instead, its plan is to design mobile computing solutions that will interface with any of the wireless or public switched networks supplied by Motorola, Ericsson, ARDIS, RAM Mobile Data, etc. IBM used to have a strategic relationship and equity participation in ARDIS, but it sold its share back to Motorola in 1995. Not having its own wireless WAN available for long-haul wireless traffic, this decision was partially motivated by a desire to align the company with multiple wireless network infrastructure and service vendors, including Ericsson's Mobitex, GTE MobileNet's CDPD, and emerging PCS players. IBM announced in 1995 a bilateral strategic relationship with GTE MobileNet to offer integrated network solutions by combining the limited reach of its ADVANTIS WAN with GTE MobileNet's CDPD and circuit-switched offering. The company's systems integration unit (ISSC) has also announced the formation of an outsourcing team with Ameritech.

- IBM's major strength as a mobile computing solution provider lies in the area of application and systems integration. TCP/IP communications software was not one of IBM's core competencies or strategies in the past, but the company does fulfill this requirement through its own version of TCP/IP, through third parties, or through its open systems center specialists. It should be noted that a majority of mobile computing projects employ TCP/IP as an OSI layer-three transport mechanism. IBM's ISSC organization is one of the world's major systems integrators building custom end-to-end solutions (it being, for example, the systems integrator in J.B. Hunt's mobile computing project).

Unlike Motorola, IBM has not thus far developed many vertical mobile computing business applications, tending instead to encourage its customers to use third-party solutions from other companies. They are, however, focusing on sales force automation as a growth area for mobile computing. With the establishment of several mobile computing competency centers in different regions of North America and Europe, however, we should see more focused mobile computing developments from IBM in 1996 and 1997. For now, the company should most certainly be considered a serious mobile computing contender in projects involving extensive integration with IBM mainframes as back-end business-system platforms. With systems integration experience in the Atlanta Olympics in 1996, where mobile computing applications were an integral part of the overall system, IBM could become a viable technology integrator in multivendor mobile computing projects as well.

14.3   Network Services Infrastructure Vendors

In this category, we shall cover ARDIS, RAM Mobile Data, QUAL-COMM, AT&T, SkyTel, and RadioMail.

14.3.1   ARDIS

Originally formed in 1990 as a joint venture between IBM and Motorola, ARDIS is the largest wireless data network provider in the United States. When IBM sold its share in 1995, it became a wholly owned independent subsidiary of Motorola. We shall confine ourselves here to a brief description of ARDIS's business and product strategy. (Refer to Chap. 8 for a more detailed description of the ARDIS network services.)

ARDIS's strategy is based on the following:

- Capitalizing on the company's early lead in the implementation of a nationwide network infrastructure by focusing on large Fortune 500 clients who need to connect field forces to IBM mainframe-based data centers (as in the case of IBM itself and Sears).

- Promoting its superior in-building signal penetration as a major technical advantage of the ARDIS wireless network.

- Upgrading its network to 19,200 bps with RD-LAP wherever dictated by business requirements and economically justifiable.

- Making the ARDIS network a core business offering while at the same time being seen as a facilitator and key organizer of marketing efforts to promote the creation and adoption of mobile-aware business applications. (ARDIS's President Purnell was elected chairman of the Worldwide Wireless Data Network Operators Group, or WWDNOG, in 1995.)

- Emphasizing its software developer program by giving third parties facilities, software, and attractive network service rates to persuade them to develop certified ARDIS-compliant software and business applications.

- Working closely with vendors such as RADIO MAIL to promote "ready-in-a-box" personal messaging services and e-mail gateways.

- Joining the Inmarsat-led initiative to enable communications between the Microsoft Exchange Server and Windows 95 applications. This initiative involves ARDIS, Inmarsat, AT&T Wireless, Vodafone, and Microsoft. It uses Microsoft's messaging application programming interface (MAPI) to access multiple wireless networks from a common e-mail in-box on Microsoft Exchange.

- Promoting PDAs like ENVOY and MARCO for their future potential, despite lackluster consumer response so far.

- Working with its partners such as Rockwell International to provide specialized solutions to the trucking industry. Rockwell provides a satellite-based wireless network service that provides excellent coverage in open rural areas but is hampered by tall buildings in downtown areas—where ARDIS has its excellent in-building coverage. The combination thus provides coverage in both rural and urban areas. The company has a similar arrangement with QUALCOMM. ARDIS works closely with IBM in many projects that involve IBM platform-based business systems integration.

14.3.2   RAM Mobile Data (RAM)

RAM Mobile Data, 49 percent owned by BellSouth, is the second largest nationwide wireless data network services provider. While the ARDIS network is based on Motorola technology, RAM's network uses Ericsson's Mobitex technology and operates in England, Belgium, and Holland as well as the United States. The company is a member of the Mobitex Operators Association (MOA), which represents Mobitex networks in 14 other countries, besides the United States.

In many ways, RAM's strategy is similar to that of ARDIS. The following points describe strategic differences:

- Not being owned and controlled by Ericsson the way ARDIS is 100 percent owned and controlled by Motorola, RAM is more open than ARDIS in its relationships with third parties. (Strictly speaking, Ericsson does control the proprietary Mobitex standard in terms of implementation. Nevertheless, it does so under the auspices of the MOA, which dictates the requirements of its customer base.)

- The RAM network operates at 8000 bps compared to ARDIS's 4800 in most places. ARDIS, however, could well leapfrog RAM's network speed to 19,200 bps over the next few years, while RAM has no plans to upgrade its speed to 16,000 bps. It could perhaps at some future date achieve a quantum speed advantage with GSM/PCS technology.

However, there are still many technical difficulties in implementing GSM/PCS for data in both Europe and the United States.

- RAM is more hierarchical than ARDIS in its network architecture topology, with the result that local traffic within the same MOX does not have to go through a central switch.

Additional network options from RAM.*   In a distinct turn from its previous strategy based on Mobitex only as infrastructure network everywhere, RAM has mapped out a new cellular strategy. In late 1995, RAM announced that it will support cellular, paging, satellite, dial-up, and private networks, in order to extend the reach of its packet network. This is an interesting recognition of the fact that no single wireless network may be able to meet customers' requirements in the future. You need hybrid network implementations, as we have suggested in our architecture discussion in Chap. 5. What this implies is that an application designed for Mobitex could and should run on other networks such as cellular, satellite, PCS, or PSTN.

RAM's new strategy calls for 96 percent coverage of all urban areas in three years, filling in gaps with modem polling services in conjunction with cellular carriers, supporting land-line connections, and acquiring more spectrum at PCS auctions. RAM is discussing with several carriers business arrangements to link cellular and CDPD networks. To allow users to run their applications on these networks through asynchronous communications services, RAM will also provide new software drivers. Figure 14.1 depicts this strategy graphically.

14.3.3   QUALCOMM

Based in San Diego, California, with $386 million in revenues in 1995, QUALCOMM has established itself as a leader in providing end-to-end mobile computing solutions to the transportation industry. By the end of 1995, the company had over 100,000 users and 80,000 trucks equipped with its OmniTRACS system (see Chap. 2) in the United States, Canada, Mexico, and Europe. Hundreds of fleet owners use QUALCOMM as the primary method of communication with their trucks. Based on a satellite-based wireless network, GPS, and custom-designed application software, OmniTRACS provides scheduling, tracking, and maintenance information to dispatchers and maintenance managers.

QUALCOMM has also become a major player in the emerging PCS market with the development of a CDMA implementation of PCS. Industry leaders, including the PCS Technology Advocacy Group

---

* Information based on *Communications Week,* Nov. 6, 1995, and discussion with the vendor.

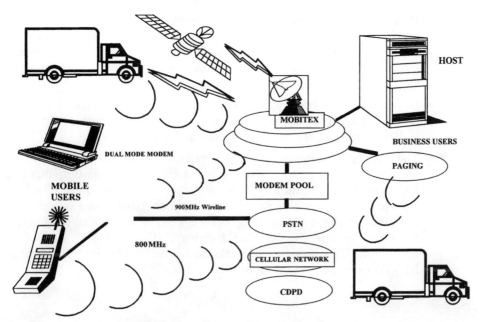

Figure 14.1   RAM Mobile Data's new network strategy. (*Source: RAM Mobile Data—as reported in* CW WEEK.)

(PCSTAG), and users alike have given their seal of approval to this technology by accepting it as an alternative standard to TDMA. The PCSTAG has recommended that IS-95 based on QUALCOMM's proprietary CDMA wireless communications technology be accepted as ANSII standard. The technology has been licensed to Northern Telecom and a Korean consortium of Goldstar, Hyundai, and Samsung. The TDMA/CDMA debate will persist, however, as the race to offer full-fledged services in North America continues (CDMA is estimated to be 6 to 12 months behind TDMA, according to some analysts). We believe that CDMA will survive this debate. It will eventually emerge as an alternative and perhaps better technology than TDMA.

QUALCOMM has extensive experience in designing and operating satellite-based networks. The company is a partner in the LEO satellite project in conjunction with Globalstar Corporation. It is also the supplier of Eudora, perhaps the most popular e-mail package ever, used on the Internet by millions.

QUALCOMM's business and product strategy is characterized by the following:

- Provision of a complete end-to-end solution to the transportation industry. This includes hardware, software, training, on-going support, and maintenance services.

- Provision of new operational applications, such as its Vehicle Information System (VIS), and enhancement of operational information applications with decision support information for fleet planning purposes, as evidenced, for example, by the development of the QUALCOMM Decision Support System (QDSS).

- Development, marketing, and licensing of its CDMA technology. QUALCOMM intends to sell directly to end user organizations as well as to manufacturers like Northern Telecom.

- The forging of relationships with other network providers to augment coverage in urban areas where satellite communication may be hampered by tall buildings. As was mentioned earlier, QUALCOMM has entered into a strategic relationship with ARDIS, for example, to combine excellent in-building coverage with equally outstanding coverage for open areas and highways. QUALCOMM has also become an IBM industry remarketer for AS/400 QTRACS and the RISC/6000 implementation of OmniTRACS.

14.3.4 AT&T*

AT&T is the largest telecommunications company in the world, with a considerable stake in computers (largely through its acquisition of NCR, which it has set up as a separate entity once again). AT&T Bell Laboratories developed much of the technology that has made wireless communications possible. The company made several innovative forays into mobile computing with EO and other offerings from 1992 to 1993 and is thus an important mobile computing vendor. In 1994, it acquired McCaw Cellular, with the result that by the end of 1995 it was able to boast five million cellular subscribers. It offers cellular services in 105 major metropolitan areas, covering 35 percent of the total U.S. population.

AT&T Wireless Services is at the forefront of the telecommunications and mobile computing industries, driving the development of wireless technology standards, encouraging cooperating alliances, and continuously implementing innovative ways to communicate. The Wireless Data Division is overlaying new data technology on existing cellular networks to transmit packets of data using CDPD technology. There are several business units:

- Cellular Voice, serving over five million customers

- Messaging, with over 800,000 customers

- Air-to-ground services, providing digital telecommunications for U.S. aircraft

---

* "AT&T—About the Company," a paper published on www.mccaw.com.

- Wireless Data, offering secure data communications

- Strategic Technology Group, laying the groundwork for the future of wireless.

In 1995, AT&T Wireless participated in FCC auctions to extend the reach of its spectrum licenses. The company purchased two national licenses in order to offer two-way messaging services nationally. At the same time, it purchased PCS licenses to expand its coverage in the United States. With 21 new major metropolitan areas, AT&T Wireless will soon be able to offer services to more than 80 percent of the U.S. population.

AT&T launched its CDPD AirData business service on April 17, 1995, in Miami, and by the end of the year had extended it to 75 percent of the company's footprint.

AT&T's strategy in deploying CDPD is marked by five key factors:

1. Ubiquitous infrastructure deployment

2. Enabling of availability of application solutions

3. Offering of cost-effective devices

4. Reseller and third-party channel development

5. Awareness of the benefits of wireless data

AT&T believes that the right applications for CDPD are transaction-based (short messaging, remote monitoring, dispatch, data entry, POS authorization, where message traffic is less than 2 KB per minute) and interactive (database access, e-mail, online access, short fax where message traffic is 2 to 20 KB per minute).

14.3.5 The CDPD vendor forum

The CDPD Forum is an independent industrywide consortium formed in 1993 with seven carriers. In 1995, it included 94 member companies comprising 17 carriers and vendors of hardware, software, and other cellular-related products and services worldwide. While it is not a single-vendor organization, it does represent CDPD vendors as a group. Major members of the CDPD Forum are GTE MobileNet, AT&T Wireless, Bell Atlantic/Nynex, Ameritech Mobile, IBM, Microsoft, Motorola, Oracle, Novell, PCSI, Racotek, etc. The Forum has the following objectives:

- The development and promotion of standards for CDPD so that infrastructure manufacturers can build products to an industry standard, thereby eventually leading to interoperability between networks.

- The building of a limited-size messaging specification for CDPD.

- The definition of a mobile end system (M-ES) certification program, thus ensuring that certified devices are compliant and interoperable within infrastructures conforming to the CDPD release 1.1 specification.

- The promotion of CDPD as an alternative to specialized packet-switched networks such as ARDIS and RAM Mobile Data.

- The promotion of CDPD technology for business applications. The forum quotes CDPD's superior coverage, higher speed (19,200 bps), voice/data integration on the same infrastructure, choice of carrier in service areas, and commitment and investment by over 100 companies as major advantages.

- Production of a CDPD deployment scorecard on a quarterly basis.

14.3.6 SkyTel*

SkyTel Corporation, a subsidiary of Mobile Telecommunications Corporation (Mtel), is a leader in the emerging market of two-way messaging. While it has a far smaller share of the overall paging market (536,000 subscribers in 1994 versus 4.4 million for market leader PageNet, according to the Yankee Group), SkyTel was awarded a Pioneers Preference license by FCC in the narrowband PCS spectrum. Pioneer's Preference licenses are granted to companies that demonstrate technological innovations likely to significantly affect the course of an industry. The granting of the license gave the company a head start in developing and implementing a network and services for SkyTel 2-Way, which it inaugurated in September 1995. At the time of this writing, the company was still in the process of debugging and stabilizing the new technology. This early experience should keep SkyTel in the future as two-way paging takes hold in the market.

SkyTel 2-Way's two-way messaging service allows people to instantly respond to and automatically acknowledge pages. People sending messages know when the message is delivered and can receive an immediate reply via the two-way network. As SkyTel expands its services, wireless messaging will expand to encompass new concepts and integrate new information technologies. The SkyTel 2-Way network enables interactions between workers and between workers and systems, relaying information back and forth so that field projects can be monitored and people in the field stay in touch with their home office.

---

* "SkyTel's 2-Way Technology Backgrounder," published on World Wide Web on Internet.

Today, the paging market has a subscriber base of 27 million people. Although SkyTel is initially targeting current paging users and mobile professionals as its primary markets, it also sees itself eventually tapping into a far larger market. Over time, as the range of two-way products SkyTel offers increases and the capabilities of the service expand, SkyTel 2-Way should attract the attention of a wider and more general audience.

SkyTel has been the first company to market many services that have helped evolve the paging and messaging industry. The following are noteworthy:

- SkyPager—a satellite-distributed nationwide paging service

- SkyWord—a nationwide text messaging service

- SkyTalk—integrated voice mail

- SkyFax—integrated fax services

- SkyMail—integrated e-mail

- SkyNews, SkyQuote—information services

SkyTel's current strategy is to focus on high-end business travelers while competitors such as PageNet concentrate on the mass market. The company intends to expand its current base of users with its two-way paging service and to extend its application base for two-way paging. SkyTel is also working with other IT leaders (Lotus Development Corporation, Microsoft Corporation, Sony Electronics, Toshiba, etc.) to expand and integrate paging services with other software applications and hardware devices. These relationships have led to solutions that integrate paging with a variety of tools that businesses and mobile professionals use to work and communicate.

SkyTel's 2-Way paging network.   In our opinion, two-way paging based on narrowband PCS will become a major mobile computing offering for affordable and high-function paging applications (though it is inherently a low-level function when compared to full function CDPD/ARDIS/RAM applications). Therefore, we are providing information on SkyTel's two-way paging network as a future trendsetter (Fig. 14.2). No doubt, subsequent incarnations will improve upon the current implementation.

SkyTel 2-Way was developed in response to two fundamental needs in today's paging market:

1. Paging users (particularly among service and support groups) want easy and automatic confirmation of message delivery, as well as integrated message-response capabilities that will allow them to increase productivity and efficiency in day-to-day operations.

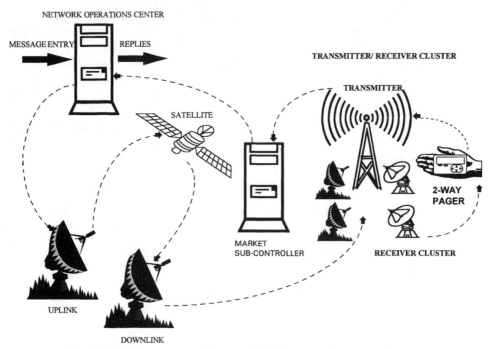

Figure 14.2 The SkyTel 2-Way network. (*Source: SkyTel Technology backgrounder paper.*)

2. Mobile professionals want a low-cost, two-way wireless messaging solution that can be integrated with corporate messaging and information systems, thereby expanding their information networks so that they can be reached wherever they are.

Both groups want a wireless messaging service that combines the advantages of a one-way paging network's nationwide and in-building coverage (using small, low-cost devices with long battery lives), with the advantages of two-way communication.

Four components comprise SkyTel's 2-Way paging network:

- The SkyTel Network
- The SkyTel Network Operations Center (NOC)
- SkyTel two-way messaging devices
- SkyTel Services

The SkyTel network is based on an asymmetrical, two-way network architecture that reduces cost and use of low-power devices. Outbound messages from the NOC use a different path to reach subscribers than do messages and confirmations. Mtel has three nationwide narrowband technology licenses at its disposal: the Pioneers Preference license granted by the FCC, and two licenses purchased in the FCC spectrum auctions. Together these licenses provide three 50-KHz forward paths and five 12.5-KHz reverse channels. These channels are all in the 901 and 940 MHz frequency bands.

The network design utilizes simulcast technology, which was a direct outgrowth of SkyTel's experience in designing and optimizing its one-way paging network. Simulcast technology uses multiple transmitters to achieve nonobtrusive overlap coverage. The use of simulcast technology, a satellite distribution system, and high-powered transmitters give SkyTel 2-Way its breadth and depth of coverage and facilitate deep in-building penetration.

The network uses Motorola's new ReFLEX 50 protocol. ReFLEX 50 was designed as a narrowband PCS protocol specifically for high-speed, two-way wireless messaging. It differs from InFLEXion, Motorola's other narrowband PCS protocol, which was designed primarily with voice services in mind. ReFLEX allows SkyTel to support various data types, from numeric, to alphanumeric, to binary files.

The protocol increases network capacity, reliability, and efficiency because it provides sophisticated error-correction features, support for frequency reuse, and automatic registration capabilities. ReFLEX 50 is also unique among Motorola's two-way FLEX protocols in that it is designed to support simulcast, critical to superior in-building reception. Messages are sent over a 50-KHz channel using a satellite-based system and high-powered transmitters. Information is transmitted at 25.6 Kbps—over 10 times faster than most of today's paging systems. Both confirmation of message delivery and responses from two-way pagers are received through a series of small receivers deployed around the transmitters. The data rate for return messages is 9600 bps, which minimizes battery drain on the pagers.

The NOC plays a critical role in SkyTel's 2-Way paging service. The NOC consists of a set of servers, gateways and tools that enable SkyTel to provide guaranteed delivery through a store-and-forward design. It is the NOC that allows customers to create new end-to-end messaging applications. The NOC is also the heart and brain of the network, serving as the control center and tracking all network activity. The NOC integrates all incoming messages, which originate through options such as touch-tone phones, operator assistance, Internet, and e-mail networks. It distributes the messages, collects confirmations and responses, and transmits them back to the message originators.

The NOC's centralized architecture eliminates the concept of roaming. The NOC automatically registers and tracks user devices, so users do not need to notify the network of their location. The NOC also maintains a single customer database.

14.3.7   RadioMail*

RadioMail Corporation, based in San Mateo, California, was founded in 1988 by Goodfellow, who conceived the original concept in 1981 when he was a researcher at SRI International. Motorola and 2M Invest of Copenhagen, Denmark, provided the initial capital. When the alphanumeric pager was introduced, Goodfellow developed software that translated and routed e-mail messages between the SRI computer and the pager network. In 1988, he used a NEXT workstation installed in his home to set up a gateway that allowed subscribers to one paging network to send one-way messages to subscribers on other networks. In 1992, RadioMail introduced the first two-way wireless e-mail service, using the RAM Mobile Data network. The service is now available on the ARDIS network as well. Wireless RadioMail service is available to 90 percent of all U.S. business locations. In November 1993, Radio-Mail extended one-way paging support to Apple's Newton PDA.

RadioMail provides public gateway services to mobile computer users. The company offers nationwide wireless two-way electronic mail delivery, an on-demand headline news service, and a service that allows pagers and paging cards to receive messages from nearly any electronic mail system.

The RadioMail strategy is to find users to whom it can deliver e-mail messages. Users of pagers and PDAs with pager cards are good targets for this purpose. The company also offers an API so that customers can develop message-enabled applications for which it provides the essential infrastructure: wireless gateways to almost any computer resource that can be used for information delivery, database queries, or transaction processing.

RadioMail has also developed a two-way wireless messaging application for a personal communicator based on General Magic's Magic Cap. RadioMail's two-way messaging service is available for DOS-based and Windows 3.1-based notebooks and MAC Powerbooks.

Figure 14.3 shows a schematic that explains how RadioMail works.

14.4   Full-Service Solutions Suppliers

Telxon, Norand, and Psion are three significant vendors in this category.

14.4.1   Telxon†

Telxon, based in Akron, Ohio, is a full-service company involved in the design, development, manufacturing, integration, service, support, and

---

* RadioMail Corporate Information available from the vendor.

† Telxon 1995 Annual Report available from the company.

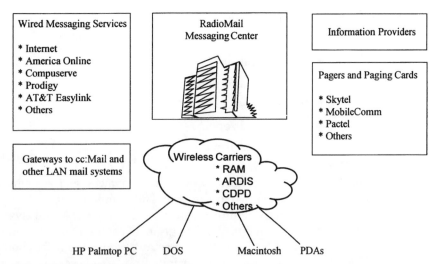

Figure 14.3   How RadioMail works.

marketing of both wireless and portable computer solutions. The company operates in Canada, Australia, England, Germany, France, Belgium, Japan, and Singapore as well as the United States. Telxon owns Aironet Wireless Communications (supplier of high-speed spread-spectrum data radios), Itronix Corporation (a manufacturer of ruggedized computers), and Metanetics (an R&D imaging company that develops 2-D barcode technology). Telxon's 1995 revenue was $379 million (28 percent higher than in 1994). During 1995, Sears Corporation ordered 14,000 Itronix (a Telxon company) computers for its field service application.

Product development and technology focus.   Telxon's product and technology development focuses on the provision of real-time data transactions. The company's core products are portable transaction computers (PTCs) and wireless communication networks. PTCs are integrated with microprocessors, memories, displays, keyboards, touch screens, OCR software, bar-code readers, printers, wireless modems, and LAN/WAN radios.

The company's proprietary application-specific integrated circuit (ASIC) technology, data radio technology, market responsiveness, and ergonomically designed packaging enable it to develop tailored solutions to meet customer needs in specific vertical markets.

Vertical market focus.   The Telxon Vertical Systems Group provides market-specific expertise and is targeted toward the following industries:

- Retail

- Industrial

- Transportation

- Insurance and finance

- Health care

- Mobile field repair

- Route accounting

Technical services, software development, and systems integration focus.
Telxon has a team of over 300 people dedicated to creating turnkey solutions, including software development and systems integration. Telxon is among a select group of companies that emphasizes this total solution concept—an important requirement for mobile computing, in our estimation. We attribute Telxon's success in large measure to this approach (besides having the right products for vertical applications).

Telxon is also a leader in the development of PenRight applications in DOS and Windows environments.

14.4.2 Norand*

Norand is a leading manufacturer and marketer of mobile computing systems and wireless data network components used in distribution and industrial settings. Typical applications include route accounting, field sales automation, and inventory database applications in manufacturing, warehouse, and retail settings. Norand systems include wireless LAN data networks as well as wireless WAN solutions for mobile applications, handheld computers using both keypad and pen/key touch input, associated peripheral equipment, application specific software, and related training and support services.

The company was established in 1968 and employs approximately 974 persons worldwide and 153 outside the United States. It has sales and service operations in Canada, Mexico, England, and Europe. 1994 sales were US$193 million. The company cites the following significant innovations:

- Developed the first handheld order-entry terminal in 1969 for use in retail applications.

- Invented the first bar-code scanner two years later. Used with order-entry terminal, data capture was greatly simplified.

---

* Company Fact Book supplied by Norand Corporation.

- Created the first computerized cash register in 1972.

- Pioneered the application of computerized handheld data collection for wholesale route distributors. Norand first manufactured the system in 1979 and now has a market leadership position of over 80 percent, representing at least 150,000 routes—more than all other competitors combined.

- Developed the first radio data network in 1985, providing real-time interaction between remote handheld terminals and a host computer. The company offered the first RF terminal and integrated barcode printer, and the first terminal with portable/mobile capability. It is a leader in supplying RF systems for LANs, with over 40,000 terminals installed (representing a market share of 21 percent).

Norand's product and services strategy is based on the following:

- Provision of turnkey solutions that include complete solutions comprising hardware, software, network components, applications, training, and support. These IT solutions are provided for two primary markets:

  1. Route accounting and mobile field applications serving bakery, beverage, dairy, parcel-delivery, and service operations.
  2. Logistics and material-handling applications (manufacturing, distribution, transportation, and retail operations)

- Becoming the best player in niche application areas such as route accounting, where, according to the company, it has 80 percent of the market.

- Sustaining growth in the future by way of product upgrades, supporting open WAN solutions, and expansion in international markets.

- Entering into strategic partnerships with other vendors aimed at expanding product and customer bases. For example, Norand manufactures RF modems for IBM, who in turn sells Norand products.

The Norand product suite includes the following major hardware and software components:

- 1000 series modular RF terminals
- The RT 3210 RF system
- The Route Commander 4000 series entry-level system for distribution automation
- The Pen*key 6100, 6200, 6300, and 6600 series (Intel architecture, DOS-based, or Windows-based) for enhanced functionality

- The ACN Communications controller (providing MCSS functionality)
- Vertical application software for route accounting

Please see App. A for additional information on Norand's products.

14.4.3   Psion

Psion is a European company that is engaged in the development, engineering, marketing, and distribution of portable computers and software, data communications equipment, and application software. Psion also offers consulting services for the development of specifications and the implementation of custom solutions. The company has offices or sales associates in the United Kingdom, Europe, the United States, and Canada.

Psion's core competency lies in the supply of handheld rugged terminals for distribution, manufacturing, and transportation applications.

14.5   Mobile Computing Component Suppliers

In this category, we have included vendors who do not necessarily participate in all aspects of mobile computing, but who supply components of a mobile computing solution in the following subcategories:

- End-user hardware
- PCMCIA cards
- Wireless LAN hardware

14.5.1   End-user hardware suppliers

The ever evolving notebook computer is well covered in trade journals and trade shows, so we shall make only a few summary comments here.

Apple, IBM, NEC, Toshiba, Compaq.   Apple, IBM, Toshiba, NEC, and Compaq are major vendors in this category. As discussed earlier, IBM provides a full range of notebooks, including semirugged devices. Toshiba, NEC, and Compaq target the general purpose business and consumer market. Their product strategies are quite similar and are highly competitive: each is attempting to increase market share by ongoing R&D into miniaturization, display technology, improved multimedia features (such as full motion video), hardware-based speech recognition, and increased battery life.

We expect all these vendors (with the possible exception of IBM, which has a wider product coverage for corporate clients) to address the consumer segment of mobile computing in an aggressive fashion during 1996 and 1997.

Apple*.   Apple has demonstrated a strong commitment to mobile computing that is unique in several important areas:

- The company offers a serious alternative to IBM-compatible PCs as mobile devices with an excellent user interface and integrated hardware design that has won many industry awards for innovation and ease of use. While it is true that Windows 95 has finally given the PC an excellent and widely accepted user interface, the Apple Macintosh is still synonymous with superior ergonomic design for an integrated hardware and software product.
- The company offers a complete range of notebooks, subnotebooks, and PDAs, along with communications software and peripheral devices that can exchange information with online services.
- Apple offers complete solutions: hardware, peripherals, operating software, communications software, and third-party applications on the remote end. Tight integration between hardware and systems software is essential for many mobile applications. In an all-Macintosh environment, Apple can provide the application servers as well.
- The Apple Powerbook holds its own when it comes to mobile computers. If Apple had the suite of business applications or development tools that the IBM-compatible PC industry has, the Powerbook would probably be the notebook of choice.
- Apple is well on its way to supporting all DOS/Windows-compatible applications on the Mac platform in addition to Mac applications, either by way of software emulation or in native mode, with the same performance levels as a 486-based PC. In this respect, PowerPC notebooks are more universally compatible than PCs, because PCs are unable to run Mac applications. Both PCs and Apple can support data interchange via diskette, however.
- Apple led the way in bringing multimedia to notebooks. Computer-based documents and presentations that integrate sound, video, and animation first became available on Powerbooks.
- Apple has changed its strategy of tight controls on hardware and software licenses by becoming more open in this area. Recent licens-

---

* "Mobile Computing—An Apple Technology Perspective," published on www.info.apple.com.

ing of Apple and MacOS technologies to IBM and Motorola will likely expand the reach of their platform.

- Apple has done more than any other vendor to introduce the PDA concept to the marketplace. A bold and risky move on the part of Apple's former chairman Scully, Newton was the first PDA on the market. While trade press and users alike have heaped considerable scorn on Newton's handwriting recognition capabilities, we should remember how difficult it is to perfect any technology (let alone something as innovative as the PDA) the first time out. Newton had nothing to emulate or improve upon. It inevitably takes several iterations to create a product that fulfills customer requirements and expectations, if for no other reason than the fact that those very requirements and expectations themselves take time to become fully apparent. It appears that the new Newton MessagePad with OS 2.0 is on the right track and is much improved over its predecessor. This effort has in fact led to the formation of a PDA software development cottage industry, with Newton as the preferred platform. Apple intends to use customer feedback for future enhancements and upgrades to the Newton PDA. It is our contention that Newton will have to incorporate personal communication features on the motherboard as the PCS industry comes into its own and more PDAs and personal communicators merge into a unified product. What will slow this down is the time for convergence of PCS standards.

- Apple has upgraded its product line in processing power, peripherals, and features to keep it competitive with the much larger PC industry. New RISC-based PowerPC chips give high-end Powerbooks the muscle they need to compete with Pentium-class PC notebooks.

- Apple is determined to pursue an aggressive pricing strategy for its product line—a well-orchestrated strategy on Apple's part to keep prices of its mobile products in line with similarly powered PC notebooks from first-tier vendors such as IBM, Compaq, Toshiba, and NEC. In early 1996, Apple decided to concentrate on high-end mobile workstations and license its low-end technology to other manufacturers.

Newton: Apple's unique mobile computing product.   Besides Apple's Powerbook notebooks and subnotebooks, Apple is unique with its groundbreaking Newton technology. Newton is not simply a low-end personal computer in a new guise, it is a whole different platform that directly complements the capabilities of personal computers, servers, and other traditional computing resources. Newton technology is optimized to work with information in much smaller, more granular chunks—paralleling the way people go through their daily routines.

Newton technology integrates extensive communications capabilities, including one-way PCMCIA paging reception, two-way wireless communication, fax using stand-alone or PCMCIA fax/modems, a built-in infrared transceiver for ad hoc information exchanges, and serial communications for networking, printing, and interaction with personal computers. A mobile worker could use a Newton device to make observations on a factory floor (or while on a sales call, or even on-site at a geology dig)—and could later transfer the information into a Macintosh application back at the office.

Newton's intelligent assistance and recognition architecture. Newton offers intelligent assistance to the user by associating information *objects* in the Newton environment. For example, after highlighting a person's name in the text of an e-mail message, a user could request Newton to schedule a meeting, find references to the person in a collection of project notes, or retrieve the person's fax number.

Newton's improved handwriting recognition capability is giving earlier critics cause to reevaluate the technology. We expect the technology to evolve further until it reaches a state where it is basically error-free and intuitive in use.

Apple's mobile computing services. Apple offers two mobile computing services:

eWorld. eWorld is a comprehensive online information and communication service that features a user interface that is both innovative and easy to work with and understand. eWorld information resources are organized using the metaphor of a community, geographically represented as a neighborhood of buildings, each containing an area of online service: a library, a professional plaza for business services, Reuters news service, a newsstand, an e-mail center, and so on. Besides its information services, eWorld provides electronic communication services, including a flexible e-mail system with gateways to other mail systems and the Internet. One of the integrated mail services is NewtonMail, which functions as an online server to client software built into the Newton MessagePad devices. eWorld supports a broad range of real-time interactive communications, including town meetings of up to 250 people, lectures, debates, and small-group discussions. Down the road, eWorld will offer platform-independent services that will include Windows 3.1 applications.

Apple Wireless Messaging Service. Apple collaborated with Mobile-Comm (a subsidiary of BellSouth) to create its Wireless Messaging Service, which provides messaging to both Newton and Powerbook users throughout the United States and Canada. Postcard-length messages can be sent to small pager-style receivers or PCMCIA receiver cards that slip into Newton devices or Powerbook models with PCMCIA modules. The messages can come from a wide variety of sources, including gateways to virtually all e-mail systems through global Internet, dial-up message servers, or even voice messages entered by operators.

In summary, Apple is an important vendor in mobile computing. At the time of writing, Apple was considering the sale of both services, eWorld and Wireless Messaging Service. During fiscal 1994, the company's worldwide revenues were $9.2 billion and it employed 14,500 employees. It invests 6 to 7 percent of its revenue in R&D.

14.5.2   PC Card (PCMCIA) adapter supplier: Xircom

Xircom characterizes itself as a company of mobile networking experts. It develops and manufactures a comprehensive set of network access solutions for mobile and remote PC users. These solutions provide mobile computer users operating from a fixed location both wired and wireless network connectivity. Additionally, Xircom's suite of ISDN network access products provides remote users with portable and desktop connectivity to corporate networks, the Internet, and other online services from a wide variety of locations such as hotels, airports, and home offices.

Its major products include:

- PC Card (PCMCIA) credit card and pocket Ethernet adapters
- Credit card Ethernet + modem adapters
- Credit card modems for PSTN and cellular dial-in
- Credit card Token-Ring adapters
- An Ethernet credit card NetWave adapter and access point
- A Netaccess multiport modem card for an RNA solution
- Netaccess basic rate ISDN adapters

By mid-1995, Xircom had shipped over 1.5 million cards. The company's strategy is to continue to enhance its line of PC Card (PCMCIA) products for wireless LANs and RNA solutions at affordable prices. Xircom intends to make all its cards compliant with the wireless LAN 802.11 standard.

The company has demonstrated a carefully orchestrated strategy to become a leader in the PC Card (PCMCIA) card market. Xircom is also pursuing the wireless LAN market aggressively with its new Netwave software products. Netwave 2.5 has enhanced roaming capabilities and faster throughput. Netwave has a unique graphical utility to assist network administrators in conducting site surveys, checking connection qualities, and determining PC card properties.

14.5.3 Wireless LAN suppliers

We shall discuss Solectek and DEC as two vendors in the wireless LAN category. Besides IBM (already described), Proxim is another well-known player in this arena.

Solectek. Solectek Corporation, based in San Diego, California, designs, develops, and markets LAN products that provide computer users with high-speed wireless links to data networks. Its business mission is to provide wireless metropolitan area network (MAN) solutions that add range and mobility to local area networks.

The company is well recognized within the wireless industry for its AIRLAN product line, which has many high-profile customers (American Express, General Motors, and IBM, to name a few). Solectek's business strategy is to specialize in the wireless LAN segment of the mobile computing market. It sells through resellers and directly to end users.

Solectek AIRLAN products offer wireless solutions for in-building LANs on a single floor or on multiple floors, and outdoor applications across multiple buildings that link LANs up to 25 miles (40 kilometers) apart under ideal conditions (the more usual range is 6 to 7 miles, or 9 to 11 kilometers). By using the following AIRLAN products, users can create blanket coverages resulting in campus-type LAN solutions:

AIRLAN/Access (wired to wireless access point) is a wireless access point that connects easily to a wired LAN and creates a 50,000-square-foot (4645-square-meter) cell of connectivity. Using AIR-LAN/PCMCIA or AIRLAN/Parallel adapters, mobile users can roam freely from Access cell to Access cell while maintaining a connection to a LAN.

AIRLAN/Bridge Ultra (long-range wireless bridging system) provides high-speed wireless Ethernet or Token-Ring bridging between LANs up to 25 miles (40 kilometers) apart. This is a spread-spectrum radio frequency device that operates in the 2.4-GHz band.

AIRLAN/Bridge Plus (multipoint wireless bridging system) uses spread-spectrum radio network technology for wireless network connectivity between buildings up to 3 miles (5 kilometers) apart, eliminating the need for expensive leased lines and underground cabling systems.

AIRLAN/PCMCIA (mobile adapter) provides mobility for notebook users. A corresponding ISA bus card is also available for desktops.

Digital Equipment Corporation (DEC): RoamAbout LAN vendor. DEC holds a strong leadership position in major elements of the networking convergence market: switching, routing, FDDI, and remote access. The company pioneered a convergence model for networking, in which hubs, routers, and switches are deployed on a single integrated platform: the DECswitch 900 MultiSwitch. According to revenue figures from International Data Corporation (IDC) and Dataquest, DEC ranks number two in the RNA market in North America. The company has networked more than 13 million users worldwide.

DEC's family of RoamAbout wireless LAN products capitalizes on the company's high-performance bridging expertise. As part of a fully manageable and scalable product set, the RoamAbout Access Point can be used as a stand-alone device, or it can be installed in DEC's Multi-Stack System, the DEChub 900 MultiSwitch, or the DEChub 90. With these products, customers can have untethered access to basic networks, as well as to sophisticated networks such as virtual LANs.

The RoamAbout product line includes wireless modems—both PC cards (formerly PCMCIA) and ISA NICs—that use the 2.4-GHz frequency band and support the two major types of wireless modem technology known as frequency hopping (FH) and direct sequence (DS). This overall capability gives customers a choice of wireless technology that best suits their mobility and application needs.

DEC's RoamAbout Access Point, engineered with a standard PC card slot for its wireless interface, was the first wireless bridge to support both the FH and DS radio technologies in the 2.4-GHz band, and the DS radio technology in the 915-MHz band. It is a full-featured, compact, easy-to-install, and SNMP-manageable wireless bridge offering protocol-independent bridging, address and protocol filtering, and downline software load capabilities. Because of its wireless bridging design, adding new wireless technologies to the RoamAbout Access Point is simply a matter of upgrading the software. Customers' investments are thus protected by the flexibility to easily add new technologies, while at the same time, network and IS managers are able to manage and support multiple wireless technologies with only one bridging platform.

Powerful roaming capabilities embedded in the RoamAbout solutions give customers unlimited mobility across multiple LAN segments—and across routed subnets with a complementary software product called RoamAbout Mobile IP. By making it possible for customers to roam while they work, these products increase employee productivity and reduce costs with continuous wireless connections to a wired network.

DEC's product strategy is to take advantage of its networking experience by seamlessly integrating remote and mobile users into customers' enterprise networks.

According to Virginia Brooks, senior research analyst at the Aberdeen Group, "Digital's philosophy is to address the needs of the mobile work-force by keeping the technology simple and making it work reliably to provide performance and robustness. Digital's approach is sound and ensures the adaptability needed to stay abreast of a multifaceted technology that is still evolving."

14.6   Mobile Communications Server/Switch and Connectivity Software Suppliers

There are two subcategories under this heading: RF switch and connectivity software suppliers and RNA vendors.

14.6.1   Wireless MCSS suppliers

We described the MCSS architecture in Chap. 5 and several product implementations in Chap. 9. However, we have not seen any consistent product strategy on the part of vendors. Usually, products are created to a particular customer's specifications and then generalized for the rest of the market. The following vendors have designed and developed MCSS solutions for different network environments:

TEKnique (see Chap. 11) has several implementations of its TX5000 switch/gateway. A low-end configuration runs on an Intel platform and a more robust and scalable configuration runs on a VME platform. The company specializes in building wireless switching hardware and software solutions interfacing with LANs, minicomputers, and IBM mainframes. It also provides software integration services.

Nettech Systems, based in Princeton, New Jersey, is a specialized mobile computing connectivity software vendor. It provides Nettech RFgate (multiplatform connectivity software for ARDIS and RAM Mobile Data networks), RFMlib (a software development tool kit with a universal API), and RFlink (a transport layer API). Nettech software runs on multiple hardware platforms. Nettech boasts of a very efficient and optimized airlink protocol. Though it has been adopted by Motorola and others, it is proprietary.

Research in Motion (RIM), based in Waterloo, Ontario, is a specialized developer of hardware and application software interfaces for Mobitex networks. See Chaps. 11, 12, and 13 for additional information on RIM products.

Many of the infrastructure vendors described earlier, including Motorola, RIM, and IBM, have their own unique MCSS implementations.

14.6.2 RNA vendors

The RNA industry was described in Chap. 10. These vendors supply PSTN/ISDN-based solutions for mobile computing. The following vendors are active in the industry:

Shiva Corporation is a leading supplier of RNA solutions for a complete range of customer configurations, from small companies to Fortune 500 corporations. Current strategy is to exploit the PSTN and ISDN dial-in market. Shiva offers connectivity support only for cellular circuit-switched wireless networks at this time. The company has formed several strategic business relationships with other vendors to offer more complete RNA solutions. One such arrangement is with Xcellnet. It is Shiva's declared strategy to eventually support CDPD and other wireless networks.

The Network 3000 family of products from Xyplex, based in Littleton, Massachusetts, offers a comprehensive set of RNA solutions in the terminal server environment. These products feature LAT support and are therefore suitable for DEC-based networks.

DEC, CISCO, US Robotics, Telco and several other vendors also offer solutions in this category. Attachmate offers remote network connectivity solutions in predominantly IBM environments (AS/400 and IBM mainframe, etc.).

14.7   Software Suppliers: Middleware and Application Software

14.7.1 Oracle Corporation

Oracle has made a conscious and deliberate decision to pursue mobile computing and Internet markets. The company has well orchestrated in the press its strong interest in an intelligent terminal/Internet appliance for remote access to information. Taking advantage of its database market leadership, they announced and delivered in 1994 *Oracle in Motion* (see Chap. 13 for more details), a software engine that acts as a store-and-forward gateway between mobile users. This was enhanced and renamed Oracle Agents in 1995. The product also offers a development platform through Oracle's fourth-generation application development tools. In our estimation, ORACLE was the first vendor to provide what we have described in this book, a mobile-aware middleware software engine. The company has received a limited acceptability in the market, primarily because of the emerging status of mobile computing.

14.7.2   Sybase

Sybase is a leading vendor of database solutions. Not to be outdone by Oracle, Sybase acquired Complex Architectures Inc. (CAI), a Gloucester, Massachusetts, company as its entry into mobile computing connectivity solutions. CAI specializes in providing transparent connectivity and network-independent communications in complex computing and telecommunications enterprise environments.

The company provides a third-generation software called Enterprise Messaging Services (EMS) version 3.0.1—a general-purpose messaging platform for wired-line and wireless networks such as ARDIS, RAM, EMBARC, cellular, asynchronous dial-up, TCP/IP, LU 6.2, and X.25 links. EMS offers seamless network-independent connectivity for CICS/MVS, Stratus VOS, Tandem Guardian, VAX VMS, UNIX System V 3.2 and 4.0, Sun OS/Solaris, SCO from MS-DOS, Windows, Macintosh, and OS/2 client applications. As far as we can tell, however, it is not mobile-aware in its current version.

Sybase/CAI has aggressive development plans (see Chap. 13).

14.8   Application Development Companies

PenRight is a developer of PenRight systems software for pen-based computing. The company has directed its corporate energies into the evolution of core software development kits for pen-based mobile applications, especially for PDAs.

Florida-based UCS has developed many pen-based business applications for the public safety and several other vertical markets.

PRC is a consulting company that has enhanced its own public safety applications for wireless networks. PRC has a major presence in the public safety market. Both UCS and PRC are business alliance members of Motorola.

14.9   Mobile Communications and Application Integration Services Vendors

We shall mention Racotek and Xcellnet as two entrepreneurial companies specializing in the mobile application integration business. Racotek seems to be strong in communications software engineering, while Xcellnet is strong on the application side.

14.9.1   Racotek

Racotek, based in Minneapolis, Minnesota, is a small but highly experienced company specializing in mobile computing products and services. Racotek provides consulting, application development, training, systems integration, and end-to-end wireless systems support. During the five-year period from 1990 to 1995, Racotek provided solutions in several vertical markets. As a result of this experience, the company has created KeyWare, a new enabling technology that comprises software components user organizations need to implement complete mobile computing solutions.

KeyWare is a distributed computing environment that uses what Racotek calls a *client/server, server/client* architecture. The term denotes a conventional client/server environment, with the addition of an intelligent agent that resides in the server but serves the client. As was discussed in Chap. 13, KeyWare includes a KeyBuilder application development tool kit that allows user interfaces to be developed with third-party extensions to PowerBuilder and Visual Basic. The communications and management functions of KeyWare include messaging services, naming services, and optimal delivery services.

Racotek's product strategy is to isolate a common functionality unique to mobile computing solutions across different vertical markets and to package it in a standard middleware that can be used to build custom business applications. This separation of middleware and business functions also helps customers in the development of their migration strategies as new network options such as CDPD become available. KeyWare supports ARDIS, RAM Mobile, SMR, cellular, CDPD, EDACS, and paging networks—a comparatively rare suite of supported networks for connectivity software. Racotek is also unique in its service strategy, providing as it does, complete end-to-end application and systems integration services.

In 1995, Racotek employed a staff of 120 professionals specializing in mobile computing solutions.

14.9.2   Xcellnet

Xcellnet, a company based in Atlanta, Georgia, is a mobile and remote network access application provider. The company offers the Communications Management System (CMS), a full-function application software engine for mobile workers. Xcellnet entered the industry with a vertical application focus on sales force automation. The company has been very successful with its products and has been growing fast. Xcellnet provides software that can run on different hardware platforms using multiple communications servers, multiple transport protocols, and multiple server operating systems. In 1995, the company entered into a cooperative marketing arrangement with Shiva Corporation to offer a bundled hardware/software solution. Shiva intends to optimize its hardware to run more efficiently with Xcellnet software.

RemoteWare Server is the heart of the CMS. Currently it runs on the OS/2 environment only, though Xcellnet's strategy is to make its software run under multivendor environments (the Microsoft NT version of RemoteWare will be available in 1996).

RemoteWare Server software acts as the control point for information exchange, providing the resource management, scheduling, communications services, information monitoring, and control services needed by the network.

RemoteWare Node runs in the background of the remote (wireless in some cases) client PC and works with RemoteWare Server.

RemoteWare Forms, RemoteWare Documents, RemoteWare Reports, and RemoteWare Mail all come with RemoteWare. These applications have APIs that can be used to integrate them with customers' operational business applications.

Xcellnet supports wireless networks such as RAM Mobile Data and ARDIS. It also has plans to support emerging networks such as CDPD.

14.10   Mobile Computing Innovation Vendor: General Magic

General Magic was incorporated in 1990 with the objective of developing the technological underpinnings of personal intelligent communications. Originally the project was conceived at Apple, but it soon became clear that in order to break new ground and not be encumbered by existing technologies, a new, independent company was needed.

To ensure that the ground-breaking technology would be woven into the very fabric of everyday life, General Magic sought to team with companies that were known for changing people's lives with their products and services. In 1991, Sony and Motorola joined Apple as investors in General Magic and licensees of its technology. AT&T, Phillips, and Matsushita joined in 1992 and 1993. By late 1995, several of the world's largest computer, telecommunications, and consumer electronics companies had either invested in General Magic or were licensing its technology. The investor list includes Nippon, Fujitsu, Toshiba, Oki Electric Industry, Mitsubishi, Sanyo, Cable & Wireless PLC, France Telecom, and Northern Telecom. The licensees include SkyTel, Radio Mail, Oracle, ex-Machina, Intuit, and GDT. In total, this is an impressive group of powerful companies who believe in the future potential of General Magic's technology.

General Magic introduced the Magic Cap and Telescript technologies in 1994. Magic Cap is a software platform that helps people keep in touch by integrating e-mail, fax, telephone, paging, infrared beaming, and other methods of communicating. Magic Cap technology is built into personal communicators and allows developers to build a new class of software and services centered on communications and user personalization.

Telescript communications technology is an object-oriented, remote programming language designed to enable the creation of electronic agents capable of navigating the world of electronic services and diverse computer systems to perform everything from sorting and organizing electronic correspondence to monitoring stock activity and shopping for goods and services.

General Magic's mission is to create personal communications products and services by developing technology and licensing it to equipment manufacturers, network information service providers, and software and entertainment companies.

General Magic has lost some of its limelight as a result of the unprecedented success of the Internet, Java, and new electronic commerce vendors. We expect General Magic to reposition itself and become a niche player. However its technology or a variation of it may be embedded in the electronic commerce software of the future in conceptual form, if not in product form.

14.11   Systems Integration Companies

While many large consulting and systems integration companies advertise their mobile computing capabilities, only a few companies can actually claim to have a core competency in the area. You should be looking for a systems integration company with the following attributes:

- Experience with several mobile computing projects
- A core group of mobile computing specialists on staff
- In-house mobile computing development centers
- Ability to demonstrate unique specialized skills

Failure to find a company with these attributes may mean dealing with a company that is learning the tricks of the mobile computing trade at your expense.

Hughes Networking Systems, EDS, Systems House (acquired by MCI in 1995), IBM, and Anderson Consulting all have varying levels of expertise in the emerging mobile computing area.

Management consulting companies such as KPMG and Deloitte & Touche are also expanding their services into the mobile computing market by leveraging their knowledge of BPR and using third parties for infrastructure skills. They would be a good fit where business process reengineering is a significant component of a project.

14.12 Mobile Computing Value-Added Resellers (VARs)

Wireless Telecom Incorporated (WTI), with headquarters in Aurora, Colorado, was founded in 1993 to provide qualified VARs with a single-source access to mobile computing and wireless data communications products, applications software, carrier and messaging gateway ser-

TABLE 14.1 Core Strategies of Mobile Computing Vendors

Apple	Developer of end-user mobile devices: Powerbooks, Newton, software, and eWorld messaging services. Promoter of third-party mobile software application development
ARDIS	Largest public shared packet radio wireless network services provider. Also, a facilitator of mobile computing applications for enterprise deployments that require superior in-building penetration and IBM mainframe connectivity
AT&T	The largest cellular network provider and an important player in CDPD. A good vendor for integrated wired and wireless network integration
DEC	An important player in wireless LAN and RNA. Provides an integrated enterprise solution for wire-line and wireless networks
Ericsson	After Motorola, the second most important wireless network infrastructure supplier. Very strong in Europe and Asia. Very aggressive in PCS and GSM
IBM	Important vendor in mobile computing. Covers many components: end-user devices, PCMCIA modems, wireless LANs, and systems integration. Expected to cover more components and become a big player in mobile computing as time goes by
General Magic	One of the most innovation-driven vendors. Has interested some of the most prestigious names in the computer, telecommunications, and consumer electronics industries. Their technology could become embedded in electronic commerce software of the future.
Northern Telecom	A very R&D-oriented manufacturer of telecommunications infrastructure. Important PCS supplier. Known for innovative design and products
MobileWare	An important middleware supplier
Oracle	Becoming an increasingly important player for database access software: Oracle Agents
QUALCOMM	Market leader in satellite-based communication for the transportation industry. Also developer of CDMA technology for PCS
Racotek	A very specialized and knowledgeable supplier with extensive expertise for end-to-end applications and systems integration. Supports all current and future wireless networks
RAM Mobile Data	Second largest packet radio wireless network supplier. Planning to support access from cellular and other non-RAM networks
SkyTel	Innovation-driven paging technology leader. First in the United States to offer two-way paging. Expected to play an important role in high-end paging applications
Solectek	A leader in the wireless LAN market
WTI	Distributor of mobile computing and wireless products. Sells through reseller channel
Xircom	An important player in the PC card (PCMCIA) market and RNA

vices, education, and technical support. It is a master distributor of mobile computing products. Its strategy is to bundle services and products from different vendors and create complete application solutions for vertical markets. It assists VARs to sell these solutions to end-user organizations. WTI has strategic arrangements with many large vendors, including ARDIS, RAM Mobile Data, AT&T, and IBM to distribute its products through reseller channels. The company had signed over 300 resellers specializing in wireless products by the end of 1995.

14.13   Vendor Products and Resources

Please refer to App. A for specific information about vendor products. Appendix B lists vendors' names, addresses, telephone numbers, and major products.

Summary

In this chapter, we have reviewed the product and marketing strategies of a few of the major vendors in mobile computing. This is an evolving and fast moving industry. Doubtless, there will be rationalization, but we believe that many of the vendors mentioned here will survive. Smaller vendors will almost certainly be absorbed by larger vendors in the process. Hopefully worthwhile technologies will endure and continue to shape the future products of their new owners. Table 14.1 features overviews of the core competencies of these vendors.

Systems Design and Integration: Gluing the Components Together

Technical Design and Ergonomic Issues in Mobile Computing

Technical design of mobile computing solutions offers unique challenges to systems professionals. Mobile users have unpredictable usage patterns, networks have more components, and ergonomic considerations in mobile offices are formidable. Poor technical design and/or a disregard for ergonomics may ultimately lead to the failure of a mobile computing project. CHANDER DHAWAN

About This Chapter

In Part 4 of this book, we discussed the various components of an overall mobile computing solution. Now we need to look at the technical and ergonomic design of an end-to-end solution. This is where systems engineering and human-factor design skills are called into play. The integration of components from different vendors imposes additional constraints that systems designers must also take into account.

15.1 Technical Design Issues

We shall discuss the following topics under this heading:

- Network design
- Capacity planning and response time calculations
- Data compression considerations
- System availability design
- Security issues

15.1.1 Network design

The design of a mobile computing solution based on a wireless network is far more complex than a corresponding wired-line solution largely because wireless networks are very often public shared networks. A packet radio network such as ARDIS or RAM Mobile Data has many intermediate links (RF and wired-line) and several switching hardware processors involved, on the one hand, while the traffic pattern of users is unpredictable, on the other. With a public shared network, not only is it necessary to obtain information about one's own workload, but it is also necessary to worry about demand peaks created by other users. And even when it is possible to logically describe the tasks that mobile users perform, how does one translate application-level workloads into network traffic?

Local area network professionals have had it easy during the last decade. For the past ten years, they have had the luxury of working with ever faster LANs—first the 10 Mbps Ethernet, then the 16 Mbps Token-Ring; now the 100 Mbps Ethernet. Consequently, they have never had to worry about translating application traffic into network traffic. They have also become accustomed to working with superfast routers with more than 20,000-packets-per-second switching capacity.

Because they do a lot more than just act as hardware switches, MCSS components for wireless networks do not offer that kind of throughput. There is no formal discipline or pool of heuristic knowledge available to aid in the design of mobile computing solutions. Experts and specialists in network suppliers' engineering groups have some know-how, but they tend to be unwilling to freely share their knowledge, which is, after all, one of their few marketing tools. How does one design wireless networks, then, in such an environment? That is the $64,000 question.

Let us adopt a pragmatic and engineering approach to this problem and ask ourselves, just what are the design issues? How can we address these issues with the engineering discipline and knowledge we have? The following paragraphs, we believe, help to answer these questions.

Wireless LAN design issues. Stand-alone wireless LAN design is relatively simple compared to the wireless WAN design that we shall discuss later. Coverage, office layout, choice of technology, and product are all important considerations. Network designers should investigate the following questions before finalizing a wireless LAN design:

- How many mobile users in total will use the wireless LAN? How many will be active during the peak period?

- Which LAN applications will they be accessing? Remember that wireless LANs operate at much slower speeds than wired LANs. Will the slower speed be acceptable to the intended users?

- Is a notebook with a wireless NIC going to be the primary end-user device? The use of notebooks as primary computers is being increasingly mandated even in fixed-location offices.

- In which areas of a building or campus will users be roaming?

- How many access points will be needed?

- Where should access points be located?

- What is the access-point range?

- What impact will construction materials like steel frames used in walls and ceilings have on signal penetration?

- Which of the two main technologies is preferable: spread spectrum or frequency hopping?

- Will there be radio frequency interference from any other devices in the office, factory, or campus?

Wireless LAN requirements should be documented and vendors should be asked to first survey the site and then propose configurations with coverage diagrams showing location of access points. Different solutions from vendors can then be compared and the best option selected. Chapter 7 should be reviewed as part of any evaluation process. For more detailed information on the subject, refer to *The Wireless Local Area Networks* by Davis-McGuffin (McGraw-Hill, 1994).

Wide area radio network design issues.   LAN/WAN design and internetworking are far more difficult to deal with than basic wireless LAN designs. When it comes to inserting a wireless WAN into a LAN/WAN internetworking schema, for example, the finer points of network design can be truly appreciated. In order to resolve mobile computing network design problems in this area, network planners must document the answers to the following questions:

- Should a private radio network be built, or can a public shared radio network be used?

- Which radio network technology among those that are available now and will be available in the future, is most appropriate for the suite of applications that will be used?

- How are user application-usage profiles matched to a given network's capability?

- Should RNA technology be integrated with a radio network infrastructure?

- How are good coverage and a minimum number of dead spots ensured?

- Should a distributed wireless network design with several MCSSs like the one suggested in Chap. 11 be used?

- Which MCSS should be selected?

- Should agent-based application-development tools like Oracle Mobile Agents be used?

- How do various network design options influence the way logical networks will be managed?

- Is network outsourcing a realistic business option to pursue? If it is, does the proposed network service provider (NSP) have the necessary capacity and capability?

Private radio networks versus public shared networks. Radio networks started out as private SMR networks for radio dispatch, paging, and public safety applications at a time when large organizations could afford them. Gradually, SMR networks shared by relatively closed groups of same-industry users emerged. Then, with the introduction of paging networks and packet radio networks like ARDIS and RAM Mobile Data, public shared networks came into their own.

Because of their cost and the inefficient use of a spectrum that is in short supply as it is, the trend now is definitely away from private radio networks. Even with available spectra, only larger organizations can afford them, especially when the required coverage is large. In certain situations, private networks do still make economic sense—such as in the case of public safety applications where municipalities and townships consolidate the requirements of the three emergency services. These organizations can justify private radio networks by pointing to the need for consistent response times, and to the fact that in situations where large numbers of transactions are made in thickly populated areas, a transaction-based public shared network could in fact be very expensive to use compared to the cost of operating a private network.

An important advantage of using a public shared network is the opportunity it affords to capitalize infrastructure costs over a shorter life cycle while still providing a suitable return to shareholders. This is particularly advantageous in light of the ongoing rapid evolution of technology. A single corporate user with a private radio network would be hard-pressed to justify frequent enhancements of the network. In a public shared network, it is easier to build a business case when costs are spread over large numbers of users.

Matching network capabilities with application profiles. Today, we have a choice of the following network technologies for our mobile computing needs:

- SMR

- ESMR

- Packet radio (e.g., ARDIS, RAM Mobile Data)
- Paging
- Cellular (circuit-switched)
- CDPD
- PCS (narrowband for now, broadband in future)
- GSM in Europe
- Satellite-based wireless networks

As was discussed in Chap. 8, wireless networks have different characteristics, so it is important to select a network that matches application requirements. Table 15.1 compares the capabilities of different networks from an application perspective.

Although individual wireless networks are initially designed for specific needs, over time, they tend to broaden their applicability, resulting in overlap from one network to another. Thus, these days there is generally more than one network appropriate for any given application, a fact that is highlighted in Table 15.2, which lists application characteristics and corresponding network options.

Network choice depends on communications software.   We would like to caution you that the choice of a network should be based as much on its inherent transportability as on several other factors, including the availability of compatible communications software and application-development tools. The final choice in selecting a network will inevitably be based on more comprehensive criteria than the previous suggestions. Table 15.2 merely attempts to highlight the more obvious choices on the assumption that other fundamental needs such as software compatibility, security, and network management are also being met.

Integration of RNA technology with radio network infrastructure.   As we stated in Chap. 10, the RNA industry currently provides mobile workers with connectivity to existing LAN-based applications through ISDN, PSTN, and now Internet connections. Ideally, from a network manager's perspective, a communications hardware and software solution should support both wireless and dial-up connections—and it should do so without the need for additional wireless network MCSS training and support. Unfortunately in practice, however, problems are still being experienced in integrating the two approaches. Wireless speeds are relatively low and many LAN applications are still not mobile-aware. As well, RNA industry communications servers do not support wireless WAN MCSS functions for the likes of ARDIS, RAM, CDPD, or ESMR. The MCSS defined in Chap. 11 is functionally richer than anything the RNA industry offers.

TABLE 15.1 Capabilities of Various Wireless Networks

Network technology	Description	Best applications	Geographical coverage	Speed and performance	Pros	Cons
SMR	Specialized radio network for dispatch	Dispatch: voice and data	Local or regional	4800 bps	Voice and data	Slow speed, regional coverage only
ESMR	Digital radio to base stations, then public wired network	Voice/data: paging and short messages for dispatch	Regional	4800 bps	Integrated voice and data; digital network	Currently offers only limited coverage and availability
Circuit-switched cellular	Data sent over circuit-switched cellular network via special modems	File transfer, long textual messages	Nationwide	4800–14,400 bps	Good coverage; economical for file transfer, fax	Expensive for short messages
Packet radio (ARDIS or RAM)	Radio network designed for packet data	Short bursty (point of sale, database queries, fleet dispatch, telemetry)	Nationwide	4800–19,200 bps (ARDIS) 8000 bps (RAM)	National coverage, strong third-party support	Expensive for long files
Narrowband PCS	New two-way PCS-based network	Short messages with brief interaction	Regional to national	25.6 kbps on SkyTel	Two-way messaging, cost attractive	Emerging; limited messaging protocol

426

Broadband PCS	Initial focus on voice; data will come later	Not clear; potential for both short messages and file transfer	Currently available only as pilot projects	No reliable information available; potential high	Potentially could replace packet radio, cellular, CDPD	Pilot testing only; heavy investment required
CDPD	Packet-based digital data over cellular infrastructure	Short bursty (similar to ARDIS and RAM)	Metropolitan areas now; national eventually	19,200 bps	More ubiquitous coverage to come; faster than RAM.	Currently, only spotty coverage available; the higher priority given to voice affects its availability for data
Satellite-based network	Data sent over low earth-orbit satellites	High-speed file transfer	National and international	9600 bps	Greatest coverage, except in downtown core	Limited availability

SOURCE: Information in *Data Communications*, March 1994, upgraded and modified by the author.

TABLE 15.2   Matching Applications with Network Types

Application type	Network choices	Key decision factors
E-mail	ARDIS or RAM Mobile Data	Economical for shorter, fewer messages per user, per business session
	Narrowband PCS	Very economical
File transfer and fax	Cellular circuit-switched	National coverage and cost are the most important considerations
	ESMR	ESMR networks are being designed for file transfer and fax services
Telemetry	ARDIS, RAM, CDPD	Short transactions in minimal quantities at predictable intervals
Point of sale transaction processing	Packet radio: ARDIS, RAM	National coverage; more expensive than CDPD
	Circuit-switched cellular	National coverage; expensive response time; takes 30 seconds to set up call
	CDPD	Limited coverage, priority of voice over data
Fleet dispatch	ESMR	Voice and data integration on the same device; one way messaging now, two-way to come; short messages
	Private packet radio	Coverage; two-way messaging
	CDPD	Limited coverage; speed; constant connection; no setup time; cost
Couriers/transportation	Switched-circuit cellular	Nationwide coverage, expensive but large customers can get discounts
	Private packet radio	Nationwide coverage, cost, short transaction
	CDPD	Limited coverage; less expensive than circuit-switched cellular and private radio; speedy response time at 19,200 bps
Public safety (police, fire, ambulance)	Spread spectrum	Limited area in municipalities; very low cost
	CDPD	Cost is less; coverage may be acceptable
	Private radio packet	Cost for constant connection and many applications; consistent response time
Database access	Spread spectrum	Cheap in campus environment
	Private packet radio	Acceptable only for mobile-aware DBMS applications
	CDPD	Coverage; response time at 19,200 could be less

SOURCE: *Data Communications,* March 1994, upgraded and modified by the author.
Note: The table is for illustrative purposes only. It does not imply that only the specified network types can be used for a given application.

Coverage issues in wireless network design. The limited coverage of many wireless networks is one of the major complaints heard from users, and therefore a very important factor to consider in determining the suitability of a network for a given application. If the coverage is too limited, or if there are many dead spots, there is no point in considering the network. If, however, there are just a few zones with poor coverage in an otherwise satisfactory network, it may be possible to negotiate with the provider for the installation of additional base stations in appropriate areas. This is especially true for large customers. UPS and

FedEx were able to negotiate the establishment of a special wireless network. The cost of microcell-type base stations has come down considerably in the last two years and will continue to do so as the technology becomes more mainstream, making it possible to improve overall coverage at a relatively small delta expense. A network provider may well be looking for a large customer to help it make that initial investment.

Hybrid networks should also be considered. For example, QUALCOMM has teamed with ARDIS to offer a hybrid solution for truckers who find that QUALCOMM's satellite-based service provides excellent coverage on highways, but poor coverage in downtown cores.

Distributed wireless network design.   In Chap. 11, we described a distributed wireless network design that utilizes existing backbone infrastructures. A better solution involves the use of the network provider's own backbone and a thick pipe into the customer's main communications center. This solution is preferable for three reasons: (1) it does away with the possibility of overloading the customer's wired-line network backbone; (2) network providers always have redundant links, with automatic switching in case of failure; (3) network providers' prices include anyplace-to-anyplace transportation of data, so why not use the service to the maximum extent?

Selecting the best MCSS.   Choice of MCSS is an extremely important decision in any mobile computing solution. We have described the MCSS as the heart and nerve center of any mobile computing system. It must be reliable and it must provide the required functionality and meet performance specifications. Network designers should thoroughly research this area. We have described various currently available options in Chap. 11. The future trends discussed in Chap. 20 should also be taken into account.

Impact of the agent-based client/server model on network design.   As was discussed in Chap. 12, implementation of the agent-based client/server model can result in considerably decreased network traffic. Its ability to consolidate many interactive tasks into a few communications sessions can significantly affect network choice. If an application follows a conventional client/server design, one of the results may be constant interaction with the LAN. Thus, for this type of application a circuit-switched cellular network would be a poor candidate because of the high cost of multiple short transactions. If an agent-based client/server model is used, however, most of the work can be done off-line, leaving the actual uploading and downloading of data to just one or two communications sessions—and thereby making the use of a circuit-switched cellular network far more acceptable.

Impact of network design on network management. It is important to be able to automatically access information about the physical status of various wireless connections from a network provider's control center. Ideally, this access should be integrated into the network management scheme (as is the case with NETVIEW/6000 for IBM-centric installations, or HP OPENVIEW for many UNIX-based installations).

Network design and outsourcing. Outsourcing is the wave of the future and it is entirely possible that network outsourcing will come to include wireless networks as well. Thus, such an option should be considered as a serious contender: it could be very cost-effective and it would leave the burden of maintaining the network to those who know the business best.

15.1.2 Capacity planning and response time calculations on wireless networks

A mobile computing application transaction traverses many hardware and software components before it reaches the destination server—and must cover the same path again in reverse to complete the trip (Fig. 15.1). There are many physical links (hops), wireless and wired-line,

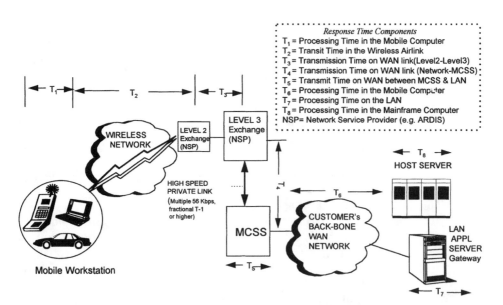

TOTAL TRANSACTION TIME=$(T_1+T_2+T_3+T_4+T_5+T_6+T_7+T_8)$ FOR INPUT MESSAGE
+ $(T_1+T_2+T_3+T_4+T_5+T_6+T_7+T_8)$ FOR OUTPUT MESSAGE

Figure 15.1 Components of transaction response time in wireless networks.

between the end user's client application software and the information server. There are also several pieces of software involved, many of which feature queuing (i.e., they are asynchronous).

Thus, there are complex rules for scheduling priorities on a network. This makes it extremely difficult to build a mathematical model to estimate response times—and therefore to plan reliable network capacities in advance. Network providers do give some estimates, using either complex queuing models or rule-of-thumb calculations based on the experience of other customers.

The greatest surprise often comes in discovering the precise contribution of software to total response times. The average TCP/IP specialist may very well not be able to even estimate how many packets will be shipped over a wireless network with a specific application using a client/server design and new middleware. While middleware may simplify application logic, there is no way of knowing just what is going on at any given moment. Many delays occur because the software configuration, the number of buffers, the number of physical ports, the number of parallel sessions, and the control blocks are not optimally tuned in the MCSS and the DBMS server.

It is important, therefore, to go back to basics. We need to turn to fundamental engineering knowledge and mathematical modeling skills to come up with an initial estimate. This initial estimate can later be refined as understanding of the application's behavior on the wireless network with the communications software of choice grows.

We cannot overemphasize the fact that the greatest single cause of inaccuracies in response time calculations is the inability of systems designers to translate application traffic into network traffic. This inability too often stems from attempts to use crude mathematical models to represent very complex networks comprising multiple servers and queues. A second major contributor to the inaccuracy of estimates is the failure of many designers to allow for the holding delays in software queues at various points. Far greater attention needs to be paid to estimating this traffic and to configuring software parameters.

One way of obtaining information about application-to-network traffic is to use a network probe inserted at an appropriate point while running the most typical transactions and recording the network data flow. There are various software tools available to assist the communications designer in this task.

Once network data flow information is available, the service provider's network designer can use a mathematical formula similar to the one that follows to calculate transit times through the wireless portion of the network:

Transit time through wireless network (TTWN) =

transfer time between the end-user device and the base station, including retries

+ queuing delay to access the channel

+ transfer time between the base station and the network controller

+ queuing delay in the network control program (NCP) buffers

+ transfer time between the NCP and the higher-level message switch (e.g., MOX for RAM or national switching center for ARDIS)

Also, please note the following simple single-server queuing model, as a guideline for rough rule-of-thumb-type calculations:

Transit time in a server = (average service time in the server)/

$$(1 - \text{server-utilization factor expressed in units of 1})$$

For example, if server-utilization factor $\mu = 0.5$ (i.e., 50 percent utilization), transit time including queuing time = $S_t/(1 - 0.5) = 2S_t$ = twice the service time.

As can be appreciated, the number of hops in the path of the network could increase the transit time considerably. Remember that every new hop in the network path adds propagation time plus link delay, however small it may be. (On any communications link, there are three response time components: queuing time, propagation delay, and transfer time.)

Note that end-to-end application transaction response times are much higher than network transfer times for packets of data. Application data might be segmented into multiple packets that have to be reassembled at the other end before the data is presented.

Thus, application response time =

Processing time in the mobile workstation

+ transit time through the wireless network

+ queuing/processing time at the MCSS

+ queuing/transfer time on the host link between the MCSS and the back-end DBMS server

+ queuing/processing time on the LAN server or mainframe

+ time for the response message to traverse all the network components described above in the reverse direction

For information on calculating the optimal number of ports and parallel sessions, please refer to James Martin's *Systems Analysis for Data Transmission* (Prentice-Hall, 1972, or later versions).

15.1.3 Data compression considerations

Wireless network bandwidth is scarce and expensive. Every possible technique should be used to get the utmost service out of this bandwidth. Compression of data is one such technique.

The most common place to compress data is in the modem. While a lot of advanced compression work has been done on PSTN, ISDN, and cellular modems, the same efficiencies have not been achieved so far in proprietary packet radio modems like Mobitex or ARDIS. Refer to Sec. 10.6 of Chap. 10 for further discussion of this topic.

It is a good idea to go beyond modem hardware in reducing the amount of traffic on wireless networks. Intelligence can be built into client application programs so that short message codes can be used to indicate common data occurrences (e.g., an item description need not come from the database resident on the host computer, but instead can be generated on the mobile computer from an item number). Forms and screens should always be stored on the mobile computer.

Interesting ideas can be gleaned from the way the two-way paging protocol (the key objective of which is to keep network traffic to a minimum) is being implemented by companies such as SkyTel. Once application designers find themselves worrying about response times, they tend to start coming up with ingenious ideas.

15.1.4 Fault-tolerant design for higher availability

Mobile computing applications range from e-mail inquiries to mission-critical ambulance or fire truck dispatch. Whatever the application (but particularly in the case of emergency service applications) the network is being relied upon to provide a service, and any lengthy downtime is unacceptable. In other words, a high degree of fault tolerance is needed.

Typically, wireless networks have redundancies built into them and network service providers guarantee an extremely high network uptime percentage, relative to downtime. Generally, base-station hardware and network controllers have hardware redundancy, and message switches are typically built on fault-tolerant platforms such as Tandem and Stratus. However, nothing should be taken for granted, and public shared network providers should be asked for details of their redundancies. Similar redundancies should also be built into private networks.

Another vital component that must have redundancy built into it is the MCSS. Find out if the method of switch-over from failing components to standby components is automatic or manual. If it is manual, is the station staffed, and how long does a switch-over take?

A useful safeguard is to have two or more smaller MCSSs at a location rather than one large one. That way, if an MCSS does fail, only a part of the network will go down. If the MCSSs have a reliability of 99 percent each, then the probability of two MCSSs failing simultaneously and losing the entire network = $1 - (0.99 \times 0.99)$ percent, or 0.0199 percent. As well, excess capacity can be built into each of the MCSSs, so that if worst comes to worst, degraded service at least can be maintained for users—which is better than no service at all.

We advocate end-to-end integration and design in this book. This means looking at all components—from end-user devices to information servers—and researching the likely effect of failure of any component. Remember, the system is only as reliable as its weakest component, unless that component is duplexed.

15.1.5 Security issues in mobile computing

Securing information from unauthorized access is a major problem for any network—a problem that has suddenly become very visible with the explosive growth in the use of the Internet as a platform for electronic communication.

Security is an even greater problem for wireless networks, since radio signals travel through the open atmosphere where they can be intercepted by individuals who are constantly on the move—and, therefore who are difficult to track down. Horror stories of hackers scanning airwaves and siphoning off cellular ID numbers for fraudulent use have become commonplace. Even pager messages are no longer safe.

Here are some examples of common security breaches:

- Interception of law enforcement data on SMR, private radio, or CDPD networks by criminal elements
- Interception of credit card authorizations over wireless networks
- Physical breach of security at unattended base stations or other communications centers
- Theft of database information
- Theft of airtime

While it may not be possible to make any system 100 percent secure, there are certain steps that can be taken to ensure that the risk of security breaches is minimized.

Sources of security leaks. Many local exchange carriers use microwave communications for their inter-LATA calls. Since the frequencies used by carriers is public information, it is not at all difficult for an intruder to intercept both voice and data transmissions (Fig. 15.2).

Even with the increasing conversion of interexchange circuits to fiber, radio and satellite transmissions are still used by many carriers. Network providers should be asked specifically what type of circuits are used for traffic back-haul from base stations, and in particular what circuits will be used in any proposed networking solution.

Tampering with cellular NAMs as a security leak. A common problem in the cellular industry is the theft of airtime by individuals who make cellular calls without paying for them. These people have found a relatively easy way to pirate the numeric assignment numbers (NAMs) of valid users. Even combinations of NAM plus MAN1 or MAN2 sequences are no longer secure.

Stealing information. It is not uncommon for individuals intent on industrial espionage to scoop up vast quantities of information by placing small scanners at appropriate locations and searching with very powerful algorithms (Fig. 15.3). Credit card numbers and bank account numbers are among the most common types of information stolen. While such an effort does require determination and planning on the part of the thieves, remember that the law breaker is often far more motivated than is the person in charge of security. Security system designers need to keep this in mind and make their security arrangements as tight as the technology and the budget will allow.

Designing for security in mobile computing. There are several steps that can be take in designing for security in mobile computing networks and applications.

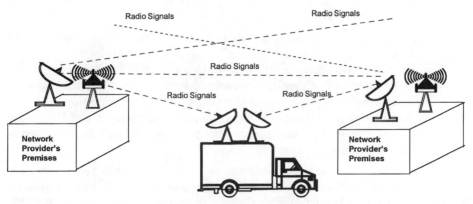

Figure 15.2 A vehicle parked in the direct/indirect path of transmission, with proper antennas, could receive the radio signals without the knowledge of the sender or receiver. Signal between the valid parties is not affected.

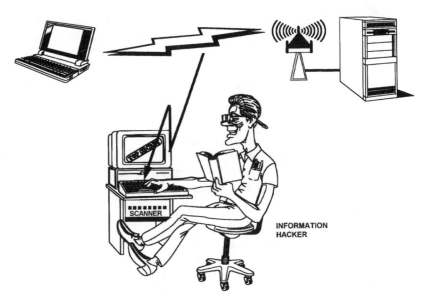

Figure 15.3   Use of scanners poses serious risks to security of information. (*Source: Adapted from* Wireless Networked Comm *by Bates.*)

Physical security, policies and procedures.   There is no point in implementing expensive hi-tech security systems while the physical security of end-user devices, base stations, and information servers is ignored. A notebook left in the back seat of an unlocked car is an obvious and only too common security violation that should be discouraged in the strongest possible terms.

This potential problem will soon be exacerbated with the advent of inexpensive PCS/PCN microcells located in small and unattended sites throughout communities.

Building security features into equipment.   There should be security features built into every piece of equipment sold. Unfortunately, vendors will start paying attention to this fact only when forced by customer demand. Certainly, security features carry a price tag—but not as hefty a price tag as the cost of stolen information. Cellular equipment manufacturers were taken by surprise in the 1980s and 1990s and are now attending to the problem. We will discuss encryption techniques separately as a subtopic.

Application and system-assisted security.   The whys and wherefores of ensuring security through user passwords and similar mechanisms has been taught in many information systems classes. We shall not dwell on these techniques here because the subject is well covered by security system specialists and there are many excellent reference books available on the market. Instead, we shall concentrate on mobile computing security issues.

Dial back as a security technique.   RNA-type mobile computing applications can incorporate dial back in business environments where users and their location are known. Many hardware-based security servers provide this feature.

Firewalls: security servers at the host.   Many specialized companies are providing security servers that can be installed at the host. Cylink, LeeMah, and Racal DataCom are a few of the companies providing such access security servers. Many RNA communications servers also provide this functionality as an integral part of the communications server.

Cylink is well known for providing RNA security products. Figure 15.4 shows a schematic of its SecureAccess System offering X.509 certificate-based security. It is designed for large Fortune 500 installations with thousands of mobile users.

Racal's Guardata WatchWord II token offers convenient alternatives to passwords based on common names, birthdays, etc. (Fig. 15.5). When using WatchWord II, critical information is never entered in clear. The operating principle is based on the challenge/response

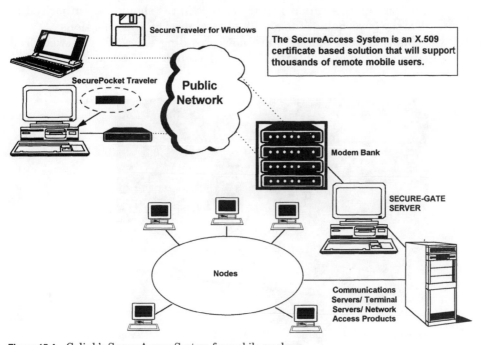

Figure 15.4   Cylink's SecureAccess System for mobile workers.

mechanism described in the ANSI X9.26 secure sign-on standard. The user enables the token by entering a PIN. The WatchWord Generate process takes a digital challenge from the host computer system entered into the token, which then generates a seven-digit response: a one-time password. The response is calculated from the challenge using the DES cryptographic process. There is a security controller or server at the host between the modem pool and the information server. It is anticipated that the next generation of security products will integrate security into the modem or communications server products.

The encryption process in mobile computing. *Encryption* involves scrambling digital information bits with mathematical algorithms and is the most potent protection available against security intrusions into wireless and wire-line communications. Different encryption schemes have been proposed and implemented. The Data Encryption Standard (DES) is one algorithm that has held sway since the 1970s. RSA, based on public key cryptography and named for the three MIT professors—Rivest, Shamir, and Adleman—who developed it, is another. Pretty Good Privacy (PGP) is a public domain implementation of RSA available for noncommercial use on the Internet in North America. The debate still rages as to whether or not the controversial Clipper chip proposed by the Clinton administration in the United States should be authorized to give the government the ability to eavesdrop on transmissions.

Many cellular carriers are now providing encryption between cell sites and the MTSO (Fig. 15.7). Unfortunately, the last segment (i.e.,

Figure 15.5   Racal's Guardata WatchWord II token. (*Courtesy of Racal DataCom.*)

Sending A Secure Message

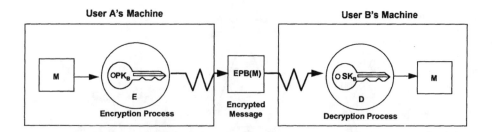

1. User A wants to send a secure message to User B. User A encrypts (E) the message (M) using User B's public key, PK$_B$.

2. The encrytped message (EPB(M)) is sent across the wire.

3. User B gets the encrypted message and decrypts (D) it using the secret key (SK$_B$), resulting in a decrypted message (M). Encryption and decryption are inverse processes.

> *Legend*
> PK$_B$ -- Public Key Of User B
> SK$_B$ -- Private (secret) key of User B
> EPB(M) - Encrypted message for B

Figure 15.6   Mobile computing security via public and private key combination. (*Source:* Open Systems Today, *October 1994.*)

between the end-user device and the cell, or base station) obviously cannot be encrypted and this is where all the theft occurs. For end-to-end security, the only answer is to build encryption/decryption capabilities into the end-user device itself. Unfortunately, this can be done only with end-user devices on digital cellular networks—and digital cellular is still not ubiquitous (with only 40 percent coverage in the United States in 1995, according to Dataquest).

Encryption key types.   There are three types of keys used in encrypting data:

1. A private key known only by the sender and the recipient

2. A private/public key combination

3. A one-time key

In private-key systems, the two parties have a secret key which they use to encrypt and decrypt data.

The private/public key combination is more secure, however. In this scheme, the recipient's public key—available to all who need it to send encrypted data—is used to encode information for transmission. The recipient uses a private key associated with the set to decode the information.

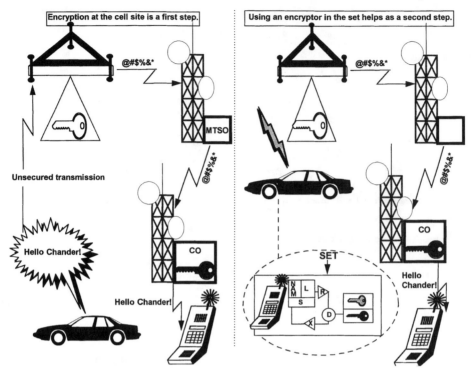

Figure 15.7   Encryption. (*Source:* Wireless Networked Comm *by Bates.*)

The one-time key method is based on the generation of a new key every time data is transmitted. A single-use key is transmitted in a secure (encoded) mode and, once used, becomes invalid. In some implementations, the central system will not issue a key for a new connection until the user supplies the previously used key.

Electronic signatures.   Electronic signatures can be used to ensure that users are who they claim to be. With the appropriate hardware and software—PenOp from Peripheral Vision in the United Kingdom—a system can literally demand a valid signature. While the primary use of such software is in contract-related applications (mortgages, loans, etc.) there is no reason why it cannot also be used as a substitute for a password.

PenOp is based on a biometrics signature-verification technique. It supports a variety of signature capture methods, ranging from low-cost digitizers attached to desktop PCs through to handheld PDAs or pen computers (Fig. 15.8).

An end-to-end encryption scheme with a constantly changing public/private key set.   While each of these encryption schemes provide a certain

Figure 15.8 Digital signatures. (*Courtesy of Peripheral Vision.*)

amount of security in and of themselves, we believe the best scheme is one based on end-to-end encryption using private and public keys, where not even the network provider's control center knows what information is being transferred. To achieve this, the client machine and the information server must each perform encryption/decryption as appropriate, depending on the direction of the transmission. Several PC (PCMCIA) cards provide encryption capabilities, and, while hardware cards are certainly the fastest way of achieving DES and RSA encryption, software-based encryption is also available. (With Pentium notebooks, software encryption with older DES algorithms takes only 10 to 25 milliseconds for a typical 500-character message.)

This approach works independently of any security that the network provides. In fact, depending on the number of mobile users involved, the cost of carrier-provided encryption may be much higher than end-to-end encryption implemented by the user organization.

15.2 Ergonomic and Logistics Design

The ergonomic and logistics design of mobile devices is extremely important for users who employ these devices for extended periods.

This is especially true in situations where the device is permanently installed in a vehicle that doubles as the user's office. Such is often the case with police officers and route salespersons. For them, convenience factors are vital. Several ergonomic considerations have already been discussed in Chap. 6. We will list them again here.

15.2.1   Form factors of end-user devices

In general, end-user devices should be highly portable. Application requirements might well affect shape, size, and weight, in which case, usability pilot testing should be carried out with different models and devices. Intended users should be involved in the selection of the device themselves. Trade-offs that may be necessary should be brought to the users' attention (e.g., functionality and ruggedness over elegance and superior ergonomic design).

15.2.2   Battery life

A device permanently installed in a vehicle can be powered by the vehicle's battery. For others, however, the device's own rechargeable battery is often the only source of power. In such cases, the duration of the charge is an important consideration in selecting the right device. While we should see continued improvements in this area, current limitations constitute an inconvenience factor that has to be endured. The habit of keeping a fully charged spare battery close at hand is a good one to cultivate.

15.2.3   Input: Keyboard, touch, pen, or speech

Whether a keyboard should be used or not depends to a large extent on the amount of data to be entered—and on the current state of the art of speech recognition. As speech recognition becomes more sophisticated and powerful, so keyboards may be needed less. Some applications are perfectly acceptable with touch input or pen-based input. Refer to Chap. 6 for further discussion of these issues.

Color screens are not effective in bright sunlight. Improvement can also be expected in this area.

15.2.4   Ruggedness

There is no doubt that mobile equipment is far more subject to "the slings and arrows of outrageous fortune" than its desktop counterpart, and as a result, public safety and industrial users have demanded military-standard ruggedization for some time now. We believe that notebooks will have to be upgraded more and more frequently as technology continues its exponential growth, and that the business-case justification for long-lived ruggedness, therefore, will increasingly be questioned. Please review the comments in Chap. 6 on this issue.

15.2.5 Health and safety issues

In post-1995 vehicles, the implementation of dual air bags put serious constraints on the permanent mounting of mobile units. Available locations are by no means necessarily convenient locations and compromise is required. In addition, there is debate over the impact of radio frequency emissions on health. You are advised to stay abreast of current research on these issues.

15.2.6 Fixed or portable

In pondering an overall mobile computing solution, a decision has to be made as to whether or not to have portable end-user devices, or end-user devices permanently mounted in vehicles, or one of the available variations on the theme. Several companies specialize in building in-vehicle mobile-device mounts. There are also portable slave units available that work in conjunction with permanently mounted devices. Another alternative is to install a portable-unit docking station in a vehicle.

Summary

In this chapter we have reviewed technical and not so technical design issues that are important to mobile applications. We noted that network design, capacity planning, and security issues should be carefully analyzed by project planners. We suggest that specialists be called in to assist project staff in evaluating various options. Most large organizations have staff members who are experts in network design, capacity planning, and security issues. If there are radio engineers in the organization, they should be brought in for help and advice. Additional help can be obtained from network service providers—but for truly independent advice, the aid of external consultants should be sought. Hopefully, this book will provide sufficient background to enable you to ask the right questions from the right people, thereby leading to truly in-depth evaluations.

The Functional Specifications of a Mobile Computing Solution

If we could design a perfect solution that was inexpensive and had lots of economic benefits with very few risks, we would no longer need systems designers and technology planners. It is the art of balancing conflicting tradeoffs and of balancing risks against potential benefits that makes our jobs so exciting.

CHANDER DHAWAN

About This Chapter

Our approach to the design of a mobile computing solution has been to focus on integrating components into a functional whole. With the discussion in Chap. 15 of technical design issues completed, a picture should now be emerging of the way the complete system will look.

In this chapter, we shall prepare a functional-requirements document. We will illustrate the process by describing functional specifications of the major components and the systems-integration tasks that need to be performed by either an internal or an external organization. Once the requirements are documented, we can shop for a specific solution.

16.1 Functional Specifications of a Mobile Computing Solution

Mobile computing projects can range from simple e-mail applications to comprehensive, enterprisewide implementations of custom-developed, mobile-aware, mission-critical applications across thousands of notebooks, PDAs, and personal communicators. In this book, we have addressed these solutions to the needs of a large enterprise. However, we have also consistently broken each problem down into its component parts and then looked at the total system. We shall take the same approach here.

In order to describe the functional requirements of a complete project, then, we shall first describe the components that make up the entire solution. Then we shall look at the integration issues that have to be addressed in order to bring the components together to form a functional system. Integration requirements can be divided into two types: system integration at the lower layers of the OSI model and application integration at the higher layers.

16.1.1 The mobile computing scenario

Because public safety applications have already been described in Chap. 2, Chap. 3 (business process engineering), and Chap. 4 (business case preparation), we will use for our example the requirements of a hypothetical state government building with an integrated public safety system for a variety of applications. Such an application suite is interesting for the following reasons:

- It demonstrates the often-times mission-critical nature of mobile computing applications and their ability to satisfy demanding response-time constraints.

- It cuts across many different organizations—law enforcement, emergency health services, fire protection agencies, driver registration databases, and the justice system (courts).

- Applications can be implemented on RF-equipped notebooks as well as on extravehicular PDAs.

- Information servers reside on different platforms, including mainframes, minicomputers, and PC LAN servers. In this regard, it is pertinent to note at this point that—as with large financial and insurance systems—a majority of state/provincial division of motor vehicle (DMV) databases reside on IBM mainframes.

- The new infrastructure will require the integration of legacy applications with mobile-aware applications.

- With the exception of narrowband PCS and circuit-switched cellular, all network implementations (SMR, ESMR, ARDIS, RAM, CDPD, and GSM) are possible for this application. While in the past, private radio networks have been common, government users are now considering public shared networks.

- Business process reengineering is an essential component of such a project.

The mobile computing scenario is illustrated in Fig. 16.1.

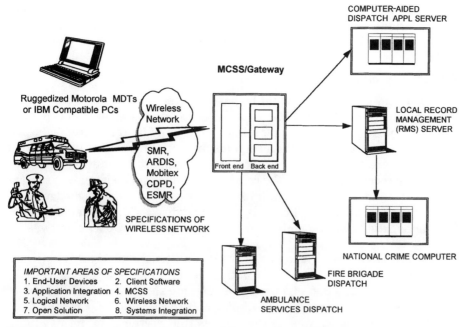

COMPUTER-AIDED
DISPATCH APPL SERVER

MCSS/Gateway

Ruggedized Motorola MDTs
or IBM Compatible PCs

Wireless
Network

SMR,
ARDIS,
Mobitex
CDPD,
ESMR

Front end   Back end

LOCAL RECORD
MANAGEMENT
(RMS) SERVER

SPECIFICATIONS OF
WIRELESS NETWORK

NATIONAL CRIME COMPUTER

FIRE BRIGADE
DISPATCH

AMBULANCE
SERVICES DISPATCH

IMPORTANT AREAS OF SPECIFICATIONS
1. End-User Devices   2. Client Software
3. Application Integration 4. MCSS
5. Logical Network   6. Wireless Network
7. Open Solution   8. Systems Integration

Figure 16.1 Functional specifications of a mobile computing solution.

The application suite comprises the following:

- Inquiry into local, state, and national offense databases (typically DEC, UNIX, Tandem, or IBM)
- Automated on-site citation production
- Inquiry into DMV (driver/motor vehicle) records, typically on an IBM mainframe
- Automated on-site accident (vehicle collision) reporting
- Safety inspection of trucks on the road
- Emergency health services interface during accidents
- Interface with court systems for fine-payment inquiries

Primary users will be police officers, ambulance drivers, and state highway control inspectors.

The requirements definition prior to a formal request for proposal (RFP). The description of the requirements in this chapter could be expanded into a formal RFP after discussion and analysis of a real-life case. Note that we use the terms *requirements* and *specifications* interchangeably. While we shall not provide details of the specifications that may be necessary for an RFP, the following headings could be used to construct a formal requirements document for internal use, or an RFP that may be sent to vendors for soliciting proposals.

16.1.2 End-user device specifications

The end-user device should be based on the platform of choice in the organization. While IBM-compatible notebooks are the de facto standard in the industry, Apple's Powerbook may be considered as a primary or an alternate platform if a substantial number of the intended users are familiar with it. The cost of supporting multiple end-user devices can be less than the cost of retraining and resistance to switchover—and the potential failure of the project that that engenders.

The following specifications should be discussed, evaluated against application requirements, and then described in the document:

- Processor speed, amount of RAM, and hard disk size. If software encryption is to be used, Pentium-based processors may be recommended. The cost of a processor upgrade should be weighed against the overall increase in flexibility and throughput that results from the upgrade, as well as the savings that accrue from not having to buy PC (PCMCIA) encryption cards. Consider also equipment turnover. It is not always necessary (or even wise) to buy the latest and the greatest equipment, but it should not become obsolete in the first year of operation, either.

- Number of PCMCIA slots.

- Type and the size of screen. Color, if a color screen suitable for outdoor operation is available, or monochrome.

- Swivel/tilt nature of the screen.

- Size and weight of the unit.

- Modem. Either integrated into the unit or, perhaps, separately mounted in a vehicle. Most mobile workers need to be able to bring their units (including modems) with them when they leave their vehicles.

- Flexible mounting brackets or docking stations in vehicles to which end-user devices can be attached.

- Ruggedness specifications. For example, resistance to vibration at high speed. NATO military specification 810 C&D should be considered, if appropriate. (See Chap. 6.)

- Ability to function in a broad range of temperatures (e.g., from $-35°$ to $40°C$). Such an ability may be required for extravehicular public safety applications, depending on geographic location. Note, however, that the price for these particular specifications can be high. The number of vendors offering suitable units is very small.

- Input devices: keyboard, touch, pen, or voice; active or passive. In-vehicle applications—where use of a mouse is totally impractical—may use pen, touch, or voice input. Speech should always be evaluated as a serious option.

- A magnetic-stripe reader attached to the unit. This should be available for applications where credit cards or driver licenses may have to be swiped.

- Specifications for any peripherals that will be used, such as portable printers, scanners, and fax machines.

Ergonomic specifications

- Size and weight of the unit.

- Type of screen.

- Easily manipulated screen controls for contrast, brightness, etc.

- Availability of a movable, tactile keyboard. The size of the keyboard and placement of function keys is important.

- Keyboard should be impervious to liquids.

- Power management and battery specifications.

- Adjustable (vertical and horizontal) position of the unit when mounted in a vehicle.

16.1.3   Mobile user client software specifications

The mobile end-user device should support the organization's environment (UNIX, DOS, Windows 3.1, Windows 95, Windows NT, OS/2, or MacOS).

Software drivers with APIs for various radio modems such as Mobidem, ARDIS, CDPD, and, soon, GSM and PCS, should be available. Communications software support under TCP/IP should be rich and compatible with different networks. Ensure that the software driver has been tested with the chosen network. Encryption software should be available on the mobile-client end as a communications utility.

Software support should be available for pen, both as an alphanumeric input device and a drawing tool. Remember that a pen is a far more convenient user interface than a keyboard in many field applications. However, a keyboard is faster when users have to enter long narratives, as, for example, in the case of an automobile accident report, an audit, or a report by an arresting police officer.

Note also that it will soon be possible (perhaps as early as 1997) to use speech as an economical and much superior user interface, so it should be factored in now as a future requirement. Because speech-recognition algorithms are processor intensive—this is one reason why IBM has implemented the technology on a PowerPC—systems designers should plan for hardware that has the necessary power and software that has the intelligence to accommodate this eventual capability.

We expect that GPS/GIS will become a more common requirement in field dispatch and vehicle location applications as time goes by. If appropriate, this capability should be also be allowed for by keeping a PC Card slot available.

16.1.4 Application-integration specifications

During the 1970s and 1980s, many organizations developed business applications that could be accessed through dumb terminals such as the IBM 3270, the VT220, or PCs emulating these terminals. A few of these applications are now available in client/server versions. Any decision to adapt these legacy applications to a mobile environment should be made in the full knowledge that such an undertaking can be a very complex and costly task. In fact, it may be difficult to economically justify application modifications, and resources may not even be available. Far better to develop new applications wherever possible, using the mobile-aware design principles described in Chap. 13. Either way, application-level connectivity presents a challenging task: be very thorough in evaluating its complexity.

Many vendors offer what they call *transparent tools* for connecting existing remote network applications by emulating the AT command set on wireless networks (a la RAD I/O from RIM of Waterloo, Ontario, in the Mobitex environment). We believe that this strategy should be considered only a stopgap arrangement. The long-term goal should be to exploit the advantages of a client-agent-server model through development engines such as Oracle Agents or Racotek's Keyware middleware.

Application-integration requirements should be described in as much detail as possible, for it is in this contentious area more than any other that misunderstandings and disagreements occur. You should answer three basic questions:

1. Does the vendor understand the applications and their requirements?

2. Are the applications themselves sufficiently well documented?

3. Does the internal development staff responsible for application integration understand the communications interfaces that the vendor will be providing?

The following additional questions must also be answered in the course of documenting application-integration specifications for the vendor:

- Do the business applications use a terminal interface?

- Is the application based on a client/server design with the server on a LAN?

- What software communication interface does the application use: terminal server, LAT, LAN MANAGER, NETWARE, IPX/SPX, NET-BIOS, etc.?

- What non-LAN server platforms need to be supported? Are connections needed to IBM mainframe-based applications?

- Are there applications that retrieve information from multiple platforms for a single transaction (or business-unit of work)? For example, in order to issue a citation, an inquiry may have to be made of a DMV registration database on an IBM mainframe, local-incidents information may have to be pulled from an Oracle offense database on a DEC/VMS platform, and an accident database under Sybase on a Netware LAN may have to be updated.

16.1.5 Mobile communications server switch (MCSS) requirements

The MCSS should meet the following specifications:

- It should provide a transport-layer (level three of the OSI model) interface to multiple wireless networks, such as private SMR, ARDIS, RAM Mobile Data, and CDPD at a minimum, but preferably ESMR and NB-PCS, as well.

- The back-end portion of the MCSS should support logical (application-level) connectivity to information-server platforms (IBM, DEC, and Tandem in our scenario).

- It should support multiple concurrent sessions from a single client application to multiple host servers, as per the single-transaction citation example above.

- It should support the following network protocols:
  - TCP/IP
  - X.25
  - SNA LU 2.0 and LU 6.2
  - NETBIOS (though not mandatory, in many cases)

- It should be scalable in terms of performance. That is, different models should support from 100 to 5000 users, with up to 500 users connected simultaneously.

- It should have a fault-tolerant design, preferably with an automatic switch-over rather than a manual switch-over.

- It should support asynchronous (unblocked two-way) logical communication between the client application and the server application under the supported transport protocols. (An unblocked two-way logical protocol is one in which either party can send messages at any time, rather than having to wait for a response to one message before sending another.)

- It should have a persistent (hard-disk-based) message queuing and recovery facility.

- It should provide standards-based (SNMP) network-management functions that can be integrated into an organization's central-site network-management platform.

16.1.6 Logical network architecture specifications

The logical network architecture represents the connection between the client and the server applications in the client/server model. In the client-agent-server model, it includes both client/agent and agent/server connections.

The logical architecture should support multiple wireless network protocols. Application design should be independent of the network layer. Middleware especially designed for mobile-aware applications should be available.

There should be as few protocol conversions in the logical connection as possible. Wherever appropriate, a common transport protocol, preferably TCP/IP, should be used. Desirability of using TCP/IP as a standard protocol over the airlink should be weighed against the advantages of an optimized and more efficient protocol, such as Nettech's instantRF, so long as it is supported by an established vendor such as Motorola.

16.1.7 Wireless network specifications

As we have seen in Chap. 8, the wireless network scene is changing fast. From an application perspective, a radio/wireless network should meet the following requirements:

- Packet-based networks (ARDIS, RAM, and CDPD) should be available for OLTP-type transactions. For long text messages, file transfer, and fax-based applications, connection-oriented networks such as cellular should be supported. In our scenario, the applications fall into the first category.

- The network should support physically continuous connections, meaning there should be no long delays in call setup, as is the case with cellular circuit-switched networks. Many public safety applications cannot tolerate long connection delays.

- The wireless network should support speeds of at least 9600 bps, with 19,200 bps the preferred speed.
- The network should support TCP/IP as the common transport protocol.
- The wireless network (or service provider, in the case of a public shared network) should provide monitoring, audit, and performance information to the logical network operations staff operating user-support hot lines.
- The effective transit time for messages on the vendor-provided portion of the network solution must not exceed the specifications shown in Table 16.1.

16.1.8 Open-solution specifications

Proprietary components in a mobile computing solution should be kept to an absolute minimum and should be used only where functionally equivalent standards-based open-solution components are not available. Ours is a pragmatic definition of an open solution. In our view, an open solution is one that is based on either a de jure or a de facto standard, where market and user acceptance of a standard is more important than its de jure nature. Based on this criterion, a mobile computer conforming to an IBM-compatible PC specification is more open than a workstation based on an ISO- or IEEE-sponsored standard if the market has not accepted such a standard.

The following questions should be answered in determining the open nature of a mobile computing solution.

- Can workstations be acquired from multiple sources?
- Can different radio modems be used in the proposed configuration?
- Does the proposed wireless network's API support multiple OS platforms (Windows, OS/2, MacOS, etc.)?
- Can the solution be adapted to a network other than the one proposed by the vendor? How much would it cost to switch from ARDIS to RAM Mobile Data, and vice versa?
- Does the MCSS run on more than one hardware platform?

TABLE 16.1   Transit Times for Messages

Inquiry transaction into DMV database (50 characters in, 400 characters out)	90 percent of all transactions in 3 seconds or less
Inquiry transaction into local-incidents database, with a one-incident response (50 characters in, 300 characters out)	90 percent of all transactions in 2 seconds or less

- Can maintenance and support services for the solution be purchased from third-party maintenance organizations?

16.1.9   Systems integration experience specifications

It is important that care be taken to prevent a situation where a vendor ends up using a project as a learning experience. Often vendors trot out experienced resources at the time of a sale, but are subsequently unable or unwilling to assign them to the project on a full-time basis. There should be a clearly written contractual agreement dealing with the supply of key technical resources, on the basis of which the vendor has sold the solution. Ideally, the vendor selected for a mobile computing project should have had prior experience in building mobile computing solutions.

16.1.10   The current technology environment

No solution should be chosen without there being a thorough understanding all around of the organization's current IT infrastructure: network, communications software, applications. The environment should be clearly described in the document so that vendors know what they are dealing with and can provide accurate estimates. By keeping this background information in mind when making recommendations, vendors are better able to interface a new solution with the existing infrastructure—and, consequently, will be less inclined to inflate time and cost estimates to cover unknowns.

16.2   Matching Specifications Against Implementation

Once requirements have been set down on paper, the next task is to match them against what is available on the market. The objective here is to get the best solution for the organization, usually by inviting proposals. Either an RFP can be sent to a wide group of vendors, or a selected group of vendors can be asked to submit proposals if a prior screening of vendors has been done.

It is our firm conviction that superior solutions will be recommended by vendors if they know that the requirements document has been meticulously researched and subsequently prepared in a careful and detailed manner. Remember that a sense of healthy competition always elicits the best response from vendors. Even a favored vendor will prepare a better proposal if it is clear that there are contenders from other quarters responding to a well-thought-out requirements document.

The overall task of arriving at an optimal solution can also often be made a great deal easier by certain givens about your technology environment and/or preselecting some of the components. For example, the current network infrastructure might dictate a particular communications server or mobile computer.

16.3   Evaluation Criteria

Comprehensive criteria should be established for evaluating different solutions and selecting the optimal one for the organization. The following factors should be included in this criteria:

- Compliance with functional specifications.
  - absolute compliance with mandatory specifications
  - compliance with as many desirable specifications as possible
  - compliance with ideal wish-list requirements
- Compliance with performance and capacity specifications.
- Compliance with design specifications (i.e., redundancy of components).
- The vendor's experience with technical design, application development, application integration, systems integration, and mobile computing project management. Customer references in the organization's vertical industry should be obtained.
- The vendor's track record in finishing projects on time and within budgets.
- The vendor's technical support and hardware-maintenance services.
- Overall costs for the proposed solution.
- Risks associated with the proposed solution.

16.4   Evaluation Methodologies
for Determining an Optimal Solution

The discussion in Chap. 4 on the preparation of business cases should be referred to in conjunction with this section.

We propose that a formal evaluation methodology be used to determine an optimal solution—a methodology based on the quantification of as many of the criteria items as possible. First, requirements should be separated into three categories: mandatory, desirable, and an ideal wish list. Start with mandatory requirements that are flat-out indispensable. Any proposed solution that does not meet these requirements is automatically discarded. Desirable specifications can be assigned dollar values that indicate their worth in terms of benefits, or the cost of acquisition from a third party, or the cost of in-house devel-

TABLE 16.2 Comparing Different Vendor Solutions

Criteria	Vendor # 1 solution			Vendor # 2 solution	Vendor # 3 solution
	Compliance or penalty cost	Weight	Score (0–10)		
1. Mandatory requirements					
End-user client device specifications	Meets	5	10		
Mobile-user client software specifications	Meets	5	10		
Application-integration requirements	Meets	15	8		
MCSS specifications	Does not meet	15	0		
Logical network architecture	Meets	5	8		
Physical wireless network specifications	Meets	10	10		
Open solution specifications	Meets	5	10		
Systems integration requirements	Meets	10	6		
2. Desirable specifications					
End-user client device specifications	Meets	5	10		
Mobile-user client software specifications	Meets	1	6		
Application-integration requirements	Meets	5	8		
MCSS specifications	Does not meet	5	0		
Logical network architecture	Meets	2	7		
Physical wireless network specifications	Meets	5	8		
Open solution specifications	Does not meet	2	7		
Systems integration requirements	Meets	5	8		
Total penalties for desirable specifications	$2.0 million	100		$0.5 million	$1.0 million
Total costs—one time	$5.0 million			$6.0 million	$7.5 million
Total operational cost for project life	$4.5 million			$3.0 million	$4.0 million
Total costs—one time, ongoing, penalties	$11.5 million			$10.5 million	$12.5 million
3. Other factors					
Vendor track record	Good			Very good	
Vendor's future mobile computing strategy	"Me too"			Leader	
Recommendation from references	OK			Good	
Major pros	Low cost			Previous exp.	
Major cons	Does not meet mandatory req.			Timing not good	
Other risk factors	Lack of exp.				
Overall rank and recommendation	*Disqualified*			*Best overall solution*	*Next best solution*

Note: This is a hypothetical sample table. A real-life table would list all the specifications.

opment. Cost penalties equivalent to their dollar value can then be levied against vendors who are unable to supply one or another of them. This will result in an apples to apples comparison of different vendors' proposals. Table 16.2 shows a partially completed template that compares competing solutions.

Summary

In this chapter we reviewed ways in which to create a requirements document for mobile computing. We discussed the high-level specifications of various components and integration tasks by using a sample scenario based on a public safety application, on the one hand, while keeping the subject matter as generic as possible, on the other. We also described evaluation criteria and a methodology for effectively using these criteria. Armed with this knowledge, you should now be in a position to select an optimal solution for your organization.

17

Implementation Plan for a Mobile Computing Project

The information technology industry can learn two important lessons from industries that have been around for a lot longer—building construction, for example. The first lesson is to standardize components, and the second lesson, using our example as a vehicle, is to rely on experienced professional contractors—rather than carpenters, brick layers, or design engineers—for project planning and management tasks. Too often, both lessons are ignored when it comes to IT projects. CHANDER DHAWAN

About This Chapter

In the last chapter we developed the functional specifications for a mobile computing solution. We also discussed evaluation criteria and a methodology for using those criteria, paving the way at last to the actual selection of an optimal solution for our mobile computing project. However, that optimal solution presently exists on paper only.

Now comes the job of implementing the chosen mobile computing solution. In order to simplify this difficult task, we will break it down into smaller implementation units and develop a comprehensive project plan showing how these units are linked together. Dependencies must be thoroughly understood and trained resources must be acquired from within or outside the organization. In other words, a comprehensive project plan is needed. In this chapter, we will discuss ways in which to develop such a plan. Before our discussion, however, we shall examine various mobile computing implementation strategies. After our discussion, we will look at a sample project plan.

17.1   Implementation Strategy for a Mobile Computing Project

The following implementation strategies should be analyzed before a detailed plan is developed. Based on prior experience with many of these issues, we feel that it is important that they be addressed right at the beginning of any project planning exercise.

17.1.1   Pilot before rollout

Because mobile computing is still very much an emerging technology, many components have not yet been fully tested in real-life environments. To complicate matters further, the vendor community continues to advocate proprietary technologies that will likely be taken over by standards-based implementations in the future. On the other hand, promising technologies introduced by small, innovative companies are often at risk of being subsumed by larger companies as a result of eventual rationalization in the industry. In other words, the mobile computing industry is rife with many unknowns.

Taking a conscious and calculated risk in order to gain a competitive advantage is a valid business strategy—but the result of such a plan may be a very shaky financial case which, with user uncertainty about proposed new concepts being introduced, may be difficult (if not impossible) to float. A pilot implementation may be the answer to this dilemma.

A technology pilot can reveal which components work in a particular scenario, and which do not. At the same time, it can expose components that do not deliver promised functionality, while serving as a training experience for the development and integration team. Modifications to the original design resulting from the pilot can then lead to the development of a final rollout plan. The one major disadvantage of a pilot, however, is that it does inevitably delay the rollout timetable.

17.1.2   Application-integration responsibility

A major issue that has to be dealt with in any mobile computing project is deciding whose responsibility it is to interface with mobile computing business applications. This is an issue that has to be resolved regardless of whether applications are acquired from external vendors, are custom built, or are mobile-aware versions of existing applications.

A review of the availability and experience of internal resources and the corporate IT strategy on outsourcing can lead to a resolution of this issue. Generally speaking, it is best to keep business-rule analysis tasks internal to the organization. Actual application-programming development and integration of applications with the proposed infrastructure can be outsourced to a systems integration vendor, provided the vendor has previously built similar applications in a mobile or wireless environment. If an external vendor uses any particular project as a learning exercise, then it does not add value to the project. In such cases, serious consideration should be given to handling the application development task from within the organization.

17.1.3 Internal versus external systems integration

The same issue arises with the overall systems integration of the project. Most organizations use external vendors for this task. Large organizations (banks, insurance companies, telecommunications carriers, automobile manufacturers, etc.) might have internal systems-integration groups. If so, they also have extensive development and support resources to assist the systems-integration team. In smaller, modern, slim, trim, and agile business organizations, in-house expertise need be built only for operational or ongoing tasks. Specialized tasks should be outsourced. If the mobile computing project is a one-time-only venture, it is not economical to acquire the necessary expertise. Faster return on human resource investment can be had by buying the expertise from outside. Even in cases where systems expertise does exist internally, management consultants and organizational experts recommend a focus on core business competencies. Obviously if outside expertise is not available, or if trained resources are currently not busy with internal projects, an exception should be made to this general recommendation.

17.1.4 Single vendor versus vendor consortium

Who should handle vendor management, if multiple vendors are involved? Should the entire project be handed over to a single systems integrator who acts as a prime contractor that manages subcontractors? Or should individual contractors be dealt with separately? IT management needs to answer these mobile computing project questions. Currently, there are very few systems-integration companies with extensive experience in mobile computing projects. There is no critical mass of such projects that justify the expense of keeping the appropriate skills in regional offices. A national vendor's experience base that is founded on mobile computing projects in other parts of the country is useful only if the same team that acquired the experience is an available resource to your project. Telephone advice and occasional audits by these experienced specialists are ways in which national vendors do their best to leverage their experience base.

Our advice is simple: if the prime vendor is experienced, has a good track record, and can assign dedicated resources with prior experience in mobile computing, give the responsibility for vendor management to that company and save headaches (and, possibly, legal problems) all around. A prime contractor is generally more experienced in dealing with subcontractors than a user organization is.

Teaming agreements that clearly lay out task responsibilities should be included as part of any master agreement with the prime contractor. There should be irrevocable penalty clauses between the prime vendor and the organization, and similar clauses should be included in teaming agreements between the prime vendor and any subcontractors.

It should also be mandatory for project leaders from subcontracting firms to attend joint project meetings where they can give status reports on their tasks direct to the organization (which, after all, is paying the final bill). Do not allow the prime contractor's project manager to act as a buffer or to filter information during project review sessions. Contractual and financial matters, however, should, of course, be left to the prime vendor to resolve with subcontractors.

17.2 Major Tasks

The following are major tasks that can have a significant impact on the successful outcome of any mobile computing project:

- Business requirements analysis and business process reengineering (BPR)
- Development of the initial technology architecture of the solution
- Technology research
- Wireless network selection
- Wireless network coverage audit
- Vendor product selection
- Prime vendor selection
- Determining staff requirements
- Project management
- Development of mobile-aware applications
- Integration of applications with chosen communications software
- Back-end server changes
- Security design
- End-to-end optimization and tuning
- Stress testing

- Training and technical support
- Implementing network management
- Resource planning

17.2.1   Business analysis and business process reengineering (BPR)

All IT projects should start with a detailed business analysis. As well, research shows that mobile computing projects that include BPR have had greater degrees of success than those that are treated as wireless infrastructure projects or application-integration projects. Even if the proposed application extends only to providing wireless e-mail access, the possibility of taking the opportunity thus presented to change current processes by which staff call in for messages should be researched. As was discussed in Chap. 3, mobile computing offers significant opportunities for BPR—opportunities that should be addressed before rollout.

Planning for BPR impacts process flow and organizational changes and therefore requires entirely different people skills. Industrial engineering expertise is required as a primary discipline rather than software development expertise. There is no reason not to implement a technology infrastructure and application-development pilot project from purely mobile computing logistics and organizational points of view, but there does come a point when BPR experts should also be brought in to analyze opportunities to change existing business processes.

17.2.2   Initial technology architecture of the solution

Every building construction starts with a conceptual design. Similarly, every IT project must start with an application model, a data model, and a technology architecture, as discussed in Chap. 5. A team of senior designers, planners, and technical managers must be assembled to develop an architecture based on technology principles already accepted by the organization.

17.2.3   Technology research

The next step is to research technologies, products, and options that may have to be evaluated as to their ability to satisfy business needs. The information in this book is sufficient to start the process only. Knowledge bases will have to be updated, vendors contacted, and courses and seminars attended to bring skills to a level where a mobile computing project can be undertaken with confidence.

17.2.4   Wireless network selection

Selecting an appropriate wireless network or PSTN (dial-up or ISDN) connection in an RNA case is an important task that was discussed in Chap. 15. Some organizations choose to preselect a network before evaluating and selecting other components. Such a component-by-component approach is acceptable, provided extensive traffic analysis is done before network selection is finalized.

Other organizations have sent the entire set of functional requirements, including network specifications, to systems-integration firms and vendor consortia.

A few network providers (ARDIS and RAM Mobile Data, for example) have a reasonable amount of information systems experience, from a planning point of view. They can make recommendations that can be compared prior to a decision.

In general, researching other users' experiences with similar applications has the potential to greatly simplify any project.

If a private wireless network is an option, proposals will have to be invited for a network design showing base stations, antennas, and controllers. Because the criteria for a radio network is so different from that of a wired-line network, in such an undertaking, radio network engineering specialists should be called in to help out.

17.2.5   Wireless network coverage

Once a network has been selected, the vendor should be asked to supply detailed coverage maps.

Note that network providers carry two types of coverage maps: one is a sales-oriented map and the other is a computer-generated engineering map showing signal strengths in greater detail. (Some network providers have maps showing actual signal strength measurements in various locations.) Obviously, the latter map is the one to review. The ultimate coverage test, of course, is to literally drive around with a measurement instrument.

There are coverage-prediction software programs available from independent consultants as well as from vendors. Mozaik is one such program from Motorola that runs on a PC under OS/2. The company also provides FactWare, complimentary DOS-based software that calculates a statistically appropriate number of map grids to validate Mozaik coverage predictions.

Vendors also sell coverage-design services.

17.2.6   Vendor product selection

A mobile computing solution comprises many component products that have to be evaluated against requirements. Either an RFP process can be used to accomplish this, or minitasks can be assigned to product specialists.

17.2.7   Prime vendor selection

Selecting a prime vendor as partner in a mobile computing project is certainly one of the most important tasks in any mobile computing undertaking—a task, therefore, that should be done by way of a formal selection process. This selection process, along with associated issues, has been described in Chap. 16.

17.2.8   Determining staff resource requirements and training

The project manager must acquire the necessary human resources from inside or outside the organization. These resources may have to be formally trained before they can be considered effective and fully productive. They are not easy to come by and the organization should be prepared to pay a premium, if necessary.

17.2.9   Project management

In a large mobile computing systems-integration project, two project managers are involved: one from the prime vendor and one from the user organization. Conventional project management techniques must be followed. The project should be broken into smaller segments, with all teams responsible for specific tasks attending regular project meetings. All individual project plans should be amalgamated in one master plan.

Serious problems can arise if misunderstandings are allowed to develop between the user organization and the vendor's organization. Such misunderstandings can not only lead to project delays, but they can also divert management attention with unproductive dialogue, discussion, and debate. Formal, well-documented work assignments to individual teams, the prime vendor, and the subcontractors (signed by participating project leaders) can go a long way to ensuring a clear understanding of individual responsibilities.

There should be a well-established change-request mechanism that is clearly understood by all parties. This mechanism should specify who is authorized to make changes, and how changes will be conveyed. The impact of any changes on costs and the schedule must be determined and conveyed to senior management.

To better foster effective communication, tools that can automate various forms, including change requests and approvals (Lotus NOTES, for example), should be used. For project monitoring and resource control, Project Manager, TimeLine, or similar tools should be used.

17.2.10   Development of mobile-aware applications

In the course of conducting the business analysis, a catalog will be compiled of proposed user applications. One decision that has to be made is whether new mobile-aware applications are going to be developed, or whether existing applications will be modified to make them mobile-aware. This is a new discipline for which there are very few tools available.

These tasks should be assigned to the best application designers available who can then use the design guidelines outlined in Chap. 13 and selected application development tools to get them to work. Oracle's Mobile Agents is a good tool to start with, if it fits the design and database environment. More mobile-aware tools can be expected in the future.

17.2.11   Integration of applications with chosen communications software

Application integration is the software glue that connects mobile client applications to back-end information servers. Thorough analysis of the chosen software is important. Preliminary analyses are available in trade magazines like *Data Communications*. More detailed evaluation of the right tool for the application and the network infrastructure should be undertaken by those who understand not only telecommunications, but also the relevant application interfaces. Be extremely careful about using middleware unless it is clear that the advantages are greater than the price. At the very least, make sure it is mobile-aware middleware that has proven itself.

17.2.12   Back-end server changes

A mobile computing project should not necessitate significant changes to back-end applications and information servers, since they serve fixed networks as well. Nonetheless, some changes are inevitable. These changes can range from terminal (or LU, in IBM SNA parlance) definition, security, and TCP/IP address definitions, to more extensive server application changes needed for a new mobile-aware design. The systems programming group at the mainframe, the database administrator at the LAN, and the host application-design group should be asked to evaluate all the changes required. Quite often, mainframe systems staff have strict control over changes and a predetermined schedule that must be observed. Allowance should be made in the plan timetable for this possibility.

17.2.13   Security design

The design and implementation of proper security should be a distinct task assigned to a team that draws upon the skills and experience of the corporate security administrator. We have discussed this topic in Chap. 15. You should review this chapter to determine detailed security requirements and how you intend to meet these requirements.

17.2.14   End-to-end optimization and tuning

Once an application has been developed and is functioning, its optimization should be undertaken in the spirit of our end-to-end design theme, with messages being traced through the various software and hardware components. Be aware, however, that this is a process that can take months rather than weeks in complex multiplatform environments, and is, therefore, a task that requires a degree of dedication.

17.2.15   Stress testing

For large rollout projects, time should be allowed for stress testing of the network, the MCSS, and the information servers' capacity to handle the workload. Should there be an absence of appropriate stress testing tools, systems engineering know-how and prediction abilities will have to be relied upon to project eventual online performance. The network service provider should be able to help in this test.

17.2.16   Training and technical support

The training of users is an extremely important task. The training organization itself should be involved as soon as possible so that it can develop a thorough understanding of the new applications and prepare user training programs. In addition to the initial training programs, technical support procedures for remote users should also be designed. (Existing technical support procedures will require changes because mobile users and telecommuters are now involved.) It is always a good idea to use context-sensitive online help wherever possible, backed up by more extensive reference manuals. These manuals should be published electronically to save costs and keep them current. In a new system, many revisions become necessary as the technical support database grows with experience.

17.2.17   Implementing network management

Network management will be discussed in Chap. 18. It may be necessary to use network management tools that are unique to mobile computing.

17.3   Resource Planning

Mobile computing projects draw upon many different disciplines in an organization. Many of these skills are available in the information technology and business operations departments of the organization. The following skilled human resources should be well represented on the project team, either on a full-time or part-time basis:

- Business users affected by the mobile computing applications analysis
- Business analysis and business process reengineering specialists
- Application-design and -development specialists
- Systems programmers for legacy applications
- Corporate infrastructure planners
- LAN/WAN network designers
- Radio engineers, if they exist in the organization
- Security administrators
- Training and technical support personnel
- Network managers

Negotiation with department heads will often be necessary to acquire appropriate human resources. The involvement of department heads at the inception of the project will help to ensure willing participation and cooperation. In some instances, it may be necessary to look outside the organization for specific resources. Any such additional costs should, of course, be entered into the business case.

17.4   A Sample Project Plan

We have included a sample project plan that can be used as a template for your mobile computing project (Table 17.1). You can get some ideas from this plan and develop one for your own project. Please note that any project plan is an evolving exercise. It will be necessary to add, delete, and upgrade as time goes by.

Summary

In this chapter, we reviewed various strategies for the systems integration of a mobile computing project and discussed the tasks involved. We indicated that a mobile computing project is a multidisciplinary undertaking, and that mobile computing applications depend on a smooth integration of existing software and hardware with the new mobile network infrastructure. In order to achieve this integration successfully, a carefully monitored and detailed project plan is necessary.

TABLE 17.1 Sample Mobile Computing Project Plan Version 1

ID	Name	Duration (days)	Scheduled start (month/day/year, A.M.)	Scheduled finish (month/day/year, P.M.)	Predecessors	Resource names
1	Mobile computing project initiation	45	7/1/96 8:00	8/31/96 5:00		
2	Project orientation	22	7/1/96 8:00	7/30/96 5:00		Project manager
3	Project requirements definition	35	7/15/96 8:00	8/31/96 5:00		Project manager
4	Organizational roles and responsibilities	24	7/15/96 8:00	8/15/95 5:00		Project manager
5						
6						
7	Project resources hiring	21	9/1/96 8:00	9/30/96 5:00		Project manager
8	Radio communication specialist	21	9/1/96 8:00	9/30/96 5:00		Project manager
9	Mobile-application-development specialists	21	9/1/96 8:00	9/30/96 5:00		
10	TCP/IP networking specialist	21	9/1/96 8:00	9/30/96 5:00		
11	Business process reengineering specialists	21	9/2/96 8:00	9/30/96 5:00		
12						
13	Business process analysis	304	10/1/96 8:00	11/28/97 5:00		BPR specialist
14	Business requirements for mobile computing	20	10/1/96 8:00	10/28/96 5:00		
15	Current process analysis	20	11/1/97 8:00	11/28/97 5:00		BPR specialist
16	Reengineered processes	20	12/1/96 8:00	12/27/96 5:00		BPR specialist
17	Convert business requirements into solutions	21	1/1/97 8:00	1/29/97 5:00		Project architect
18						
19	Business case preparation	1	6/30/98 8:00	6/30/98 5:00		
20	Analyze preliminary benefits and costs for business case	1	6/30/98 8:00	6/30/98 5:00		
21						
22	Mobile computing technology research through RFI	110	9/1/96 8:00	1/31/97 5:00		
23	Technology review—preliminary	20	9/1/96 8:00	9/27/96 5:00		Technical architect
24	RFI preparation and issuance	22	10/1/96 8:00	10/30/96 5:00	4	Senior project leader
25	RFI response evaluation	12	12/15/96 8:00	12/31/96 5:00		Project manager
26	RFI vendor presentations	10	1/6/97 8:00	1/17/97 5:00		Project manager, project director

Note: This project plan lists only some of the tasks in a mobile computing project. Dates and task dependencies are purely arbitrary.

469

TABLE 17.1   Sample Mobile Computing Project Plan Version 1 (*Continued*)

ID	Name	Duration (days)	Scheduled start (month/day/year, A.M.)	Scheduled finish (month/day/year, P.M.)	Predecessors	Resource names
27	RFI evaluation report	1	1/31/97 8:00	1/31/97 5:00	26	Project manager
28						
29	Project application-development support	130	9/2/96 8:00	2/28/97 5:00		
30	New application requirement definition	87	9/2/96 8:00	12/31/96 5:00		Senior business analyst (20% of resource)
31	Development tools selection	20	12/16/96 8:00	1/12/97 5:00		Senior dev analyst
32	Investigate wireless test lab setup	10	12/1/96 8:00	12/15/96 5:00		Comm analyst (10% of resource), dev analyst (10% of resource)
33	Create wireless dev/test lab	43	1/1/97 8:00	2/28/97 5:00		Vendor, dev analyst, comm analyst
34						
35	MC technology architecture development	40	9/1/96 8:00	10/25/96 5:00		
36	Application architecture review	5	9/1/96 8:00	9/6/96 5:00		Project leader
37	Data architecture review	5	9/6/96 8:00	9/12/96 5:00	36	Project manager
38	Technology architecture	30	9/15/96 8:00	10/25/96 5:00		Steering committee
39	End-user device selection	10	9/15/96 8:00	9/27/96 5:00		Project manager
40	Software architecture	20	9/15/96 8:00	10/11/96 5:00		Project manager
41	Wireless network architecture	30	9/15/96 8:00	10/25/96 5:00	39	
42	Matching wireless network to application characteristic	10	9/15/96 8:00	9/27/96 5:00		
43	MCSS—back-end server interface architecture	10	10/1/96 8:00	10/14/96 5:00		
44						
45	Mobile computing RFP—build and issue	63	1/2/97 8:00	3/31/97 5:00		
46	RFP draft creation	22	1/2/97 8:00	2/1/97 5:00		Project leader
47	RFP review by user group and steering committee	1	2/7/97 8:00	2/7/97 5:00	46	Project manager
48	RFP issued to vendors	1	2/21/97 8:00	2/21/97 5:00		Project manager
49	Vendor briefing	1	2/28/97 8:00	2/28/97 5:00	48	Project manager
50	Vendor responses received	1	3/31/97 8:00	3/31/97 5:00		
51						
52	RFP evaluation	65	3/1/97 8:00	5/30/97 5:00		

53	Develop evaluation methodology	10	3/1/97 8:00	3/15/97 5:00		Project leader, project manager
54	Pass 1—mandatory evaluation	11	4/1/97 8:00	4/15/97 5:00		valuation team project manager
55	Pass 2 evaluation—desirables, risk, etc.	5	4/16/97 8:00	4/22/97 5:00	54	Evaluation team
56	RFP benchmarking	4	5/10/97 8:00	5/15/97 5:00	55	Project leader, technical architect
57	Reference checks	4	4/23/97 8:00	4/28/97 5:00	56	Evaluation team
58	Evaluation report	12	4/22/97 8:00	5/7/97 5:00	57	Evaluation team, project manager
59	RFP evaluation review by management	0	5/10/97 8:00	5/10/97 5:00		Steering committee
60	RFP AWARD recommendations	1	5/15/97 8:00	5/15/97 5:00	59	Project manager
61	Contract negotiations with prime vendor	9	5/20/97 8:00	5/30/97 5:00		Project manager, legal team
62						
63	Communications H/W and S/W changes	273	10/15/96 8:00	10/30/97 5:00		
64	Requirements/analysis	24	10/15/96 8:00	11/15/96 5:00		Comm analyst
65	Mainframe communications hardware/software	20	10/15/96 8:00	11/11/96 5:00		
66	NETWARE LAN software changes	24	10/15/96 8:00	11/15/96 5:00		LAN analyst
67						
68	Detail design	65	11/16/96 8:00	2/14/97 5:00		
69	Mainframe comm. software—preliminary design	20	11/16/96 8:00	12/15/96 5:00		Mainframe comm analyst
70	Mainframe comm. S/W—prelim. design review	1	12/20/96 8:00	12/20/96 5:00		Comm analyst
71	Mainframe comm. S/W—detailed design	29	12/21/96 8:00	1/30/97 5:00		Comm analyst
72	"Mobile-aware" design review	10	2/1/97 8:00	2/14/97 5:00		
73						
74	Code/unit test	96	2/15/97 8:00	6/30/97 5:00		
75	Mainframe comm. S/W code	54	2/15/97 8:00	5/1/97 5:00		Comm analyst
76	Sales automation client appl. code	75	2/15/97 8:00	6/1/97 5:00		
77	Application walk-through	13	4/15/97 8:00	5/1/97 5:00		Appl dev team, users

Note: This project plan lists only some of the tasks in a mobile computing project. Dates and task dependencies are purely arbitrary.

TABLE 17.1 Sample Mobile Computing Project Plan Version 1 (*Continued*)

ID	Name	Duration (days)	Scheduled start (month/day/year, A.M.)	Scheduled finish (month/day/year, P.M.)	Predecessors	Resource names
78	Application test plan	9	5/5/97 8:00	5/15/97 5:00		Test team, appl dev analyst
79	Client application unit test	11	6/15/97 8:00	6/30/97 5:00	75	Appl dev analyst
80	Mainframe comm. S/W unit test	21	6/1/97 8:00	6/30/97 5:00	76	
81						
82	Integration test with wireless and MCSS	65	8/1/97 8:00	10/30/97 5:00		
83	Create end-to-end test environment	22	8/1/97 8:00	9/1/97 5:00		Vendor
84	Mainframe system definition (SYSGEN)	11	8/15/97 8:00	8/29/97 5:00		
85	SFA client appl. testing	10	9/2/97 8:00	9/15/97 5:00	79	Appl dev team, users
86	Mainframe server testing	10	9/16/97 8:00	9/29/97 5:00	80	
87	End-to-end application/server testing	22	10/1/97 8:00	10/30/97 5:00		
88	Sign off by systems assurance group	33	8/15/97 8:00	9/30/97 5:00		
89	Review of hardware/software by SA team	11	8/15/97 8:00	8/31/97 5:00		SA team
90	Acceptance by SA group	1	9/30/97 8:00	9/30/97 5:00		SA team
91						
92	TCP/IP-based comm middleware	99	4/1/97 8:00	8/15/97 5:00		
93	Analysis	11	4/1/97 8:00	4/15/97 5:00		
94	Selection of middleware product	22	4/16/97 8:00	5/15/97 5:00	93	Comm analyst, dev analyst
95	Design of appl. interface	21	5/16/97 8:00	6/15/97 5:00	94	Comm analyst, dev analyst
96	Code	22	6/16/97 8:00	7/15/97 5:00	95	Comm analyst, dev analyst
97	Unit test	23.74	7/16/97 8:00	7/30/97 5:00	96	Comm analyst, dev analyst
98	Integrate application with middleware	10	8/1/97 8:00	8/15/97 5:00		Comm analyst, dev analyst
99						
100	"Mobile-aware" sales automation application	356	11/1/96 8:00	3/13/98 5:00		
101	Requirements	96	11/1/96 8:00	3/16/97 5:00		
102	Requirements gathering	61	11/1/96 8:00	1/24/97 5:00		Appl analysts
103	Requirements analysis	21	1/25/97 8:00	2/24/97 5:00		Appl analysts

ID	Task name	Duration	Start	Finish	Pred.	Resources
104	Requirements document creation	10	3/1/97 8:00	3/15/97 5:00		Appl analysts
105	Review requirements document	0	3/16/97 8:00	3/16/97 5:00		Appl analysts
106						
107	Analysis	30	2/1/97 8:00	3/15/97 5:00		Appl analysts
108	Preliminary analysis and design	20	2/1/97 8:00	2/28/97 5:00		Appl analysts
109	Preliminary analysis and design doc.	5	3/1/97 8:00	3/7/97 5:00		Appl analysts
110	Review requirements and design/process doc.	0	3/8/97 8:00	3/8/97 5:00		Appl analysts
111	Reengineered process document	10	3/1/97 8:00	3/15/97 5:00		Appl analysts
112						
113	Application prototype development	25	3/16/97 8:00	4/20/97 5:00		Appl analysts, users
114	Prototype Ver 1.0 development	10	3/16/97 8:00	3/30/97 5:00		Appl analysts, users
115	Review prototype V. 1.0 with user	4	4/1/97 8:00	4/5/97 5:00		Appl analysts, users
116	Prototype V. 2.0 development	7	4/6/97 8:00	4/15/97 5:00		Appl analysts, users
117	Review V. 2.0 prototype	3	4/16/97 8:00	4/20/97 5:00		Appl analysts, users
118						
119	Detail design	50	4/1/97 8:00	6/9/97 5:00		Appl analysts
120	Sales force automation application	33	4/1/97 8:00	5/15/97 5:00		Appl analysts
121	Order entry	23	4/15/97 8:00	5/15/97 5:00		Appl analysts
122	Inventory inquiry	23	4/15/97 8:00	5/15/97 5:00		Appl analysts
123	Integration with e-mail and communication	11	5/1/97 8:00	5/15/97 5:00		Appl analysts
124	Detail design document—draft	12	5/15/97 8:00	5/30/97 5:00		Appl analysts
125	Detailed design document—final	5	6/2/97 8:00	6/7/97 5:00		Appl analysts
126	Detailed design and code walk-thru—final review	1	6/9/97 8:00	6/9/97 5:00		Appl analysts
127						
128	Development—code/unit test	54	5/16/97 8:00	7/30/97 5:00	120	Appl analysts
129	Sales force automation application	32	5/16/97 8:00	6/30/97 5:00	121	Appl analysts
130	Order entry	32	5/16/97 8:00	6/30/97 5:00	122	Appl analysts
131	Inventory inquiry	32	5/16/97 8:00	6/30/97 5:00	123	Appl analysts
132	Integration with e-mail	30	5/16/97 8:00	6/26/97 5:00		Appl analysts
133	Develop a test plan	5	7/1/97 8:00	7/7/97 5:00		Appl analysts
134	Test plan review	1	7/8/97 8:00	7/8/97 5:00		Appl analysts
135	Unit test—do it	15	7/10/97 8:00	7/30/97 5:00	133	Appl analysts
136						
137	Sales force automation test with wireless	32	8/1/97 8:00	9/15/97 5:00		
138	Sales force automation appl wireless test	10	8/1/97 8:00	8/15/97 5:00		Appl analysts

Note: This project plan lists only some of the tasks in a mobile computing project. Dates and task dependencies are purely arbitrary.

TABLE 17.1 Sample Mobile Computing Project Plan Version 1 (Continued)

ID	Name	Duration (days)	Scheduled start (month/day/year, A.M.)	Scheduled finish (month/day/year, P.M.)	Predecessors	Resource names
139	Concurrent application access test	10	8/18/97 8:00	8/30/97 5:00		Appl analysts
140	Test by dev staff in regional office	11	9/1/97 8:00	9/15/97 5:00		Appl analysts
141						
142	User acceptance test	67	9/18/97 8:00	12/20/97 5:00		
143	Create user test scenarios and test data	4	9/18/97 8:00	9/23/97 5:00		Users
144	Testing in car with wireless and MCSS	5	9/24/97 8:00	9/30/97 5:00		Users
145	User test	22	10/1/97 8:00	10/30/97 5:00		Users
146	User signoff—go ahead for implementation	10	11/1/97 8:00	11/14/97 5:00		Users
147						
148	Documentation	35	11/1/97 8:00	12/20/97 5:00		Technical writer
149	Overview	10	11/1/97 8:00	11/15/97 5:00		Technical writer
150	User guide	25	11/15/97 8:00	12/20/97 5:00		Technical writer
151						
152	Policy and procedures	20	11/1/97 8:00	11/28/97 5:00		User rep
153	Develop new SFA procedures	20	11/1/97 8:00	11/28/97 5:00		
154						
155	User training	85	6/1/97 8:00	9/26/97 5:00		
156	Prepare detailed plan	5	6/1/97 8:00	6/6/97 5:00		
157	Prepare master course	20	7/1/97 8:00	7/28/97 5:00		
158	Schedule training for users	5	8/1/97 8:00	8/7/97 5:00		
159	Conduct training	20	9/1/97 8:00	9/26/97 5:00		
160						
161	SFA implementation in field—1st phase	75	12/1/97 8:00	3/13/98 5:00		
162	Northeast region	20	12/1/97 8:00	12/26/97 5:00		Field install team
163	Far east region	15	1/2/98 8:00	1/22/98 5:00		Field install team
164	New York area	30	2/1/98 8:00	3/13/98 5:00		Field install team
165						
166	Mobile computing infrastructure planning	86	5/1/97 8:00	8/28/97 5:00		
167	Wireless network coverage testing	20	5/1/97 8:00	5/28/97 5:00		Vendor
168	Wireless network design	20	6/1/97 8:00	6/27/97 5:00		Comm analyst
169	Mobile application network traffic and capacity planning	10	6/1/97 8:00	6/13/97 5:00		Project architect
170	Order WAN circuits from MCSS to mainframe	5	7/1/97 8:00	7/7/97 5:00		Project manager

#	Task	Duration	Start	Finish	Resource
171	Install and test above circuits	5	7/15/97 8:00	7/21/97 5:00	Bell
172	MCSS installation and testing	15	7/15/97 8:00	8/4/97 5:00	Vendor
173	Install, configure, and test MCSS software	10	8/15/97 8:00	8/28/97 5:00	Vendor
174					
175	Mobile computing vendor installation plan	99	9/1/97 8:00	1/15/98 5:00	Vendor
176	Vendor deliverable one—Northeast region	55	9/1/97 8:00	11/14/97 5:00	Vendor
177	Install radio modems in mobile workstations	10	11/1/97 8:00	11/14/97 5:00	Vendor
178	Install and test wireless network upgrade	20	9/1/97 8:00	9/26/97 5:00	Vendor
179	Test network coverage	10	10/1/97 8:00	10/14/97 5:00	Vendor
180	Install software and test application	10	11/1/97 8:00	11/14/97 5:00	Vendor
181	Vendor deliverable two—far east region	53	10/1/97 8:00	12/12/97 5:00	Vendor
182	Install radio modems in mobile workstations	10	12/1/97 8:00	12/12/97 5:00	Vendor
183	Install and test wireless network upgrade	20	10/1/97 8:00	10/28/97 5:00	Vendor
184	Test network coverage	10	11/1/97 8:00	11/14/97 5:00	Vendor
185	Install software and test application	10	12/1/97 8:00	12/12/97 5:00	Vendor
186	Vendor deliverable three—New York	54	11/1/97 8:00	1/15/98 5:00	Vendor
187	Install radio modems in mobile workstations	10	1/2/98 8:00	1/15/98 5:00	Vendor
188	Install and test wireless network upgrade	20	11/1/97 8:00	11/28/97 5:00	Vendor
189	Test network coverage	10	12/1/97 8:00	12/12/97 5:00	Vendor
190	Install software and test application	10	1/2/98 8:00	1/15/98 5:00	Vendor
191					
192	Vendor acceptance testing	64	3/1/98 8:00	5/28/98 5:00	Vendor, project team
193	Acceptance testing for deliverable one	10	3/1/98 8:00	3/13/98 5:00	Vendor, project team
194	Acceptance testing for deliverable two	10	3/16/98 8:00	3/27/98 5:00	Vendor, project team

Note: This project plan lists only some of the tasks in a mobile computing project. Dates and task dependencies are purely arbitrary.

TABLE 17.1 Sample Mobile Computing Project Plan Version 1 (*Continued*)

ID	Name	Duration (days)	Scheduled start (month/day/year, A.M.)	Scheduled finish (month/day/year, P.M.)	Predecessors	Resource names
195	Acceptance testing for deliverable three	10	4/1/98 8:00	4/14/98 5:00		Vendor, project team
196	System acceptance test (all vendor deliverables)	20	5/1/98 8:00	5/28/98 5:00		Vendor, project team
197						
198	Mobile computing project management	522	7/1/96 8:00	6/30/98 5:00		
199	Prepare preliminary plan—major tasks and resources	11	9/1/97 8:00	9/15/97 5:00		Project manager
200	Prepare detailed project plan	23	1/1/97 8:00	1/31/97 5:00	199	Project manager
201	Project monitoring	522	7/1/96 8:00	6/30/98 5:00		Project manager
202	Project status monthly meetings	522	7/1/96 8:00	6/30/98 5:00		Project manager, project director
203						
204	Mobile computing operational management	87	3/1/98 8:00	6/30/98 5:00		
205	Network operations organization	10	4/1/98 8:00	4/14/98 5:00		Project manager
206	Network management implementation, e.g., SNMP	10	3/1/98 8:00	3/13/98 5:00		Project team, network manager
207	Network management procedure development	20	6/3/98 8:00	6/30/98 5:00		Comm analyst
208	Integration with central network procedures	10	5/1/98 8:00	5/14/98 5:00		Comm analyst, central network manager
209	Training of network operations staff by project staff	10	6/1/98 8:00	6/12/98 5:00		Project team
210						
211	Mobile computing project completion	0	6/1/98 8:00	6/1/98 8:00 A.M.		
212	Mobile computing—first phase operational	0	6/1/98 8:00	6/1/98 8:00 A.M.		

Note: This project plan lists only some of the tasks in a mobile computing project. Dates and task dependencies are purely arbitrary.

18

Operational Management of Mobile Computing Networks

We are constantly amazed by emerging technologies. We compliment profusely those who implement them because they are the pioneers. However, it is not until the new technology is operating successfully that the pioneering organization actually begins to reap any rewards. If those who keep the wheels running did not operate these systems successfully, we would always be in the red.
CHANDER DHAWAN

About This Chapter

In this chapter, we shall discuss mobile computing network management requirements and tools. Since many of the tools in question are still under development, we shall concentrate our discussion on operational management requirements and the work of standards-setting bodies such as the Mobile Management Task Force (MMTF).

18.1   Mobile Computing Operational Management Requirements

In that objectives and systems administration tasks are essentially the same, the operational management requirements of a mobile computing application are in many respects no different from any other IT application. However, when it comes to network management techniques, mobile users and wireless networks do have certain unique requirements.

Apart from anything else, the use of different technologies (some of them still in a nascent stage) and connections that are temporary and often tenuous makes mobile computing networks intrinsically more complex, on the one hand, and less robust, on the other. Constantly mobile users are often not easy to locate or contact, their cellular connections busy transmitting data. Additionally, security is a far bigger problem with mobile-user devices than with fixed desktop PCs.

Unfortunately, to make matters worse, mobile workers are also among the most demanding of users and expect everything to work properly 100 percent of the time—especially in front of customers or at night in hotel rooms when there is only minimal (if any) support staff on duty. To top it all, this expectation is often exacerbated by a tendency on the part of these same workers to tinker with hardware and software configurations themselves.

There are several tasks that operational staff have to perform to keep a system operating smoothly, including:

- Network problem management
- Network asset management
- Network change management
- Network software upgrades
- Network performance monitoring
- Application status monitoring

18.1.1 Network problem management

In large organizations, network managers are currently preoccupied with multivendor internetworking problems, and with the integration of different network management schemes (increasingly based on the SNMP standard) into single enterprisewide solutions. Now, in the midst of all this preoccupation, users are starting to demand Remote Network Access and mobile computing network management. These demands can quickly—and only too often do, judging by the experiences of early implementers—become major stumbling blocks for network managers as they struggle to cope with emerging technologies and vendors who concentrate their own efforts on providing functionality first and standards-based network management second. (Or third, or fourth!) In the absence of generally accepted industry standards, vendors tend to provide only a limited amount of proprietary implementations of network management capabilities.

Problem management deals not only with recording problems, but also with tracing the recorded problems to specific hardware or software failures. Table 18.1 describes a typical set of problem-resolution requests that a hot line receives from the mobile workers.

In order to diagnose the cause of the problem, the network help desk staff members need tools to monitor the status of various components. The first challenge they face is that they do not have monitoring capability or status information about the health of their wireless network.

TABLE 18.1   Typical Problems/Requests from Mobile Users

Component	Type of problem or change request
Mobile-device-related	Setting up new user IDs and passwords Installing new software drivers and software applications Recovering and downloading mistakenly erased data files Hardware failures Out-of-memory messages Virus alerts
Operating-software-related	Missing drivers
Application- or database-related	Application hangs Older versions of application software encountered Data files corrupted Problems running multiple applications
Communications-related	Communication sessions interrupted Failures during file transfer Slow network (poor response time) Weak coverage (poor signal strength) Network modem failures
Server-related	Server hangs Backup/recovery problems Databases down

In most cases, it is a public shared network. The second challenge to the network staff comes from the mobile workers' tendency to experiment with their machines more often on the road than in the office. Many enterprising users try to be self-reliant in computer networking, despite not having had the necessary training to make these changes on their own. Such well meant—but ill-advised—experimentation often leads to more problems, and ultimately to more calls for help.

18.1.2   Network asset management

How do you manage network assets when user locations are not known? How do you ascertain user configurations when the users themselves have not been seen for days or weeks? How do you upgrade user software when there is no way of ascertaining available disk space? Which management software do you use at the central site when there are thousands of mobile users? Is the software scalable? Is there a system administrator robot that constantly polls remote mobile users to retrieve stored status information, even if the MMTF's mobile management information base (MIB) is implemented across the enterprise? These are some of the open-ended questions that have to be answered if accurate information is to be kept about mobile users' computer configurations.

Whenever possible, the same asset-management software should be used as is used by the organization for the wired network, especially if it is SNMP-compatible. Once a mobile MIB is available for end-user devices, it can be included in the network asset-management software.

18.1.3 Network change management

Mobile computing networks are very dynamic, with new users being added every day and existing users constantly adding and changing hardware.

When new users become a part of the network, they must be assigned network addresses and given user IDs and initial passwords. As well, it is often necessary to create custom boot diskettes for new users containing information that cannot be sent over the airlink because of security policies.

When existing mobile users upgrade hardware or send hardware in for repair, additional demand is created for new user IDs and passwords for replacement hardware.

All this in turn results in a need to constantly update the configuration database.

18.1.4 Network performance monitoring

One of the most common complaints from mobile users is that wireless networks are slow compared to wired LANs, and that network performance is not consistent in different geographical areas, especially in public shared networks. Understandably, network managers prefer to be thoroughly conversant with network performance in a proactive manner prior to any onslaught of user complaints.

To be able to discuss problems with a network provider, a network performance-monitoring capability is needed so that network transit times can be identified in various components (see Fig. 15.1), starting with application processing times in the client, network, and information-server components.

As far as we can tell, no tools are currently available for measuring transit times in wireless networks, but sample transactions could be time-stamped and measurement data collected. This data should be used to prepare service-level reports on a regular basis. Unlike a private WAN where you can predict load characteristics more predictably, a shared network is more volatile and you may never know how many other users the network service provider has started servicing.

18.1.5 Software upgrade requirements

Users are constantly changing their operating systems software. Today they might be using Windows 3.11; tomorrow, Windows 95. As users upgrade operating system software, they need help in upgrading software drivers and OS-dependent configuration files. While self-loading diskettes or CD-ROMs can and should be issued for larger files, the fastest method of upgrading software is via dial-up transmissions from a central site (or via wireless networks in some cases, where you have very small changes) provided the upgrades are not extensive.

At the present juncture, the use of wireless communication bandwidth for upgrades should be restricted to transfers of application data. Online systems software upgrades should be made only via faster and more reliable dial-up links using V.34 modems.

18.2   The Mobile Help Desk

The center of action for all these activities is the customer service call center, otherwise known as the *help desk*. This is where all requests for help with problems, upgrades, and changes go. Help-desk staff have access to operational staff, database administrators, and network specialists—the second-line technical support staff.

It is the responsibility of the help desk to record all problems and monitor them until they are satisfactorily resolved.

18.2.1   Do we have mobile computing management solutions?

There is a patchwork of solutions that address specific problem areas, but no single tool set capable of effectively addressing all of mobile computing's many unique requirements. A few vendors are creating a standard so that the industry can build standards-based solutions, chief among them being the MMTF's SNMP-based MIB, which we shall mention along with a few other tools in Sec. 18.3.

18.3   Mobile Network Management Tools

We shall describe the following initiatives, products, and approaches to the development of standards and tools for mobile network management.

- The MMTF initiative
- Remote-control software
- Symbol's Spectrum24 Netvision
- Plug-and-Play with Windows 95
- Individual hardware vendor approaches
- Individual software vendor approaches
- The MCSS as a network management hub

18.3.1 The MMTF initiative*

Recognizing mobile computing's unique network management problems, several vendors, headed by Epilogue Technology and Xircom, have joined together to create the Mobile Management Task Force. Epilogue Technology is a pioneer in providing SNMP-based network management development software to hardware, software, and chip vendors. Xircom is a leading vendor of PC Card–based solutions for wireless LANs and RNA. Other 1995 members were Compaq, Fujitsu, IBM, LanAire, Motorola, National Semiconductor, and Zenith Data Systems.

The basic terms and concepts of SNMP. While we do not intend to enter into a detailed discussion of SNMP in this book, we will nevertheless explain a few terms and basic concepts. For the source of a more detailed insight into the topic, refer to *Network Management Standards* by Uyless Black (McGraw-Hill, 1994).

SNMP is the most popular network management protocol for multivendor environments. Just as TCP/IP has been adopted as a de facto transmission protocol standard, SNMP has assumed a similar role in network management. SNMP architecture was designed with simplicity in mind to support the remote management of network resources to the fullest, independently of host computers or gateways.

SNMP has three components: the network element, the network management (NM) station, and an SNMP agent. The *network element* is the resource that is being managed (mobile computer, PDA, information server, etc.). The *NM station* is where management control information is requested and received. The *SNMP agent* is the software that collects control information on behalf of the NM station from the network entity.

The agent may in fact be a proxy agent in situations where the network element cannot be reached by conventional management protocols (low function devices such as modems and bridges, for example, cannot interact with the NM station). Proxy agents can serve many useful features: (1) the managed entity does not have to concern itself with network management; (2) it can perform protocol conversions; and (3) it can provide security.

Management control information is described and stored in a MIB, which resides in the managed entity. The SNMP protocol defines the way an NM station accesses information in MIBs using one of the three methods shown in Fig. 18.1.

An SNMP manager can be programmed to send periodic polling messages to the managed devices, the intervals established through an SNMP MIB. This facility is important because of the need with wire-

---

* Information available on Xircom and Epilogue World Wide Web home pages.

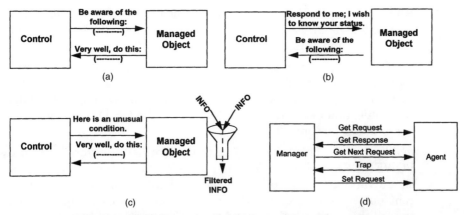

Figure 18.1 SNMP operations: (*a*) SNMP conventional interrupt; (*b*) SNMP conventional polling; (*c*) SNMP trap; (*d*) SNMP protocol data units (PDUs). (*Source:* NM Standards *by Uyless Black.*)

less networks to keep management traffic to an absolute minimum. Managed objects can send interrupts under certain conditions as well. Interrupts are not preferred by designers, however, because of the unpredictability of traffic.

Finally, there are SNMP traps, based on restricted parameters. A *trap* is an interrupt handled by a filter that conserves precious network bandwidth by reporting back only if a certain threshold value has been reached, or a combination of different parameters reaches a specified value (when message queues and the number of transmission errors combined exceed a threshold value, for example).

In the past three years, equipment vendors have made significant progress in implementing the first version (v1) of the SNMP standard, as well as its enhancements (v2), in a few cases. Conformance to the SNMP standard is now a common requirement in networking RFPs.

The Epilogue/Xircom mobile management solution. Xircom has developed the initial structure for a mobile MIB extension to the current SNMP v2 specification. Comments and contributions from industry vendors, network administrators, and end users are being solicited. These will be refined and submitted as part of a draft MIB document. Epilogue Technology will publish an informational request for comment (RFC) and will present the draft MIB to the Internet Engineering Task Force (IETF) for consideration.

Proposed mobile MIB extension. The following is the initial structure for the mobile MIB extension being proposed by Xircom and Epilogue:

mobilePlatformInformation. This is an object that identifies the mobile platform being used, including the manufacturer and model,

BIOS date, operating system (including vendor and version number), and PCMCIA client driver identification data (including driver ID, vendor, and version number).

mobileLocale.   Although it is not clear where the information will come from, this object is designed to give the network administrator local access information, specifically the country, city, local time, time zone, and in the case of dial-up users, the local telephone number of a dial-up connection.

mobilePower.   This is an object that identifies the version of the POWER.EXE driver resident on the user's computer, the version of the Advanced Power Management (APM) ROM BIOS, and the current power-saving configuration (none, low, medium, high, or the time of the last critical-suspend event). Traps related to power management could include SUSPEND and RESUME.

mobileResourceAllocation.   This object includes information about memory, IO, IRQ, and related parameters used by mobile devices, and could be expanded to track the availability of various resources, such as adapters, slots, bus extensions, and ports.

mobileCardServices.   This is an object that identifies the version of Card Services.

mobileSocketServices.   This is an object that identifies the version of Socket Services.

mobileBattery.   This is an integer that would determine whether a device is running on AC or battery, and, if using battery, determine the number of minutes of battery life remaining. It is set to 0 if unknown.

mobileConnectionMedia.   This is an integer that identifies the connection media being used: 0 = unknown; 1 = serial port; 2 = Ethernet; 3 = Token Ring; 4 = Wireless; 5 = 100 MBit.

mobileConnectionSpeed.   This is an object that identifies the connection speed of the media being used (9600, 38,400, 10 MB, etc.).

mobileConnectionMechanism.   This is an object that identifies the connection mechanism being used. (Examples include pcAnywhere, CarbonCopy, and NetWareConnect.)

Other traps being considered for inclusion in the Mobile MIB extension include insert/extract PCMCIA devices and docking/undocking.

Types of MIBs for mobile computing.   MMTF group has defined four MIBs that define the objects for mobile computing. The *system MIB* provides information about the Mobile software drivers, card and socket services, power status, and related system conditions. The *adapter MIB* gives information about the network adapter. The *link MIB* provides information about the Mobile network link to help network managers troubleshoot and optimize the connection. The *extended system MIB* contains information about system extensions. This is shown graphically in Fig. 18.2.

Table 18.2 shows three scenarios where Mobile MIB may be used.

Progress with implementation of Mobile MIB. Xircom is shipping SNMP Mobile MIB with the network management software included in its PCMCIA network adapters. Epilogue has incorporated the MIB as an extension of its Emissary MIB compiler software. Shiva Corporation has also announced plans to support Mobile MIB.

18.3.2 Remote-control software

The following remote-control software is often used for problem diagnosis:

- CarbonCopy
- NetModem from Shiva Corporation
- pcAnywhere
- ReachOut from Ocean Isle Software, Vero Beach, Florida
- IBM's Netfinity 3.0

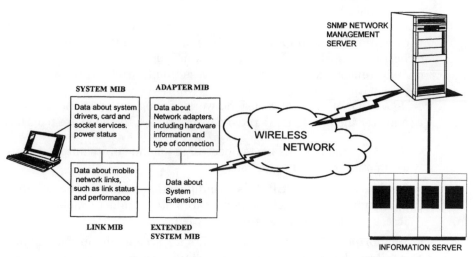

Figure 18.2 MMTF's mobile MIBs. (*Source:* Communications Week, *Jan. 1996 and MMTF page on Internet.*)

TABLE 18.2   Scenarios Where Mobile MIB Could Be Used

Problem scenario	Mobile MIB solution
Company wants to standardize on one brand of PC (PCMCIA) cards because older nonstandard card and socket services cause intermittent problems.	With the support of a specialized Mobile MIB extension residing on mobile users' laptops, the network manager can identify and upgrade the CS/SS to a newer version known to fix the problem.
Different users connect to the network using many different methods, making it difficult to standardize on one method.	By monitoring connection methods, the network manager can identify which is the most popular.
Company's auditors want an inventory of laptops.	SNMP provides a means of tracking and controlling capital equipment.

There are other utilities, such as Novell's Netware utilities, Frye's utilities for Netware, and Intel's management utilities that enable the diagnosis of notebook-related problems.

Of course, problem management and software upgrades take on a whole new dimension if the software has to be upgraded in hundreds and thousands of notebooks, and there is no way of scheduling the transfer of data from a central site. The most pragmatic solution is to have users pick up upgrade diskettes from the nearest convenient location. This solution works only if the server software can accommodate multiple versions of client software.

18.3.3   Symbol Technology's NetVISION Tools

Symbol Technology's NetVISION Wireless LAN Management Tools simplify the configuration and management of a wireless LAN. This easy-to-install program immediately initializes and configures the Spectrum24 (Symbol's wireless LAN) radio networking environment with minimal user intervention. Once installed on the network server or host system, NetVISION uses autoconfigured defaults to boot Spectrum24 remote nodes, including access points, radio terminals, and devices with wireless LAN adapters.

NetVISION's GUI provides network managers with the tools to administer the Spectrum24 radio network nodes. Included are tools for:

- *Booting:*   NetVISION boots Spectrum24 mobile units automatically with no intervention required.

- *Configuring:*   NetVISION is a configuration manager. It configures Spectrum24's range of network parameters, remote mobile units, access points, and access to security features.

- *Monitoring:*   NetVISION sets limits for acceptable operations and monitors the network for noncompliance, issuing exception reports for departures.

- *Diagnostics:* NetVISION tests the status and operation of network nodes. Run-time system errors are monitored via NetVISION diagnostics and are logged into an error file located on the host machine.

18.3.4 Plug-and-Play with Windows 95

Notebook manufacturers have started using the Plug-and-Play technology in Windows 95—along with the widespread integration of the PCI bus standard in docking stations—to provide IT managers with help in keeping track of laptops used on LANs. Unfortunately, this still does not address the wider problem of mobile users who do not visit their offices for days and weeks. However, it is expected that the PCI bus standard will soon be implemented in notebooks themselves. Central network management systems will then be able to retrieve information directly from mobile computers.

18.3.5 Individual hardware vendor approaches

MMTF's MIB implementation by the industry at large will take several years. Meanwhile, a few vendors are providing their own utilities to provide partial answers to the bigger problem of overall network management for mobile computing. Intel is providing management utilities to provide system information remotely. Similarly, Compaq includes intelligent manageability utilities in some of its notebooks, notably the Contura and the LTE Elite.

18.3.6 Individual software vendor approaches

In our limited research, we found that Xcellnet's RemoteWare product offers good practical solutions to some of the network management issues that we have discussed in this chapter, including:

- Reasonably effective but inexpensive security control where full-fledged encryption is not warranted
- Automated data/file distribution and synchronization in case of failures
- Store-and-forward transmission of large files
- Comprehensive history and control of changes
- Remote control, with system activity alerts (e.g., software updates)
- Application-configuration management

If the mobile solution is based on RNA using PSTN or ISDN only, Xcellnet's RemoteWare software bundled with Shiva's hardware platform could be a very effective combination from functionality and network management perspectives. Shiva has already announced its intent to implement MMTF's Mobile MIB in future products.

18.3.7 An MCSS as a network management hub

We have focused on remote devices so far. We also need network management information about the wireless network, the physical and logical network connections, the TCP/IP sockets, and the logical flow of data in the network. One possible source of this information is the MCSS hardware.

TEKnique's TX5000 wireless network switch makes this type of network management information available—either locally to the network staff, or remotely through a dial-up connection to the switch supplier's support staff, who can diagnose problems and apply software patches, or upgrade complete software modules through an asynchronous connection to the switch. We would like to see all switch suppliers provide an SNMP v2 MIB that can be integrated with enterprise network management software, such as IBM's NETVIEW or HP's OPENVIEW. Network planners should ask for this type of capability in their requests for proposals. At the time of print, IBM had announced such a capability with ARTour MCSS product to be available in late 1996. Graphical information on signal strength and status of sessions can be retrieved by the network staff at the central site through NETVIEW.

18.4   The Integrated Network Management of a Mobile Infrastructure

Mobile computing solutions consist of many components, and individual component manufacturers provide network management information about their products. There is not a single vendor that we know of, however, who provides an integrated, systemwide view of the various components in a mobile computing solution. The closest anyone ever comes is the occasional systems integrator who builds a limited capability for a specific customer.

We propose that network managers ask for this integration of information at the central site. We believe that the centralized network management of an enterprise is the most pragmatic approach for two reasons. First, very high skills are required for mobile computing network management—skills that are also very expensive to acquire. Secondly, for logical network management (application status), the necessary information is likely to be available only at the central site.

With present-day call-center telecommunications networks increasingly being centralized, or even outsourced, we can learn from the experiences of the airline industry. In the case of airline reservations systems, it does not matter where a call is physically serviced—the call-center representatives know the status of the network elements wherever they are.

Of course, simple asset management or server-status information can be provided by regional network administrators, if such individuals are available at these sites.

We make the following suggestions for the enterprisewide network management of mobile computing systems:

- All network elements should have SNMP v2 compatible MIBs.

- For mobile devices, MMTF's Mobile MIB should be requested, if it is available. If it is not available, the vendor should be asked to indicate its intentions regarding a future implementation.

- Wireless network providers should be asked to provide basic network management information to the central network management staff through a logical port into their network management system.

- The remote network diagnostic capability of the MCSS should be available.

- There should be standardization on remote diagnostic software such as CarbonCopy, pcAnywhere, or any other similar package selected by the organization for the RNA solution (see Chap. 10).

- For the sake of overall integration, we should use whatever network management solution the enterprise has chosen as a standard. In most cases, a new standard for mobile computing is not warranted.

18.5   Wireless Network Management Lags Behind RNA

Elemental, nonintegrated, and proprietary, network management in the wireless network industry appears to be evolving very much as an afterthought, lagging behind the RNA industry, which has made substantially more progress with network management software. There are two main reasons for this. First, network management is more easily addressed in the RNA industry. Second, there are not many vendors who are interested in a standards-based network management solution for the enterprise or in promoting and licensing a de facto standard.

We have started seeing some coordination and a movement toward standards through the National Portable Computing Professionals Association (NPCPA). However, only when there is concerted demand from all users for standards-based and integrated network management will real progress be forthcoming.

18.6   Policy and Procedures to Support
Technology-Based Network Management

There is no magic wand for network management. No amount of network management automation will solve problems created by a lack of standards and well-observed procedures. We strongly recommend, then, that network management tools be augmented by appropriate policies, procedures, and standards. The following suggestions are but a few of the many procedures that can be institutionalized in an organization:

- Standardize on a brand/model of notebook used on the network.
- Standardize on a base of permissible software configurations.
- Private software packages should be restricted and allowed only if they do not violate software-configuration standards imposed by a central organization.
- For a tightly controlled organization, consideration should be given to purging renegade software after checking hard disk directories.
- Insist on regular virus scans, backups, and software refresh procedures.

18.7   Who Should Be Responsible
for Mobile Computer Management?

Network management can be performed by either a centralized group, or it can be distributed at a departmental or regional level. This decision involves both technical and organizational issues. While centralized network management is more cost-effective, a hybrid solution may be more responsive from a users' perspective. Under such a scheme, responsibility for simple network management tasks can be given to department-level systems administrators, and more complex problems can be assigned to a skilled group in the central network organization. The user should see only a single problem-management interface with automatic and transparent switching of the call between the departmental and the central organization. The current support-organization structure should be considered as a model. In most cases, no new support organization is warranted for mobile computing.

Summary

In this chapter, we have examined network management tools and related issues for mobile computing. We stated that network management in mobile computing is in the early stages of development, with individual vendors creating proprietary and narrowly focused approaches. We discussed the MMTF's work on creating SNMP standards-based MIB. We should see a more unified, integrated, and standards-based approach in the years to come.

Chances are, however, that a single network management tool is still a long way off. Rather, it may be necessary to use a combination of tools from various vendors. For now, the more important network diagnosis, asset management, and standardization issues unique to mobile computing should be the focus of attention. For other issues, current network approaches should be used.

Reference

Black, Uyless, *Network Management Standards,* McGraw-Hill, New York, 1994. This is an excellent reference book for understanding SNMP, MIBs, and network management in general.

Mobile Computing Challenges, Opportunities, and Trends

19

The Challenges and Opportunities in Mobile Computing

In life, every change introduces a challenge as well as an opportunity. Mobile computing changes the paradigm of work in many ways—and therefore presents many challenges. To fully exploit the opportunities that mobile computing also offers, we must overcome these challenges and make it a mainstream technology.

CHANDER DHAWAN

About This Chapter

In Chap. 18, we discussed the management issues associated with the operation of mobile computing networks. With that discussion, we completed our coverage of the life cycle of a project: from conception to completion and operation. However, before we conclude this book, it will also be helpful to discuss the challenges that the industry faces. This discussion will serve two purposes: (1) it will encourage vendors and users to continue to look for long-term industrywide solutions; (2) it will assist users in developing strategies to overcome these obstacles while undertaking mobile computing projects.

Our focus in this book has been on user organizations' perspectives rather than on vendors' perspectives, and accordingly we have looked at some notable large-scale successes. UPS and Federal Express make excellent business-school and IT case studies. At the time of writing, Merrill Lynch's 1995–1996 $500 million information infrastructure rejuvenation program is centered around mobile computing applications. Sears' decision to equip 14,000 service vehicles is another example of the use of mobile computing to improve customer service—and a company's bottom line. And by discarding paper and pencil, the New York and Chicago stock exchanges are forcing reluctant traders to switch to portable computers connected through wireless networks.

These are but a few of the more visible examples of the ongoing implementation of mobile computing. Yet, despite these examples—and predictions from many distinguished visionaries and market research firms since 1993 of tremendous growth in the mobile computing industry—critics and wireless gurus like Bill Frezza still claim that mobile computing's time has not yet come. When such remarks are interpreted to mean that the time has not yet come for mobile computing to enter the mainstream, we have to agree, particularly in light of the acknowledged spotty and modest growth of the industry to date.

But at the same time we also strongly support the notion of a bright future role for mobile computing in our business and personal lives. Fundamental rationales for the adoption of mobile computing technology are strong. What is holding us back are technical and business challenges that the industry must resolve before mobile computing can really thrive.

We believe that the following 10 challenges have prevented the mobile computing revolution from becoming a reality in the time that many industry forecasters have predicted it would:

1. A lack of fast, reliable, and cost-effective wireless networks

2. An absence of standards-based mobile computing products and services

3. No universal communications interface for information servers

4. An absence of BPR strategies in mobile computing projects

5. The fact that few currently available business applications are mobile-aware

6. An absence of mobile-aware application development tools

7. A failure of resellers to provide a "cookie cutter" installation approach

8. The low priority of mobile computing projects in the total IT project mix

9. The complexity of systems integration

10. A lack of trained resources

19.1   Challenge 1: Lack of Fast, Reliable, and Cost-Effective Wireless Networks

Fixed LANs and high-speed WANs—with their substantially increased bandwidths—offer a reliable medium for the transmission of digital data within well-defined areas and fixed locations. The same is not true of wireless networks. Unfortunately, the wireless network industry did itself a great disservice early on by selling to unsuspecting users the idea that they could simply implement retrofitted LAN, minicomputer, or mainframe applications on wireless networks, before the infrastructure could deliver the speed, coverage, and reliability. While a few mobile computing applications are new, most are old applications with new communications drivers. And since user expectations have not been tempered by frank disclosure, users now expect to receive from these applications the same performance benefits on wireless networks as they have hitherto received on LANs or WANs, despite the fact that the coverage, bandwidth, and reliability are just not there to the same degree. The bottom line is that, unlike LANs and WANs, the smooth operation of applications cannot be taken for granted when the applications are migrated to wireless networks. Rather, it is essential that wireless applications be carefully chosen and matched with network capabilities.

The most obvious difference between mobile computing and fixed computing is the fact that mobile end users move around and expect coverage almost everywhere they go. The fact is, wireless networks just do not have the ubiquitous coverage that many mobile applications require. Presently, only packet radio networks such as ARDIS and RAM Mobile Data are national in scope with the promise of similar coverage in due course from CDPD carriers. While circuit-switched cellular networks have a better coverage, they are not suitable for transaction-oriented (OLTP) applications that require continuous connection and use many short messages.

As for speed, a major portion of the ARDIS network still runs at 4800 bps (albeit with upgrades to 19,200 RDLAP where economics can be justified). Mobitex operates at 8000 bps, with no plans as of yet to upgrade to the 16,000 bps of which it is capable. For a variety of reasons, users are rarely able to run circuit-switched cellular connections consistently at 14,400 bps.

CDPD carriers are promising a link speed of 19,200 bps, but CDPD has to overcome two major problems first: (1) presently it is available in metropolitan areas only; and (2) in CDPD networks, voice is given a higher priority over data. Thus no empirical real-life data exists that gives concerned systems engineers any confidence that response-time-sensitive mission-critical applications can be run successfully on CDPD networks without dedicated channels.

Presently, none of the wireless WANs can match the 28,800 bps speed of PSTN dial-up or 2×64 Kbps speed of ISDN connections—two types of links in the RNA industry. Even RNA speeds are much lower than LAN and WAN speeds of fixed computing.

The broadband PCS network industry may offer some relief in the future, but it will address voice requirements first. It will be at least two to three years before there is any engineering data available on which to base data application planning advice to users. The new ESMR networks will initially address dispatch applications in isolated vertical markets.

When it comes to reliability, wireless networks are generally more prone to error than fixed networks, with handoffs and temporary connection losses not at all uncommon. To make matters worse, currently available communications software is very sensitive to such errors.

More important, the cost of wireless network bandwidth is still very high—many times the cost of conventional fixed WAN bandwidths. As a rough rule of thumb, when conventional public shared (LAN/WAN) networks are used, the cost of network usage accounts for about 10 to 15 percent of total IT costs. When wireless networks are used, this percentage may run as high as 30 to 50 percent, assuming that data-flow traffic is optimized.

For all these reasons and more, it can be asserted with confidence that wireless networks will not become as ubiquitous, reliable, and inexpensive as their fixed network cousins anywhere in the near future.

Such is the engineering/economic state of the wireless network industry. A gloomy picture? Not really. Certainly not as gloomy as some would have us believe, though it does highlight the need to treat fixed and mobile networks differently from an application design and developmental perspective.

The industry is working hard to find solutions. In the meantime, we shall discuss in the second part of this chapter what we can do to use wireless networks effectively.

19.2   Challenge 2: An Absence of Standards-Based Mobile Computing Products and Services

In any emerging industry, vendors develop unique proprietary solutions initially. It is not until an industry leader creates a de facto standard, or users demand standards-based solutions that vendors start using standards. The mobile computing industry is no different in this respect. While there is an abundance of end-user mobile computing products available from many different vendors, wireless networks, communications software, and many other mobile computing components are only too frequently not interoperable.

In App. C, we have listed a selection of the standards that are emerging in the mobile computing area. Most notable are the wireless LAN IEEE 802.11 standard, the TCP/IP transport standard, the MMTF's SNMP-compatible MIB, and the cellular CDPD. However, adherence to these standards does not alone provide the interoperability that users want. We need far more pervasive standards, especially in the communications software and API areas.

Network-independent and platform-independent solutions should be designed at all three levels: client, agent, and server. In order to achieve such independence, a mobile-aware middleware standard is needed that provides true network independence for application development. There is no such standard currently available today, though National Portable Computer Professionals Association (NPCPA) and others have started working towards such a standard.

It is vital that middleware be mobile-aware (or, to put it another way, wireless-aware), because without such an awareness, network costs can skyrocket out of control. Middleware (and the transport stacks supported by middleware) should shield developers of mobile applications from the specific characteristics of wireless networks. Only then can they make applications so mobile-aware that they will perform well on slower, less reliable, and less ubiquitous wireless networks.

Without this awareness, we shall always expect technology to compensate for the natural tendency of application programmers to write code without worrying about wireless network traffic flow.

19.3   Challenge 3: No Universal Communications Interface for Information Servers

We have emphasized several times in this book the important role that the MCSS and its software play in mobile computing project infrastructures. We believe that the implementation of a universal MCSS on an open platform (RISC hardware under UNIX or Windows NT) with a well-supported middleware would be a major step in the establishment of a wireless infrastructure for mobile applications. Unfortunately, no large mobile computing vendor offers such an implementation presently. IBM's ARTour has the underpinnings of one, if multiplatform and third-party software developer support can be rallied by the vendor.

One of the reasons for this is that the market is small relative to the amount of software development involved. Oracle's Mobile Agents and Sybase's EMS implementation are strong candidates for the communications middleware segment of such a universal MCSS.

However, what we really need is the bundling of the functionality that RNA communications servers offer with mobile-aware middleware and software drivers for various wireless networks on a single open hardware platform. We hope that more users will ask network suppliers to create an open MCSS that will attract third-party application developers. If this does not happen, unique custom solutions will continue to prevail until a vendor sees a business opportunity in this type of MCSS.

19.4   Challenge 4: Absence of BPR Strategies in Mobile Computing Projects

One common element we have seen in successful mobile computing projects is BPR as an integral part of the automation of a mobile application. Far greater economic benefits are realized when mobile computing solutions replace old processes with new and more efficient ones. While this may appear to be a deceptively simple observation, the fact remains that BPR does indeed carry with it the potential to change the nature of a project very drastically.

Of course, the issue is not quite so simple from an organizational perspective. When BPR starts involving senior business and user management, a mobile computing project can take on a whole new complexity, organizationally as well as technically. Many IT professionals try to avoid this complexity and visibility. The users legitimately take over BPR projects, with mobile computing and other technologies becoming enabling tools. The fact is, however, that it is just such a combination of user and technical organizations that is essential to the success of a mobile computing project.

19.5   Challenge 5: Few Currently Available Business Applications Are Mobile-Aware

Because the mobile-aware concept is new to our industry, the current suite of LAN business applications are not suited to the needs of wireless networks; nor are legacy applications. In fact, the only applications that feature elements of mobile-aware design are certain public safety and dispatch applications built by companies such as Motorola and MDI (which was acquired by Motorola in the 1980s). However, those were built with 1980s technology. We are talking about mobile-aware design in the client-agent-server context.

Yet, vendors continue to paint rosy pictures that technology is simply not able to deliver and the industry continues to disappoint users with wireless applications that do not match the performance provided on LANs. Until we develop techniques and tools to make the current set of applications mobile-aware, there will always be a gap between promises and delivery.

One way of making applications mobile-aware is to use filtering or agent software. We talked about agents in Chap. 12. The concept of a filter or an agent involves an optimized data traffic flow designed specifically for wireless networks that is invoked as soon as an application finds that a client is mobile.

19.6   Challenge 6: Absence of Mobile-Aware Application Development Tools

Application development pundits propound middleware for transparent application development across the multiple platforms that are inevitable with mobile applications. While middleware might solve the problem for conventional networks, these tools must have a unique sensitivity to the limitations of wireless networks. Middleware vendors, however, are concentrating on the bigger problem of generic cross-platform application development. It will be a while before they can create unique implementations for mobile applications.

Application development purists may even question the wisdom of unique mobile computing applications, suggesting instead that wireless networks match the performance characteristics of wire-line networks. As has already been stated, this is not a realistic expectation in the near foreseeable future.

19.7   Challenge 7: Failure of Resellers to Provide a "Cookie Cutter" Installation Approach

During the 1980s, which can be called the decade of PCs and LANs, the IT industry created a massive second-tier vendor group called value-added resellers (VARs). These companies, assisted by manufacturers, break down complex implementation processes into simple tasks. It is this army of VARs that has installed several hundred thousand LANs during the past ten years. They use a "cookie cutter" approach that relies on simplified production techniques to create unique business solutions for specific vertical groups.

A similar approach is needed to bring mobile computing applications into the mainstream. Unfortunately, we have yet to arrive at the point where third-party developers and VARs take up in earnest the challenge of the mass production of mobile solutions, perhaps because the "cookie cutter" approach works only in the right circumstances.

In other circumstances, we can learn from the example of vendors such as Telxon and Norand, which are creating manufacturing, distribution, and warehousing turn-key solutions. It is when the process of creating solutions moves from custom design to a standard solution that VARs can take over from manufacturers and expand the market.

In due course, the industry will develop a group of VARs who know how to convert costly custom solutions to inexpensive mass-customized solutions for different vertical industries. In 1993, Wireless Telecom

Incorporated (WTI), a distributor based in Aurora, Colorado, started assembling a VAR group specializing in mobile computing. As this nucleus of VARs expands, we shall see more user organizations adopting mobile computing applications.

19.8   Challenge 8: Low Priority of Mobile Computing Projects in the Total IT Projects Mix

We need to remember that in any organization there are many competing projects for the same human and capital resources. There may be higher priority IT projects with a greater rate of return on IT investment than a mobile computing project. When a project involves complex technologies, trained resources tend to be scarce and the risks are higher, making it harder to sell these projects to senior management. Thus, it is important to build a sound business case and to recruit internal champions in support of mobile computing projects. A project's stature will also increase if mobile applications are associated with BPR that users understand.

19.9   Challenge 9: Complexity of Systems Integration

Providing two-way paging or an e-mail application may be relatively simple, on the one hand, whereas the systems integration of a large mobile computing project for enterprisewide deployment can be extremely complex, on the other. The number of components involved, the need to follow complete life-cycle application development processes, and the emerging nature of the technology all contribute to the complexity. Complexity also implies custom implementation by a few select systems integration companies that specialize in the technology, which in turn implies that a lack of competition in the field will lead to higher costs.

Such a business environment can act as a deterrent for small companies who want to implement mobile solutions, unless shrink-wrapped and easy-to-implement vertical application solutions exist. Industry and user organizations must take steps to reduce this complexity. (See also Chap. 14 and Sec. 19.7.)

19.10   Challenge 10: Lack of Trained Resources

There is no doubt that one of the factors that determines the priority of new IT projects is the availability (or otherwise) of skilled and trained resources, either within the organization or externally. We see a tremendous shortage of skilled and trained resources in user organizations that have not previously undertaken mobile computing projects. Those organizations that have implemented mobile solutions have had to acquire education and training the hard way—first time on the job while implementing their first projects.

Select universities such as Washington, Purdue, and Rutgers have initiated mobile computing courses, but generally speaking there are very few formal courses available. It seems training organizations are waiting for the critical mass to grow—which means we shall probably see more activity during 1996 and 1997 in this area. In the meantime, vendors continue to participate in mobile computing seminars and industry expositions in order to generate awareness. Education follows this awareness.

19.11 Converting Challenges into Opportunities

We have discussed the challenges holding the industry back. Now we can briefly mention the opportunities these challenges provide. Mobile computing will move forward as vendors and user organizations begin to exploit these opportunities.

Opportunity 1: Increase the speed of wireless networks, make them ubiquitous, and price them competitively with PSTN solutions. While users will certainly pay a slight premium for wireless networks, the current high prices can only act as a deterrent against the mass adoption of wireless applications. The industry must invest for the future. The infrastructure vendors may not be able to get quick payback on wireless data dollars.

Opportunity 2: Let vendors create standards-based solutions. User organizations should join forces to push vendors in this direction. The market will expand substantially once standards-based solutions are created.

Opportunity 3: Let software vendors create a universal communications interface for all networks, wired and wireless. TCP/IP appears to be the wireless universal transport protocol. But the current version is not optimized for wireless. We should push for mobile-aware TCP/IP that is more than the mobile IP that IETF is investigating. Let third-party software houses with wireless expertise (Oracle, TEKnique, Racotek, Nettech, etc.) build software drivers and optimized communications stacks that will work with industry-standard SDKs.

Let there be no specialty SDKs that end-user organizations are encouraged to use for wireless applications. Wireless SDKs should be intended for those who create high-level application development tools.

Opportunity 4: Let BPR be a part of every mobile computing application, wherever possible.

Opportunity 5: Build mobile-aware applications that take into account the limitations of wireless networks and the needs of mobile workers. Do not simply port current applications to wireless networks. They will not give you the performance that you expect.

Opportunity 6: Use development tools designed for mobile applications, distinguishing in the process between PSTN/ISDN and wireless WANs. Use middleware only if it is mobile-aware.

Opportunity 7: Minimize complexity by using standard solutions. Do not build custom solutions, if you can help it. Encourage systems integrators, specialized vendors, and VARs with prior experience in mobile applications.

Opportunity 8: Use mass customization techniques to implement solutions through a "cookie-cutter" approach. Use the same ingredients and change the shape of the "cutter" (configuration) as needed.

Opportunity 9: Build a sound business case, thereby raising the mobile computing project's status. Use professional management consulting organizations to develop specific business-case templates for mobile applications in different industries.

Opportunity 10: Developing training programs for mobile computing. Build development centers for third-party software houses and large Fortune 500 customers.

Summary

In this chapter, we discussed many of the challenges that face the mobile computing industry. The inherent limitations of wireless networks are a major obstacle for the industry to overcome. A lack of application development tools that recognize these limitations is another problem. We also mentioned that a lack of standards-based solutions is interfering with the widespread adoption of mobile solutions. The absence of industry standards forces users in the direction of custom solutions, which ultimately only add greater complexity to already complex mobile computing projects.

Only when we overcome these challenges will the mobile computing industry move forward and become mainstream.

Reference

Foreman, George and John Zahorjan, "Challenge of Mobile Computing," *IEEE Computer Magazine,* Apr. 1994.

20

Technology Trends Affecting Mobile Computing Progress

When evaluating options for emerging technologies, you must be long-sighted. Not only is it important to see what is out there now, but it is also important to consider what is on the horizon.

CHANDER DHAWAN

About This Chapter

Planning for emerging technologies is a risky exercise because many technologies introduced during the first phases of their evolution are eventually replaced by more durable ones. An entrepreneur develops a hot new idea and tries it on the market. The big players wait in the wings and observe its success or failure. If it is successful, they then either move in with their own variations of the same idea or buy the entrepreneur out.

In the early stages of a technology evolution, vendors play many business games. They float trial balloons and wait for user and trade press reactions. In an environment where there are no technology standards and durable technologies, it is wise to watch the trends carefully while planning for the future. In this final chapter, we shall discuss technology trends that are important to watch while making mobile computing plans.

The technology trends described in this chapter are based on our personal observations of where the mobile computing industry is headed. They represent our analysis of the industry, as well as a consolidation of views and opinions expressed by industry watchers, the trade press, and vendors.

No scientific survey techniques were employed to arrive at these conclusions. We have, however, applied a series of tests in order to include these trends in the book. The first test was to check for supporting evidence to a specific trend. The second test was to give more weight to user requirements in the shaping of technology, rather than to the vendor's view of what technology the user needs. The third test was to check for support of a trend by fundamental consumer and business needs. The tests offer a reasonable basis for an industry that is primarily market driven rather than vendor driven. We shall find out in a few years to what extent we were right.

20.1   Trend 1: More Network Options Now but Convergence in Future

According to the trade press, telecommunications carriers are making significant investments in wireless infrastructures. Overall investment in wireless networks, especially in CDPD and PCS networks, is in the tens of billions of dollars range in North America alone.

While voice is the primary target market for broadband PCS, data capabilities are inherent in a digital computer-switch-based network. Cellular carriers are not sitting idly by while PCS takes over this market segment. CDPD implementation is moving along at a sustained pace. According to AT&T Wireless, CDPD service is now available in most areas of its cellular coverage. We believe that traditional packet radio networks (ARDIS and RAM Mobile Data) will survive and will also continue to build additional capacity and provide more coverage. Each of these companies will offer more network options. RAM Mobile Data has already expressed its intent in this respect, as described in Chap. 14. The reality is that these packet data carriers have greater prior experience in creating complete solutions involving carrier services and communications software.

Current networks will grow and expand and more network types will become available. CDPD will give ARDIS and RAM Mobile Data a healthy run for their money. In Chap. 8, we discussed many different network options. Most of these networks will survive in the short- and medium-term (five to seven years) future.

Having more options will make the selection of the right network for a mobile application increasingly more important. What might not be an attractive, economic network option today could well become one tomorrow. It is important, then, to develop applications and select MCSSs with communications software that have the greatest potential now and in future.

20.2   Trend 2: The Transmission Speeds of Wireless Networks Will Increase

The current 4800 to 19,200 bps range of transmission speeds on wireless networks will increase during the next one to three years. ARDIS is already upgrading its network speed. Although RAM Mobile Data does not yet have similar plans, it will be forced to match ARDIS's performance advantage soon enough. CDPD will start at 19,200 bps, a speed we consider sufficient for most e-mail, mainframe database queries, and mobile-aware OLTP-based applications. Business needs of mission-critical applications will force CDPD carriers to offer a premium service with dedicated channels for data, with the frequency hopping technique over voice channels offered to their regular data customers. Data capabilities of PCS will take at least three years to come to initial fruition. PCS (under Q-CDMA implementation) will start in the 14.4 Kbps range but later deliver 76.8 Kbps.

We do not feel a justifiable business case can be made for bandwidth-intensive applications like wireless multimedia image transfer, except in a few cases where the critical nature of the requirement supersedes cost considerations (as in the case of news transmissions, for example). Rather, in the vast majority of cases, PSTN/ISDN/ADSL wire-line links at 28.8/128/1540 Kbps will remain the medium of choice for bulk data transfers.

Although wireless network speeds will continue to rise, the next quantum leap in their bandwidth will not take place for at least three to five years, when replacement of the current generation of base stations comes due. At that point, the infrastructure will have to be completely reengineered if ISDN-like speeds from mobile devices to base stations are to be supported. The biggest challenge will be the redesigning of radio modems in PCMCIA form factor and base station front-end components. Silicon and software technologies already exist for the back end of base stations and could, in fact, easily be borrowed from existing wire-line switches.

20.3   Trend 3: Network Prices Will Continue to Fall in the Short-Term Future

Increased network capacities, coupled with the slower rate at which applications will be implemented to take advantage of this additional capacity, will lead to lower network usage prices. CDPD prices will have to be more attractive than ARDIS and RAM Mobile Data's prices. Pressure will also come from narrowband PCS two-way paging networks contending for inexpensive messaging applications. Wireless network prices have decreased by 20 to 25 percent every year for the past three years. We expect this trend to continue. In another two to three years, the availability of broadband PCS will further fuel the downward spiral.

PSTN/ISDN-link prices will increase slightly, while still remaining lower by an order of magnitude—up to 50 to 100 times lower than cellular, proportionately speaking. (Compare 1 to 2 cents per minute for ISDN at 128 Kbps to 20 to 25 cents per minute for circuit-switched cellular at 14.4 to 28.8 Kbps.)

20.4   Trend 4: Global Coverage
for Wireless Networks

The wireless data industry is striving hard to match widespread global voice communication capabilities (which have been a reality for 30 to 40 years), and now, through roaming and satellite solutions such as IRIDIUM, wireless networks will indeed truly become global.

The reach of local wireless LANs and WANs will have substantially increased geographically by the year 2000 as wireless LANs become metropolitan/campus area networks (MANs/CANs) and regional wireless WANs interconnect to provide national coverage.

When full global coverage is achieved, vital data and information will be only as far away as a traveler's notebook computer. As globalization of business continues, there will be a need to stay in touch with your corporate information anywhere on the world.

20.5   Trend 5: RNA Industry Support
for Wireless Networks

Presently, the functionality of the RNA technology discussed in Chap. 10 is comparatively narrow in scope. RNA hardware/software solutions enable access to LAN applications and databases resident on LAN platforms—but the network links supported by most RNA vendors are limited to dial-in links through public switched network and ISDN. Very few RNA vendors support wireless network links, except circuit-switched cellular.

From a user's perspective, the same applications should be accessible whether via wireless or wire-line connections. We feel that it makes eminent technological and business sense to upgrade RNA communications servers to support wireless networks. Indeed, Shiva has already expressed its intent to support CDPD. When support for ARDIS, RAM Mobile Data, SMR, and PCS networks follows, RNA communications servers will be in a position to take on the role of an MCSS, as described in Chap. 11. An integration such as this of RNA and wireless solutions through a single MCSS would reduce costs, simplify solutions, and provide single points of contact for network management.

20.6   Trend 6: More Mobile-Aware
Middleware for Application Development

The mobile-aware design concept is an important one, as those who have tried to adapt existing applications to wireless networks will attest. We feel that special consideration for mobile applications should be limited to two areas only: the handling of wireless network traffic and the design of the user interface. One of the main attractions of Oracle's implementation of mobile-aware middleware, for example, is the ability it provides to use the same front-end application development tools as are used for nonmobile applications. However, the Oracle Mobile Agents solution should support different front-end application development tools such as PowerBuilder, Visual Basic, Delphi, and Visual Age. We encourage Oracle to do that.

We expect to see a few more vendors provide mobile-aware middleware similar to Oracle Agents. Racotek's KeyWare, for which Power-Builder can be used as a front-end development tool, is one such example.

20.7   Trend 7: Client-Agent-Server Will Become the Predominant Application Model

The client/server model has taken the IT industry by storm. If we ignore the inevitable trials and tribulations experienced by early client/server implementations, it appears that this model truly serves the needs of most modern applications.

However, the new client-agent-server model discussed in Chap. 12 is in fact a better model for mobile computing applications. The main advantage of this model is its ability to free the mobile user to carry on with other business activities. It recognizes the fact that with the wireless spectrum being as scarce as it is, there is no time for the chit-chat that normally goes on in a session-oriented application.

We feel, therefore, that client-agent-server will become the predominant model for mobile computing applications.

20.8   Trend 8: Speech Recognition as the Primary Form of Input for Mobile Applications

We have depended on the keyboard as an information input device for too long—almost since the typewriter was invented.

Pen-based computing, which has been around for several years now and is a viable alternative for many applications, nevertheless still has many obstacles to overcome.

In our opinion, speech recognition, which has come a long way during the past three years, is a far superior and more intuitive method of input for mobile applications. With the powerful microprocessors being produced today, it is only a matter of time (three to five years) before speech recognition becomes a primary input method for mobile computing.

20.9   Trend 9: Consolidation of End-User Devices

Today, many different devices from several vendors all jostle for position as the primary mobile applications device—and now IBM and

Oracle are floating the idea of a network-centric device for the Internet. We do not feel that Internet appliance is an ideal and full-function device for the mobile user, since we believe that such a user should be equipped with as much information as possible on a stand-alone basis. Nor do we expect Web servers to become mobile-aware in the near future, except to support wireless messaging.

Assuming that our clothing style does not change radically, mobile office workers need only two devices: one carried in a briefcase and the other carried in a pocket.

For the first category, a PC- or MAC-based notebook, with essential features varying only in form and style from model to model, is the undisputed candidate.

For the second, PDAs, PCAs, and PCS digital telephones will ultimately converge into a single device that will feature all the basic functions of a pager, a telephone, an e-mail receiver, an answering machine, a personal organizer, and an Internet appliance. It will not happen in 1996, but in two to three years after that.

20.10   Trend 10: Mobile End-User Devices Will Become More Powerful

The race toward ever more powerful and capable devices will continue, driven by higher profit margins in the notebook industry resulting from reduced competition and continuous innovation.

According to industry executives themselves, Pentium-based notebooks with 12-inch (30-centimeter) screens will become commonplace. Even larger 13-inch (33-centimeter) screens will follow in 1997–1998. Displays will have XGA and Super XGA resolution, and will be thinner as well. Hard disk storage will increase to 1.6 GB in 1996—with still more capacity to come in later years. Multimedia capabilities will be embedded in the motherboard. Despite slow progress, batteries, too, will eventually improve and be made to last longer.

PDAs will certainly improve in memory and speed as well. The struggle to find a standard operating system will not be a concern to users, because PDAs will be application specific. Microsoft's PEGASUS operating system will become an important environment for customized application development.

20.11   Trend 11: The Internet and Electronic Commerce Will Greatly Assist Mobile Computing

Spurred by the prospect of electronic commerce, several companies are already offering wireless connections to the Internet. It may not happen for a few years yet, but eventually it will be possible to make airline and hotel reservations from the car or theater bookings while socializing with friends. Once an electronic commerce infrastructure is fully implemented, the need to do such things only from fixed locations will be eliminated, thanks to the wonders of Magic Cap/Telescript and Java—and their inevitable successors.

20.12   Trend 12: Mobile Computing Will Become Mainstream in Two Years

Mobile computing is at last being seriously considered by all types of organizations. Although a few large projects have been executed and other pilot projects have been started, most organizations, however, are still in the research stages, studying the technology to find out if it is ready for them. They are also preoccupied with RNA solutions that are easier to implement. Thus, although the industry is almost ready for applications with a sound business case, it will be another two years yet before mobile computing applications become mainstream.

The Last Word

I recently attended a major seminar (held by DCI, Digital Consulting Incorporated) on mobile computing and Remote Network Access in the Silicon Valley. The seminar was appropriately called "Network Unplugged." I met several well-known independent and vendor experts in the field. I discussed some of the ideas contained in this book with them. I was pleasantly surprised that most of the experts agreed with the state of the industry described in this book. They also concurred with the need for an end-to-end integration approach to mobile computing. Therefore, I would like to summarize the key points made in this seminar and in the book you have just finished reading.

The time for planning mobile computing projects has come. The fundamentals for mobile computing solutions for business process reengineering are strong. Technology infrastructure for many business applications is either available now or will be available shortly. If the networks are not operational today, they will be by the time you are ready to roll out. Not every application can be justified on wireless networks with the current cost structure. However, if you do have a strong business case, you should start *planning* for the implementation now. You may go slow but you should start experimentation now. Remember that it is the early bird that catches the worms.

Learn from the experience of early implementers. We listed many vertical and a few horizontal applications in Chap. 2. While vertical applications in distribution, transportation, and service dispatch have been more successful than horizontal ones, the early implementers have already paid a price and are benefiting now. You should start with an application which is relatively easy, gives you the maximum rate of return, and provides a training ground for your people for more difficult applications.

Include business process reengineering in your mobile computing projects. Take this opportunity to reengineer your business processes for the rollout. Build the business case with the new processes but implement them in a phased fashion. Do not make the pilot project extremely complex unless senior management provides the resources and understands the risks in a complex project. Once potential benefits and possible risks are explained to management, they will support you even when you face problems along the way.

Develop a sound technology architecture for mobile computing. You must develop a long-term technology architecture for mobile computing if it is an important part of your future IT strategy. Create this architecture in the context of your nonmobile application, data, and technology architectures. Develop a migration plan from where you are and where you want to get to.

Select technology components carefully because confusion will abound with network and software product choices in the next few years. Many new mobile computing products have been introduced recently. Many more will be introduced in the next couple of years. Be careful in selecting the products for your solution, because many of them will become obsolete or the vendors will go out of business as a result of inevitable rationalization in the marketplace.

Select open and standards-based solution from a vendor with mobile computing expertise and staying power. You must protect your application and infrastructure investments by selecting an open and standards-based solution from a vendor who has staying power and who will be able to provide migration tools for future networks.

Select an optimal MCSS. MCSS is the heart of the mobile computing solution. The success of your project may depend on this component.

Use wireless networks efficiently by careful application design. Wireless networks will always be slower by an order of magnitude, less reliable, and more expensive than terrestrial LANs and WANs. Such is the state of physics. You can not create more spectrum in the same geographical area by laying more fiber. To recognize these limitations, you must invest in making your applications mobile-aware.

Client-agent-server paradigm holds the best promise for mobile computing solutions. Invest time in understanding the client-agent-server application design paradigm. It is through the agent software that you can insulate your legacy applications from changes and make your applications mobile-aware. Building these changes in the information server will introduce more complexity and affect stability of your mission-critical operational applications—a risky exercise, especially when these applications might be undergoing redevelopment under client/server architecture.

Design your solution with future technology trends in mind. Keep track of technology trends in the mobile computing arena. Design your future solutions with these trends in mind. Allow for changing your future course of action as these trends come true or false and new trends take hold. Identify your biggest investments and ask yourself, "Will my solution still be optimal if the technology changes as some experts predict." Give all these trends, predictions, and promises a reality check.

Chander Dhawan
October 1996

Appendices

Mobile Computing Products and Services

Since it is not possible to cover all mobile computing products and services in a book of this size, we shall cover only a small subset to give you an idea of what is available in the market. In some cases, we have mentioned specific products. In other cases, we have mentioned the names of those vendors who supply a family of products in a given category. Moreover, the limited space does not allow us to describe the features and functions of the listed products in any detail. However, we have described some of the products in the body of the book to illustrate certain technical concepts.

Please note that this industry is evolving at a fast pace. Accordingly, the products are changing regularly. The industry will go through a rationalization process during the next few years. Therefore, you should contact the vendors or refer to trade magazines to get more current and additional information.

For the sake of simplicity, we have divided the products into the following categories:

- Infrastructure products, including modems

- End-user products, including PC (PCMCIA) Card peripherals

- MCSS products

- Software: operating systems and connectivity related

- Software: application development and integration related

- Miscellaneous software: horizontal applications, e.g., e-mail and groupware

- Software: application packages

- SFA
- Public safety
- Services

A.1   Infrastructure Products

Motorola, Ericsson, and Northern Telecom are three major suppliers of wireless network infrastructure products. Their products are used for installing private or public shared radio networks operated by the end-user organizations, carriers, and network service providers.

A.1.1   Ericsson BRU3901 base radio unit

BRU3901 is a new *mini* base station for Mobitex networks. Major features of BRU3901 include:

- Outdoor/indoor installation, small size (13 × 17 × 7 inches), low weight (40 pounds), approximately 5 kilometers coverage
- Simplicity of installation, with up to 3 years of operation without on-site maintenance
- Three-watt output
- Frequency range Rx 896 to 902 MHz, Tx 935 to 941 MHz
- Modified nonpersistent CSMA media access control
- One system channel, 1500 maximum users

A.1.2   Ericsson M2000 Mobidem

M2000 is a radio modem that can be built into PCs or other equipment. It has no power source of its own. It does not have its own antenna, since one must be designed specifically for the host equipment. It has rated data transfer rates of 1200 to 9600 bps. It supports Mobitex MACS, AT, and X.28 protocols. There are three versions of M2000:

M2050	425 to 460 MHz	United Kingdom
M2060	410 to 432 MHz	Rest of Europe
M2090	896 to 941 MHz	United States and Canada

A.1.3   Ericsson C710/M radio modem

- It is a radio and data modem combined into a single unit.
- It is designed to run on Mobitex networks.
- It can transmit data at 8000 bps.
- It uses the GMSK modulation technique.

A.1.4 Ericsson Mobidem M1090
portable wireless modem

Mobidem M1090 is an external wireless modem that provides wireless data communications connectivity for palmtop, notebook, and laptop computers as well as handheld terminals. Utilizing Mobitex networks, the Mobidem supports seamless, nationwide roaming. It is connected to the computer device through an RS-232 serial port. M1090 operates at 8000 bps. It weighs approximately one pound, including rechargeable battery pack, and its dimensions are $1.3 \times 2.68 \times 7.87$ inches.

A.1.5 Ericsson's PCMCIA versions
of wireless modems

There are several PCMCIA versions of the Mobidem modem from Ericsson (M2190), IBM, and RIM.

A.1.6 Motorola APCO (public safety) private
radio network solutions

Motorola provides a full suite of products, under the ASTRO family for APCO-25-compliant (see Chap. 2 for additional information on APCO 25) private radio network standards. The ASTRO product line is being upgraded during the 1996 to 1998 time period to provide increased functionality and compatibility to the standard. The product suite consists of portable units, base stations, repeaters, radio network controllers, gateways, and network management consoles. APCO common air interface is based on FDMA, 12.5-KHz (6.25-KHz digital, in the future) channel, 9600 signal data rate, QPSK-c modulation, and the IMBE Voice Coder.

A.1.7 Other APCO-25-compliant products

E.F. Johnson, Bendix King, Stanilite Electronics, and Transcrypt International are other vendors providing APCO-25 compliant products.

A.1.8 Motorola's conventional
and trunking products

Motorola supplies hardware and software for both conventional and trunking networks. These products serve a single site as well as wide area networks. Wide area solutions are based on simulcast as well as SmartZone technologies.

Simulcast broadcasts identical information on the same carrier frequency from multiple, geographically separated sites. Therefore, it is used to extend a system's coverage beyond a single site. It is analogous to the cell reuse concept implemented in cellular networks. *SmartZone* connects individual sites to form one large, seamless radio network. Motorola M6809 is a trunking central controller that coordinates between sites.

A.1.9   Motorola's wireless modems

Motorola supplies wireless modems for ARDIS, Mobitex, and other private radio networks. Both desktop and PC Card (PCMCIA) versions are available. ARDIS RDLAP radio modem operates at 19,200 bps.

A.1.10   Motorola's BitSurfer Pro ISDN modem

BitSurfer is a high-speed terminal adapter containing two analog ports. Any two analog devices, such as telephones, faxes, or modems, can be connected at the same time, since ISDN provides two simultaneous lines. It includes a built-in NT1 interface, ISDN cables, and a multilink PPP.

A.1.11   Northern Telecom's wireless infrastructure products

Northern Telecom provides several wireless infrastructure products which are sold to carriers and other network service providers. We shall give a very brief description of these products.

The Nortel PCS 1900 system consists of a base station subsystem, a network switching subsystem, advanced antennas, terminals, and smart cards—essentially everything that a network service provider might need to install a GSM-compatible PCS network. The Base Station Subsystem (BSS) interfaces to the mobile stations and the network. The Base Station Controller (BSC) and Base Transceiver Station(BTS) are primary components of the BSS. The Network Switching system is based on Nortel's DMS SuperNode architecture.

Nortel also manufactures, sells, and supports IS-41-compatible CDMA network solutions for PCS networks. With DualMode (e.g., CDPD and voice) microcell architecture, Nortel products can increase overall capacity of the network and reduce coverage problems, especially at traffic intersections, tunnels, and other locations where other base stations located farther away may not be able to penetrate. The Nortel product line offers conventional AMPS-, TDMA-, and CDPD-based services from the same radio and antenna platform.

A.1.12   IBM wireless and PSTN modems

IBM manufactures and sells the following types of modems for mobile users:

- IBM ARDIS and Mobitex modems
- IBM Cellular/CDPD modems
- IBM V.34 dial-in modems for PSTN access

A.1.13 Wireless LAN products

Please see Chap. 7 for a discussion of the wireless LANs. Solectek, Xircom, IBM, Proxim, and DEC are major suppliers of the wireless LANs.

Solectek products.   Solectek supplies a complete line of wireless LAN products under the AIRLAN trade name. The product family consists of the following products:

- AIRLAN/Access Point
- AIRLAN/PCMCIA adapter
- AIRLAN/Parallel self-powered mobile adapter
- AIRLAN/Internal ISA card for servers and desktop workstations
- AIRLAN Bridge Ultra
- AIRLAN/Bridge Plus

Xircom products. Xircom markets the following products for the mobile computing marketplace:

- Creditcard Netwave Adapter
- Netwave Access Point for Ethernet

IBM wireless LAN products.   IBM markets a full complement of wireless LAN products: access points, wireless LAN adapters, and bridges. See Chap. 14 for additional details.

DEC wireless LAN products.   DEC markets both frequency-hopping and spread-spectrum wireless LAN products. See Chap. 14 for more details.

A.2 End-User Devices: Notebooks, PDAs, and Pagers

We have described the architecture of notebooks in Chap. 6. We also mentioned there the leading vendors in this product segment. Trade magazines, such as *Mobile Office* and *Pen Computing,* review these products on a regular basis. Compaq, IBM, NEC, and Toshiba are major vendors for general purpose notebooks. IBM also makes a semirugged notebook called 730T. *You should obtain current information from the vendors and refer to trade magazines for comparison of features.*

A.2.1 Two-way pagers

One-way pagers are now commodity items, available from paging network service providers. As for more modern two-way pagers, Motorola's Tango product is currently offered by SkyTel. If your organization is considering PDAs for the field force, you should consider installing PC (PCMCIA) pager cards in these PDAs. By the time this book goes to press, you will see new two-way pagers from AT&T, Ericsson, and others.

A.2.2 PDAs and PCA

Apple's Newton, Motorola's Marco, BellSouth's Simon, and other similar products are reviewed regularly in *Mobile Office* and *Pen Computing* magazines. We have described features and functions of these in Chap. 6.

A.2.3 Telxon's handheld computers

Telxon sells turnkey solutions, primarily for the manufacturing, distribution, retail, warehousing, and transportation industries. It has a very comprehensive product line ranging from inexpensive low-end bar-code scanner handheld computers to high-end pen-based 486 computers. Its Aironet subsidiary manufactures wireless LANs. Table A.1 gives an overview of Telxon's product line.

A.2.4 Norand's handheld products

Norand is also a developer and supplier of turnkey mobile computing solutions for distribution and industrial settings. Its products are extremely modular, allowing migration to more powerful components without replacing the entire unit. Table A.2 gives an overview of its product line.

A.2.5 Symbol Technology scanners

Symbol Technology also supplies PTC products and industrial scanners, used in many airline applications.

A.2.6 Psion's PTC products

HC R800/900 is a rugged handheld terminal with a 160 by 80 monochrome LCD display. It has a 53-key pad. HC R800/900 also accepts input through a bar-code reader or magnetic-card reader. It has an integrated RPM-405i modem for ARDIS network. Wireless applications are supported through an API from SEMI.

TABLE A.1   Products by Telxon

Product	Description
PTC-600	Bar-code scanning handheld microcomputer (low-end)
	Telxon's enhanced version of MS-DOS, with Telxon Common Application Language
PTC-610	Portable teletransaction microcomputer
	For sales order entry, inventory control, marketing data collection, asset management
PTC-710	Portable teletransaction computer with bar-code scanner—higher capacity than PTC-600; up to 2 MB RAM; optional two-way communication 2400 bps
PTC-860	Portable teletransaction computer, 128 K to 4 MB RAM; two-way communication
PTC-860-ES	Same as above but environmentally sealed
PTC-860-RF	Same as above but with RF communication for real-time communication
	Features wireless communication over Telxon DATASPAN 2000 spread-spectrum LAN and wide area such as ARDIS, RAM Mobile Data, and cellular
PTC-860-IM	Same as above but ruggedized for industrial environment
PTC-870-IM	For heavy, rugged, environmentally harsh conditions; 80,486 at 25 MHz; 1–12 MB
PTC-960-RF	Bar-code scanning and RF-integrated
POXpress	Portable computer for retailers on the move
PTC-910	Integrated CD scanning microcomputer for batch data communication
PTC-910L	Laser scanning handheld computer
PTC-912	Laser scanning handheld RF computer
PTC-1140	Wireless pen-based computer, one PC Card (PCMCIA) slot
PTC-1134 and 1144	High-performance 486 pen-based computer; 1144 is ruggedized model.
XC-6000 (Itronix)	Rugged mobile computer for field application (Sears field service application)
ARLAN 630 and 631	Aironet's Ethernet, Token-Ring access points respectively
ARLAN 640	Ethernet wireless bridge
ARLAN 655 and 656	Wireless LAN client and server cards respectively
ARLAN 690	PCMCIA wireless LAN adapter
RP-100	Portable printer
ZFP-80	Compact, mobile impact printer
GCS-SNA	Intelligent communications controller for handling IBM terminal emulation (3270 and 5250) sessions for portable handheld devices
TRIPS	Telxon's RFX Software Tool Kit for Interpretive Prompting System (real-time)
RAMsaver	Telxon's Micro Footprint version of MS-DOS 3.2 for PTCs (portable teletransaction computer)
RFXpress	System integration tool for RF—speeds the development of real-time systems that utilize RF PTCs
PTC-TCP/GCS-TCP	Provides connectivity between PTC and a host on Ethernet or Token-Ring LAN
SCS-OPEN	TCP/IP-based connectivity based on terminal emulation or client/server design between PTCs and UNIX or Windows NT servers

SOURCE: Telxon product literature.

A.3 Communications Servers and MCSS Products

Within this category, there are two subcategories: RNA products and MCSS products for wireless.

A.3.1 RNA products

Please refer to Chap. 10 for a general overview of RNA. The following vendors (a partial list) have a series of products from the low end to the high end:

TABLE A.2   Products from Norand

Product	Description
RT1000 Radio Terminal	Pocket terminal with 4 line by 16 character LCD display, 47 keys, standard UHF radio, and optional integrated scanner
DT1100 Data Terminal	Same as RT1000 without RF, optional scanner
RT1100 Radio Data Terminal	Same as base DT1100 with optional UHF radio or SST radio module
DT1700 Data Terminal	Handheld TM1700 terminal module, optional integrated scanner
RT1700 Radio Data Terminal	Same as above with RF module
RT3210	Handheld computer with RF acts as a real-time terminal to a host PC at 4800–9600 bps
Route Commander 4300 Series	Specialized lightweight (22 oz.) distribution route computer: base unit; DOS compatible, optional internal modem
Route Commander 4310	Allows future growth with plug-in memory cards
Route Commander 4400	Same as 4300 but faster processor and memory
Route Commander 4410	Maximum memory and multiline display
Norand 4810 Printer	40-column printer for PTC applications
Norand 4815/4820 Printers	Full 80-column wide printers to work with Norand PTCs
Norand 4980 and 4985 Network Controllers	Collects information from up to 72 Norand handheld PTCs. 4980 has 768 K memory, while 4985 has only 512 K memory
Pen*Key 6100	Pen-based, AM386 33 MHz, 1–8 MB RAM, 2 PC Card (PCMCIA) slots and IrDA
RT5940 Forklift Terminal	PTC with large display and large keys for continuous usage
Pen*Key 6200	Pen-based 386SXLV, 2–4 MB, MS-DOS
Pen*Key 6300	Pen-based 486SLC2/50 MHz, 4–12 MHz RAM. 1MB RAM. 1MB Flash, 2 Type II PC Card (PCMCIA), IrDA
Pen*6600	486DX2 50 MHz, 4–16 MB RAM, 8 MB Flash, two Type II PC Card (PCMCIA), IrDA

SOURCE: Norand product literature.

- CISCO: remote-access features in routers

- DEC: Digital NetRider/NetServer family

- IBM: LANDistance

- Shiva: Leader in this market, with full complement of products

- US Robotics: NETServer/2, 8, 16 ports and Total Control NET-Server/24, 48, 64 ports

- 3COM: Remote-access product family

- Xyplex: Network 3000 family

- Xircom: Multiport modem and ISDN adapter for the low-end communications servers

- Xylogics: Remote Annex 2000

A.3.2   MCSS products

Please refer to Chap. 11 for an overview of several MCSS products in the market. The following products are available:

- *RNA vendors' (e.g., Shiva, Xyplex) communications servers.* Most servers do not support wireless networks.
- *IBM's ARTour.*
- *RIM's RIMGATE Gateway.*
- *TEKnique's TX5000 product.*
- *Motorola's RNC 6000 Controller.*
- *Nettech's and Racotek's Software.* These are implemented on multiple platforms.

A.4 Software: Operating Systems/Connectivity-Related

Under this category are included those products that relate to the operating system, software drivers, and connectivity. Middleware software is also included here. These topics were discussed in Chap. 12.

A.4.1 Mobile computing notebook software drivers

Various MCSS and communications software suppliers provide software drivers for DOS, Windows 3.1, Windows 95, or OS/2 clients, such as TEKnique, a switch supplier from Chicago that supplies the Windows 3.1 network software driver for the Mobitex network.

MDD, available from Teli Mobile Systems in Sweden, is a Mobitex driver for DOS.

MobiWin, available from B&M System, Sweden, is a Mobitex driver for Windows 3.1.

A.4.2 Mobile computing connectivity software

The following products are available in this category.

Motorola's AirMobile product: a communications middleware. Please refer to Chap. 13 for a description of this software.

Nettech's communications software suite for ARDIS. Nettech's software products include Instant RF workbench family for developers and a family of Mobile applications for end users. Motorola's AirMobile is a Nettech product under the covers.

RFgate is an OS/2-based gateway and API that provides a high-performance X.25 connection to wireless and wire-line products. An important feature of RFgate is its support for intelligent agents.

RFgate also provides full customer configuration and network monitoring capabilities.

RFMlib is a network-layer API that makes it easy to send and receive data over wireless networks via radio modems. RFMlib offers a common, simple API across all networks that shields developers from differences in modem and network protocols.

RFlink is a transport layer API. It is a connectionless API.

Client applications can be in DOS or Windows 3.1.

Aironet's FieldNet. FieldNet (now marketed by Novell) allows wireless clients to access Netware servers over the ARDIS and RAM packet radio networks. FieldNet client is implemented as an ODI driver, plugging in below the standard, off-the-shelf network client protocol software. The FieldNet gateway runs as a Network Loadable Module on Netware. IPX/SPX protocols are supported. FieldNet optimizes the IPX/SPX headers and compresses the packets across the wireless link. FieldNet gateway acts as a Netware Multi Protocol Router.

Oracle Mobile Agents and Sybase's EMS communications middleware products. These products are listed in Sec. A.5, also, because the core functionality of Oracle Agents and EMS is application integration. Comparatively speaking, Oracle's Mobile Agents software is more mobile-aware than EMS. These products are described in Chap. 13.

AirSoft's Powerburst for remote access by Mobile Workers. AirSoft's Powerburst software is designed for remote access of your desktop PC or LAN server resources in the same way as you use them at the office. The major difference between conventional RNA access software and Powerburst software is that it has been designed as a mobile-aware package. It uses advanced techniques, including block-level caching, prefetching, data compression, and differential writeback to reduce the amount of traffic on the dial-up or wireless link. It allows you to continue to work even when you are disconnected. It does this by replicating the remote server's directory and file system in the portable. The package consists of a client software that executes on the remote machine and an agent software that runs on a dedicated DOS machine on the LAN. *It should be noted that Powerburst speeds up only those applications that are file-based.*

The performance gains from using Powerburst are quite impressive, based on the feedback from the users. One user quoted Microsoft mail to be six times faster with Powerburst than with a conventional RNA solution.

Powerburst is also transparent to remote node hardware and software, since it works with CISCO and Shiva hardware and software.

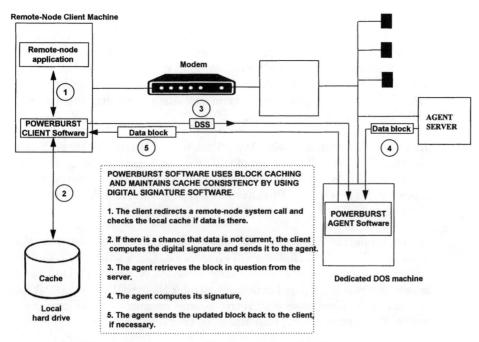

Figure A.1   AirSoft's Powerburst remote file access and synchronization software. (*Source:* Data Communication.)

TeleWorker software (Novell and Ericsson).   TeleWorker Network products are cobranded products of Novell and Ericsson for accessing Novell's Netware-based LANs from a remote client. The package consists of software and modem hardware to provide a complete solution. Initially only PSTN/ISDN solutions are available. Wireless WAN solutions are under development and will be available in 1996–1997 timeframe.

File transfer and synchronization software.   There are several products in this category. Laplink from Traveling Software is the most well known. Symantec Corporation has introduced Act Mobile Link 2.0 software for synchronization of data on the notebooks and LANs. Other packages such as RemoteWare from Xcellnet, MobileWare, and Powerburst also provide this capability.

Netware Mobile from Novell.   This is a client software that enables mobile users to access their NetWare networks. They can transfer selected network data to mobile computers and then automatically update and synchronize files when they reconnect to the network. The software is very easy to use and allows for connections by using different network configurations. It supports file replication and synchronization.

A.4.3   PDA connectivity software

Most of the third-party software developers are building products for Apple's Newton MessagePad PDA and Tandy's Zoomer. Other PDA manufacturers have a smaller following of the developers. We are listing only a subset of the products available. You may refer to *Mobile Office* and *Pen Computing* magazines or surf the Internet to get more information on these products.

Newton Connection Kit (NCK 3.0). This links Newton MessagePAD to desktop PCs. NCK 3.0 is a set of utilities that allows software developers to link PIMs, spreadsheets, and word processors to Newton PDAs. Apple has released an API that allows C programs to poll Newton's file directory.

ARDIS support of Mail-on-the-Run for notebooks and PDAs. Mail-on-the-Run from River Run Software of Greenwich, Connecticut, is a mobile-aware mail package that allows any notebook or Newton PDA user to access Microsoft Mail or Lotus cc:Mail over the ARDIS network. It has a filter that allows users to retrieve only certain types of mail messages. It supports encryption and data encryption.

InterAgent (formerly CMPump) for Newton. InterAgent (formerly CMPump for Newton) from Ives & Company, a U.K. firm, is a data routing and synchronization utility that allows the easy integration of data in heterogeneous database systems including Apple Newton. Other databases supported are Lotus NOTES, Microsoft FoxPro, Microsoft Access, Microsoft SQL Server, Oracle, and other ODBC data sources.

AllPen's NetHopper for Newton. This allows Newton OS 2.0 users to browse the World Wide Web through any ISP that supports point-to-point protocol (PPP) or serial line interface protocol (SLIP) connections.
Also see Wayfarer's connectivity software, mentioned in Sec. A.5.

IBM's Jetski connectivity software for handheld PDAs. Jetski is connectivity software, expected to be released in late 1996, for connecting PDAs running under IBM's Tricord operating system to OS/2 and NT servers. It will support the C++ tool kit in order to provide connectivity across dial-up and wireless connections.

A.5   Software: Application Development, LAN Application Access, and Integration-Related Tools

Most of the software tools are for Windows. PDAs have a different operating system and require their own unique tools. Therefore, there are two categories: one for Windows-based mobile computers and one for PDAs.

A.5.1 Application development for Windows

The following tools will be described:

Pen-based software development kits. The following SDKs are representative of what is available in the market currently:

- *PenDOS SDK* (from CIC (415) 802-7888). Provides direct access to CIC's handwriting recognition engine

- *Menuet/CPP* (by Autumn Hill Software (303) 494-8865). Development environment in PenDOS, Windows, and OS/2 for Pen for forms-based applications; has interface designer, icon and font editor, and C++ libraries

- *Pen Developer* (from Clarion Software (305) 785-4585). Full-featured development system for database application in Clarion environment; supports dBase, FoxPro, Paradox, Btrieve, Clipper Files

- *Power Pen Pal* (from Pen Pal Associates (415) 462-4888). Based on PenDOS and PenRight, a powerful integrated pen development environment

- *padBase* (from R2Z (510) 792-1477). Library of Clipper routines for pen

- *PenRight Pro* (from PenRight (510) 249-6900). Powerful cross-platform development tool being described further in this section

- *PenOp* (from Peripheral Vision (212) 262-1588). Available in PenDOS and Pen for Windows; secure capture and management of hand-written signatures. Has C++ interface

RF Query, RF File, and RF Chat. Dynamic Mobile Data's RF utilities allow developers to add wireless capabilities to Windows 3.1 applications. It supports ARDIS, RAM Mobile Data, and CDPD networks. RF Query works in conjunction with Nettech's InstantRF middleware connectivity software. RF File and RF Chat are VBXs (Visual Basic extensions) that add wireless capability to Windows applications such as Word, Excel, Lotus 123, or native Windows applications developed using Visual Basic.

Xcellnet's RemoteWare. Please refer to Chap. 12 for a description of this software.

Racotek's Keyware software. Please refer to Chap. 12 for a description of this software.

MobileWare software.   MobileWare is a client/server solution for remote access of e-mail, file transfer, remote fax, and remote print. It also supports LAN applications such as cc:Mail, Microsoft Mail, and Lotus NOTES by providing corresponding clients: MobileWare cc:Mail client, MobileWare MS mail client, and MobileWare Lotus NOTES client. It runs as an NLM and under Windows NT. It supports TCP networks, including CDPD. A schematic of MobileWare is shown in Fig. A.2. MobileWare's Transaction Management System provides the following features:

- Dynamic reconnection
- Check point
- Guaranteed data integrity
- Dynamic compression
- Encryption and authentication
- Queuing
- End-to-end acknowledgment

Technology Development Systems' (Hoffman Estates II) WorldLink 2.2. This product is in a functionality class similar to Xcellnet's Remote-

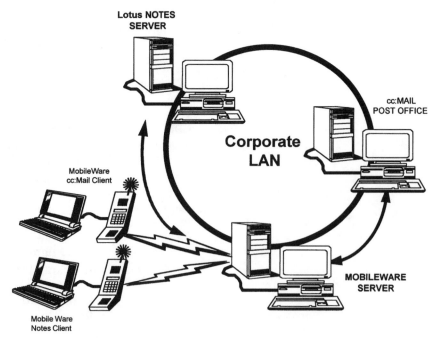

Figure A.2   MobileWare software for remote access. (*Source: MobileWare brochure.*)

Ware and MobileWare described in this section. There are of course differences in feature.

Oracle's Mobile Agents.   Please refer to Chap. 13 for a description of this software.

Sybase's (communications architecture) EMS middleware software. Please refer to Chap. 13 for a description of this software.

Mobitex software API tools

Acutus, from AU-Systems of Sweden, is a development platform that you can use to link existing or new UNIX applications to the Mobitex network.

IBM LU6.2 Radio Program, available from IBM Sweden, is a powerful gateway solution that connects mobile units with the IBM mainframe using SNA LU6.2 application-to-application protocol.

MDS/400, available from SYSteam, Sweden, is a platform for mobile application development on AS/400.

MobiLib-Plus API, available from RIM in Waterloo, Ontario, is an API in the form of Microsoft C library that can be linked with custom applications for use over the Mobitex network.

PSILink, available from PSI, Reston, Virginia, is an electronic messaging application for the Internet.

TransRmail, available from Teli Mobile, Sweden, is a complete package for application development, Mobitex communications, and access to databases. C-ISAM, Ingress, Informix, Oracle, and Sybase databases are supported.

MobiBase SQL, from Telesoft Uppsala, Sweden, is an SQL interface through a Mobitex network. *MobiGate* is a communications product from the same company. It runs on RISC 6000 and supports communication with IBM mainframes using SNA LU6.2 and APPC.

Mobitex AT command set, from RIM in Waterloo, Ontario, allows standard modem communication software to use a Mobitex network without communication. Please see Chap. 12 for additional information.

MOBI/3270, from Telepartners of Farmington, Connecticut, is an SNA 3270 software that provides a wireless link from notebooks to IBM mainframes via a Mobitex network. SIMWARE of Ottawa, Canada, has a similar package called SIMWARE 3270 for Mobitex.

Please note, more information on these products is available in the Mobitex Software directory available from Ericsson.

A.5.2 Communications connectivity
software for PDA

Only one product is being mentioned here.

Middleware-based development kit for accessing enterprise server for PDAs.
Wayfarer Corporation of Mountain View, California, has released a
messaging middleware/development tool kit for integrating PDAs into
enterprise networks and legacy systems. Please see Chap. 13 for additional information.

A.6   Miscellaneous Software:
Horizontal Applications

This category includes messaging, e-mail, groupware, etc.

A.6.1   AT&T PersonaLink

This uses Motorola's Envoy Personal Wireless Communicator to access
the electronic marketplace. This is based on General Magic's Magic Cap
and Telescript engine described in Chap. 12. Intelligent assistants travel
from Envoy communicator into the PersonaLink community, where they
can carry out the customer's instructions to find, buy, and sell things.
They can filter information to locate exactly what the customer requested
and save the customer time by performing complex transactions.

A.6.2   Software packages for e-mail
support with a pager feature

Almost all network providers such as ARDIS, RAM Mobile Data, and
CDPD carriers provide native e-mail service or interfaces to e-mail
applications on corporate networks. Both cc:Mail and Microsoft Mail
have mobile-aware versions, designed to reduce traffic on the network.
In fact, e-mail is the most popular horizontal mobile application,
according to the surveys done by market research firms.

A.6.3   Groupware applications
and mobile computing

Lotus NOTES version 4, released in 1996, has enhanced support for
mobile users.

A.7   Software: Application Packages
for Horizontal and Vertical Industries

Many horizontal industry application packages for sales force automation and vertical industries such as public safety are available in the
market. We are giving a representative sample here:

A.7.1 Sales force automation

Please see Chap. 2 for a description of these applications and products. We shall mention the following products.

KPMG's Sales Mate. Sales Mate is sales force automation software available from Motorola on the Marco wireless communicator. It allows customization and integrates with SAP's R/3 systems. Examples of functionality that Sales Mate supports for real-time processing include:

- Customer contact review and management
- Customer credit history
- Order entry, review/status verification
- Product pricing and availability checking
- Inventory counting
- Product updates and promotional inquiries
- Returns processing

SkyDispatch from SkyNotes. This application is available from Sky-Notes and allows workers in the field using a Marco personal communicator to report time logs, job completion reports, and troubleshooting information to a Lotus NOTES database that runs in the office.

Sales automation packages. Table A.3 shows a subset of SFA application packages that are either popular in the industry or provide wireless network support.

A.7.2 Public safety

The following products represent a subset of the products in this vertical industry.

PRC's law enforcement system. PRC, a consulting and systems integration company with headquarters in McLean, Virginia, is a leading supplier of public safety systems that support mobile computers in vehicles. It offers a full set of systems integration services including hardware, software, and application customization. Its application software called *Altaris* runs on DEC- and LAN-based servers under OS/2, Windows/NT, and UNIX.

UCS's pen-based solutions for public safety and EMS. UCS, a consulting and application development company based in Ft. Lauderdale,

TABLE A.3 Sales Force Automation Products

Product name	Vendor	Functionality	Databases supported	Operating system	Wireless network support
SalesTrak	Aurem Software, Santa Clara, Calif.	Contact manager, call and lead tracking, e-mail, communications, forecasting, sales analysis, sales reporting, database updating	Client: SQLBase Server: Sybase and Oracle	Windows	No
TakeControl	Brock Control Atlanta, Ga.	Contact manager, call and lead tracking, e-mail, communications, forecasting, sales analysis, sales reporting, order entry, database updating, commission tracking	Client: Windows for Workgroup 3.11 SQL Server, Oracle, Informix	Windows, UNIX, NT, LAN Server	Yes, ARDIS, RAM Mobile Data
Maximizer	Modatech, Vancouver, British Columbia, Canada	Contact manager, call and lead tracking, e-mail, communications, forecasting, salesanalysis, sales reporting, database updating, commission tracking	N/A	Windows, Netware, LAN Server, LANtastic	No
SalesKit	SalesKit Technologies, St. Louis, Mo.	Contact manager, call and lead tracking, e-mail, communications, forecasting, sales analysis, sales reporting, order entry, database updating	SQL Server	Windows, Windows for Workgroups, NT	Yes, ARDIS, RAM
SNAP/Virtual Office	Sales Technologies Atlanta, Ga.	Contact manager, call and lead tracking, e-mail, communications, forecasting, sales analysis, sales reporting, database updating	SQL Server, Sybase, Oracle, SQLBase	Windows, OS/2, NT, Netware, LAN Manager, LAN Server	RAM, other radio and cellular networks
RemoteWare	Xcellnet Atlanta, Ga.	Communications; for custom sales force automation; provides basic SFA modules. Also can use one of the above packages	SQL Server, Sybase, Oracle, DB2, SQLBase	DOS, Windows, OS/2, Macintosh, Netware, LAN Server	ARDIS, RAM, CDPD, dial-up

SOURCE: *Communications Week*, 1994 Mobile Computing Special.

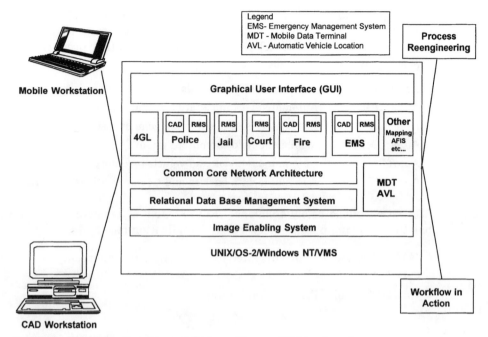

Figure A.3 PRC's Altaris system architecture. (*Source: PRC brochure.*)

Florida, has developed several pen-based applications for public safety and EMS (emergency medical services). UCS is a Motorola Alliance partner. UCS application products provide off-the-shelf pen-based products for the following applications:

- Offense/incident reporting
- Accident reporting, including diagramming
- Crime scene diagramming
- Property/evidence recording
- Fire safety inspection
- Building permit inspection
- Vehicle inspection by insurance adjusters
- EMS reporting
- City code and bylaw enforcement

Enforcement technology's automatic citation issuance system. The AutoCITE, a product from Enforcement Technology in Irvine, California, is a complete hardware/software solution to issue citations automatically for parking and other law/regulatory infractions. It consists of a rugged handheld computer with a 24-column display, 45-position keyboard, and a 42-column printer. It replaces handwritten citations with computer-generated tickets. The company has other products for law enforcement:

AutoPOLICE	Automated police records and CAD systems
AutoPARK	Automated parking citation management system
AutoALARM	Automated false alarm management system
AutoCOMP	Mobile and field computer system
AutoTRAK	Automated hotbook and vehicle inventory
AutoNET	Automated data backup and communications network

ROADSOFT public safety solutions. ROADSOFT of Longeuil, Quebec, has the following public safety–related application packages:

- *CAD.*   Computer-aided dispatch for police, fire, and emergency health services

- *GeoBase.*   For GPS-based mapping

A.7.3   FieldNotes

FieldNotes is a GIS (Geographical Information System) for laptops and pen-based computers. It has the following features and applications:

Features

- Displays and edits drawings and images such as houses and hydro poles
- Integrates database with inquiries
- Customizes database forms for specific application
- Uses Windows 3.1 on pentops, laptops, and desktops

Applications

- Inventory management
- Work-order processing
- Field mapping
- Modification of engineering drawings
- Field data collection

A.8   Services

Several vendors provide mobile computing services. The following list represents a subset.

A.8.1   ARDIS

ARDIS provides packet radio for messaging and wireless applications. Please see Chap. 8 for more details.

A.8.2 RAM Mobile Data

This provides packet radio for messaging and wireless applications. There are different Mobitex network operators in different countries. Please see Chap. 8 for more details.

A.8.3 RadioMail

RadioMail provides e-mail gateway services. Supports ARDIS and RAM Mobile Data users directly. Please see Chap. 14 for more details.

A.8.4 Metricom Network

This offers a wireless spread-spectrum network in the Bay area in California. Plans to offer it in more metropolitan areas.

A.8.5 SkyTel

SkyTel provides nationwide paging network, including two-way paging for low-end messaging and transaction-based applications. Please see Chap. 14 for more details.

A.8.6 Notable

Notable provides a variety of wireless data communications services, the most "notable" being the AirNote messaging (e-mail and telephone) service on your pager. It also offers the Septor financial information service (time-sensitive news and stock information).

A.8.7 ARDISLink

It is a bundled offering from ARDIS that includes the following:

- Physical connection from a client computer to the ARDIS message switch
- Leased circuits and ports on the ARDIS switch
- Use of ARDIS's private communications network, ARDISNET
- End-to-end network management, including availability and disaster recovery

Mobile Computing Vendors

Vendor name and address	Telephone number	Products and services
Air Communications Inc. 274 San Geronimo Way Sunnyvale, CA 94086	408-749-8089	AirTrue, a high-performance protocol for circuit-switched cellular (AMPS) networks (see Chap. 8)
AIRONET Wireless Communications, Inc. 3330 West Market Street P.O. Box 5292 Akron, OH 44334-0292	1-800-800-8016	ARLAN wireless LAN products (NICs, Access Points, etc.)
AirSoft Inc. 1900 Embarcadero Road, Suite 204 Palo Alto, CA 94303	800-708-4243 415-354-8123	Powerburst remote access software (see App. A)
AllPen Software, Inc. 51 University Avenue, Suite J Los Gatos, CA 95030	408-399-8800	Pen-based applications for PDAs and pen computers
ALPS 3553 North First Street San Jose, CA 95134	408-432-6000	RadioPort Plus wireless LAN products
Ameritech 2000 West Ameritech Center Drive Hoffman Estates, IL 60195-5000	800-662-4531	Circuit-switched cellular, CDPD, and enhanced paging services
Apple Corporation 1 Infinite Loop Cupertino, CA 95014	800-776-2333	Powerbooks, Newton PDA, Newton SDK, e-MAIL Servers, eWORLD messaging service
ARDIS 300 Knightsbridge Parkway Lincolnshire, IL 60069	700-913-1215 700-913-1453	Wireless Packet Data Network services, wireless modems, mobile computing integration services
AT&T Wireless Group 1700 South Patterson Boulevard Dayton, OH 45479	206-990-4481	Full Service Cellular and CDPD network supplier
Bell Atlantic Mobile 180 Washington Valley Road Bedminster, NJ 07921	908-306-7520	Circuit-Switched Cellular and CDPD services AirBridge wireless LAN bridge

Vendor name and address	Telephone number	Products and services
Bell Mobility Canada 20 Carlson Court Etobicoke, Ontario, Canada M9W 6V4	416-674-2220	Cellular, paging, and packet data network services (ARDIS, Canada), wireless modems, mobile hardware, and integration service
CDPD Forum Liz Durham, Marketing Manager 401 N. Michigan Avenue Chicago, IL 60611-4267	1-800-335-CDPD	Vendor consortium for CDPD network service providers
Cellular One 250 Australian Avenue South West Palm Beach, FL 33401	407-653-7044 (Phone) 407-655-7403 (Fax)	Cellular and CDPD services
Clearnet 1305 Pickering Parkway, Suite 1300 Pickering, Ontario, Canada L1V 3P2	905-831-6222	ESMR, PCS, paging network service provider
Cylink 910 Hermosa Court Sunnyvale, CA 94086	408-735-5800	Security products; also AirLink spread-spectrum microwave radios at 2 Mbps in point-to-point configurations
Dataradio 6160 Peachtree Dunwoody Road Suite C-200 Atlanta, GA 30328	404-392-0002 (voice) 404-392-9199 (fax) e-mail: info@dataradio.com	Radio modems and private radio network products
E.F. Johnson Corporate Headquarters 11095 Viking Drive Minneapolis, MN 55344-7292	1-800-328-3911 ext. 380	Radio network infrastructure, public safety APCO-compliant network products
E.F. Johnson Canada, Inc. 633 Granite Court Pickering, Ontario, Canada L1W 3K1	1-800-263-4634	Same as U.S. company listed above
ERICSSON GE Mobile Communications Inc. (a division of Telefon AB LM) Ericsson Wireless Computing 15 East Midland Avenue Paramus, NJ 07652	1-800-223-6336 201-890-3600 201-265-6600 201-265-9115 (Fax)	Complete suite of mobile computing infrastructure products (Mobitex packet, PCS, GSM, and cellular), including modems, etc.
Ex-Machina, Inc. 11 East 26th Street, 16th Floor, New York, NY 10010-1402	800-238-4738	Software between PCs and servers to pagers, e.g., NOTIFY, broadcast of software to notebooks
Extended Systems, Inc. 6123 North Meeker Avenue Boise, ID 83704	1-800-235-7576 208-322-7575 (Phone) 406-587-7575 (Phone) 208-377-1906 (Fax)	JetEye, HP SIR infrared compatible printer and desktop computer interface
General Magic 420 North Mary Avenue Sunnyvale, CA 94086	408-774-4000	Magic Cap client and Telescript server software engines for PDAs, for mobile and electronic marketplace applications
GTE MobileNet 245 Perimeter Center Parkway Atlanta, GA 30346	770-391-800-GTE-MOBL	Cellular and CDPD network services, modem pool access, wireless voice-mail, paging

Vendor name and address	Telephone number	Products and services
IBM Corporation Wireless Network Group 700 Park Office Drive, Building 662 Dept. DKMA Research Triangle Park, NC 27709	800-IBM-CALL	Wireless modems, ARTour MCSS gateway, ThinkPad notebooks, systems integration
InfraLAN Technologies 12 Craig Road Acton, MA 01720	508-266-1500 (Phone) 508-635-0806 (Fax)	InfraLAN 16 Mbps, infrared, 80 feet, IBM Token-Ring bridge, wireless LAN
Infrared Data Association (IrDA) e-mail(John LaRoche): jlaroche@netcom.com	510-943-6546 (Phone) 510-934-5241 (Fax)	Joint infrared proposal from Apple, HP, IBM, and Connexus
Iridium Inc. 1401 H Street NW, Suite 800 Washington, DC 20005	202-326-5600	Overall coordinating organization for IRIDIUM satellite-based wireless network
Laser Communications, Inc. 1848 Charter Lane, Suite F Lancaster, PA 17605-0066	1-800-527-3740 717-394-8634 (Phone) 717-396-9831 (Fax)	LACE family of products for short-haul optical transmissions in campus environments
LOTUS Corporation Cambridge, MA, USA (Offices in most large cities)	800-205-9333	cc:Mail mobile, paging-based messaging, and NOTES groupware software
MAPINFO One Global View Troy, NY 12180	1-800-327-8627	Mapping software with GPS
Metricom 980 University Avenue Los Gatos, CA 95030	408-399-8200 (Phone) 408-354-1024 (Fax)	Network service provider (see Chap. 8)
MMTF Task Force mmtf-request@www.epilogue.com	505-271-9933	Mobile Management Task Force for SNMP-compatible MIB standards
MobileWare Corporation 2425 North Central Expressway, #1001 Dallas, TX 75080-2748	214-952-1200 (Phone) 214-690-6185 (Fax)	MobileWare Software for accessing LAN applications from remote users
Motorola 50 East Commerce Drive, Suite M-4 Schaumberg, IL 60173	1-800-624-8999 ext. 105 708-576-8213 (Phone) 708-576-8940 (Fax)	Full line of mobile computing products network infrastructure, APCO-25 network products, ALTAIR wireless LAN, radio/ISDN modems, pagers, Marco communicator, Forte rugged PCs, software, integration, and more
NCR Wireless Communications & **Network Division** 1700 South Patterson Boulevard Dayton, OH 45479	1-800-CALL-NCR 513-445-5000 (Phone) 513-445-4184 (Fax)	WaveLAN wireless LAN, PCMCIA, ISA, MCA WaveLAN/ISA, WaveLAN/ISA, WaveLAN/MicroChannel, WaveLAN/ Campus Bridge, WaveLAN/ PCMCIA 2 Mbps, spread spectrum 902–928 MHz, 200–800 feet, DES encryption
NEC San Francisco, CA	1-800-366-9782	Notebooks
Nomadic Software	415-335-4310	SmartSynch 2.0 software (synchronizes files between remote systems)

Vendor name and address	Telephone number	Products and services
Norand 500 Second Street S.E. Cedar Rapids, IA 52401	319-369-3100	Ruggedized handheld portable computer systems with RF capability, route accounting software and integration services for manufacturing, distribution, and warehousing industries
NORCOM 3650 131st Avenue S.E., Suite 510 Bellevue, WA 98006	1-800-676-9951	High-speed X.25 packet data network for integrated land- and satellite-based mobile communication
Northern Telecom P.O. Box 833858 Richardson, TX 75083-3858	1-800-4-NORTEL	In-building wireless, digital cellular (CDPD), PCS, and wireless access network infrastructure
Notable Technologies Inc. 1221 Broadway, 20th Floor Oakland, CA 94612	800-732-9900	AirNote messaging and financial information services on pagers
Novell Inc. 122 East 1700 South Provo, UT 84606-6194 (Office in every major city)	1-800-258-5408	Novell Netware NOS, Netware Connect (RNA solution), and Netware Mobile
O'Neill Connectivities, Inc. 607 Horsham Road Horsham, PA 19044	1-800-624-5296 215-957-5408 (Phone) 215-957-6633 (Fax)	RS-LAWN, PC-LAWN, LAWN, RS-232 300 bps–19.2 kbps, 38.4 Kbps, 902 MHz–928 MHz, 20 mW, spread spectrum, 200–500 ft, wireless LAN
Palm Computing Inc. Calif.	415-949-9560	PIM software for Zoomer devices, Palm Connect (desktop PC connection utility)
PCSI 9645 Scranton Road San Diego, CA 92121	619-535-9500	Chip sets for dual-mode (AMPS/TDMA) phones, Mobile data base stations (MDBSs) for CDPD and narrowband PCS
PenRight 47358 Fremont Boulevard Fremont, CA 94538	510-249-6900	Pen-based software development kit
Persoft Inc. 465 Science Drive Madison, WI 53711	1-800-368-5283 608-273-6000 (Phone) 608-273-8227 (Fax)	Intersect NCR WaveLAN OEM bridges up to 3 miles, omni/didirectional antenna, wireless LAN
Photonics Corp. 2940 North First Street San Jose, CA 95134	408-955-7930 (Phone) 408-955-7950 (Fax)	Wireless LAN diffuses infrared, 40 feet, 1 Mbps AppleTalk, LocalTalk, ISA, parallel port
Proxim, Inc. 295 North Bernardo Avenue Mountainview, CA 94043	1-800-229-1630 415-960-1630 (Phone) 415-964-5181 (Fax)	Wireless LAN products (902–928 MHz and 2.400–2.483 GHz)
QUALCOMM 10555 Sorrento Valley Road San Diego, CA 92121	619-587-1121 e-mail: info@qualcomm.com	OmniTRACS satellite data radio, CDMA cellular, PCS
RACOTEK 7301 Ohms Lane, Suite 200 Minneapolis, MN 55429	612-832-9800 612-832-9383	Mobile software for application and network integration, supports most wireless networks, experience in distribution and manufacturing applications

Vendor name and address	Telephone number	Products and services
RadioLan Inc.	408-526-9170 (Phone) 408-526-9174 (Fax)	Wireless LAN system claimed to operate at 10 Mbits/s
RadioMail 2600 Campus Drive San Mateo, CA 94403	415-286-7800	RadioMail services (see Chap. 14)
RAM Mobile Data 10 Woodbridge Center Drive, Suite 950 Woodbridge, NJ 07095 email: airmail@ram.com	1-800-736-9666 908-602-5500 908-602-1262	Wireless WAN packet radio wireless network provider in United States and United Kingdom
Research in Motion (RIM) 295 Phillip Street Waterloo, Ontario, Canada W2L 3W8 email: latham@rmotion.com	519-888-7465 519-888-6906	Mobitex gateway, SDK, PC Card adapters, point-of-sale devices, and two-way paging devices
River Run Software Group 8 Greenwich Office Park Greenwich, CT 06831	203-861-0090	Application development tools for mobile applications (see App. A)
Roger Cantel Canada Inc. 10 York Mills Road North York, Ontario, Canada M2P 2C9	416-229-2000	Operator of Mobitex network in Canada, provides nationwide cellular and paging services
SEA, Inc. 7030 220th SW Mountlake Terrace, WA 98043	206-771-2182	Narrowband 220 MHz land mobile radio equipment, complete mobile radio systems for 220–222 MHz radio service, voice and/or data systems at 220 MHz
Sharp Electronics Corporation 5700 N.E. Pacific Rim Boulevard, MS 20 Camas, WA 98607-9489	206-834-8948 (Phone) 206-834-8903 (Fax)	Sharp serial infrared used on Wizard organizers and Apple Newton MessagePad
Shiva Corporation Northwest Park 63 Third Avenue Burlington, MA 01803	617-270-8320	Remote Network Access products: software and hardware, including remote Web server
SkyTel (a division of Mobile Telecommunications Technologies—Mtel) 1350 I Street NW, Suite 1100 Washington, DC 20005	1-800-SKY-USER 1-800-759-8737 1-800-456-3333 202-408-7444	SkyPager, SkyWord, SkyTalk, SkyStream, SkyFax regional, national, and international paging services, including two-way paging service
Socket Communications 6500 Kaiser Drive Fremont, CA 94555-3613	1-800-552-3300 510-744-2700 (Phone) 510-744-2727 (Fax)	PCMCIA paging card with alphanumeric display, GPS PCMCIA Cards
Solectek Corporation 6370 Nancy Ridge Drive, Suite 109 San Diego, CA 92121	619-450-9000	One of the leaders in wireless LAN products and LAN-to-LAN bridges
South Hills Datacomm 760 Beechnut Drive Pittsburgh, PA 15205	412-921-9000	Wireless NICs 902–928 MHz
Spectrix Corp. 906 University Place Evanston, IL 60201	708-491-4534	Diffuse infrared wireless LAN products

Vendor name and address	Telephone number	Products and services
Symbol Technologies Bohemia, NY	1-800-927-9626 516-563-2400	Laser Radio Terminal 3800 Token-Ring, 60.6 Kbps, 902 MHz, wireless LAN
Technology Development Systems 2300 North Barrington Road, Suite 603 Hoffman Estates, IL 60195	708-781-1800	WorldLink 2.2: database query, file synchronization
Telxon 3330 West Market Street Akron, OH 44333	1-800-800-8001	Full complement of mobile end-user devices, software, and integration services
Travelling Software 18702 North Creek Parkway Bethwell, WA 98011	1-800-343-8080 206-483-8088 206-487-1284 (Fax)	Laplink wireless software, popular for small offices, and mobile laptops; user file synchronization
Trimble Communications 645 North Mary Avenue Sunnyvale, CA 94088-3642	408-481-8000 800-874-6253 (Phone) 408-481-7744 (Fax)	Mobile GPS Receiver
U.S. Paging Corporation 1680 Route 23 North Wayne, NJ 07470	1-800-473-0845 201-305-6000 201-305-1462	North American pager network, e-mail, LAN, palmtop, notebook, laptop support
U.S. Robotics 8100 North McCormick Boulevard Skokie, IL 60076-2920 email: usr@www.usr.com	1-800-342-5877 (Phone) 847-982-0823 (Fax)	Modems, remote-access products
Windata Inc. 10 Bearfoot Road Northboro, MA 01532-1506	508-393-3330	Parallel port wireless LAN NICs (2.400–2.484 GHz, 5.746–5.830 GHz)
Worthington Products 3004 Mission Street, Suite 220 Santa Cruz, CA 95060	408-458-9938 1-800-345-4220	RF bar-code readers
Xetron Corporation 460 West Crescentville Road Cincinnati, OH 45246	513-881-3264 (Phone) 513-881-3379 (Fax)	Wireless spread-spectrum (SS) transceivers, standard RF and SS products
Xircom 2300 Corporate Drive Thousand Oaks, CA 91320-1420	Sales: 1-800-438-4526 805-376-9300 (Phone) 805-376-9311 (Fax)	Full line of PC Card (PCMCIA) and wireless LAN products

Mobile Computing Standards

Standard	Description
ANSI X9.26	Secure sign-on standard (see Chap. 15)
CORBA Middleware	Common Object Record Brokers Association's Middleware Software Standard
DCS 1800	GSM for mobile communications and digital cellular at 1800 MHz
FPLMTS/IMT-2000	Future Public Land Mobile Telecommunications Systems/International Mobile Telecommunications—2000. Next generation mobile systems expected to be introduced after year 2000
IEEE 802.11	Wireless LAN standard (see Chap. 7 and *Data Communications,* Sept. 1995)
IrDA (Infrared Data Association)	Standard for radio, wireless communication between one device to another (PC to PC, PC to printers, and PC to cellular phones, or PC to 2-way pagers) Windows 95 IrDA driver shipped by Microsoft in Nov. 1995; Beta IrDA driver for OS/2 Warp shipped in 1995
IS-41	TIA's cellular intersystems operations standard for nationwide roaming, with intersystem handoff, call delivery, validation, and authentication
IS-52	Describes allowable dialing patterns in North American cellular systems conforming to North American Numbering Plan (NANP)
IS-54/TDMA	CTIA's interim 800-MHz digital cellular air interface standard developed for the introduction of TDMA in conjunction with FDMA. Triples the capacity of previous analog systems
IS-93	Describes a number of interfaces between mobile switching centers and other telephone systems
IS-95/CDMA	800-MHz digital air interface standard based on CDMA, digital handsets, and spread-spectrum radio technology. It offers high capacity and is being enhanced for 2-GHz PCS systems
IS-124	A protocol for the online exchange of call records between mobile switching centers and network nodes designed to process these records for billing, fraud detection, and other purposes
IS-136/TDMA	PCS standard that adds digital control channel to North American TDMA standard, enables enhanced services, such as calling line identification

Standard	Description
LSM (Limited Size Messaging)	An AT&T proposal to CDPD Forum for a protocol based on Message Center concept that links any IP-based messaging service such as e-mail to CDPD services. Messaging SIG in CDPD Forum is considering it
MMAP	Mobile Management Application Protocol—supports mobility via various air interfaces on existing wired-line switched networks
MMTF MIB	Mobile Management Task Force MIB for SNMP (see Chap. 18)
NATO MIL Spec 810 C&D	NATO military standard for ruggedness of portable and mobile computers (see Chap. 6)
NDIS/Windows Extensions for Wireless	PCCA working on draft specifications for NDIS extensions for wireless
ODI Extensions for Wireless	PCCA also working on draft specifications for ODI extensions for wireless
PCS 1900/DCS 1900	The GSM-based air interface and network interface standards adopted to meet North American PCS requirements at 2 GHz
PCS 2000	Composite TDMA/CDMA spread-spectrum air interface standard proposed by Omnipoint Corporation
SNMP	Simple Network Management Protocol standard
TCP/IP	De facto standard for transport layer between remote nodes
UPT	Universal Personal Telecommunication standard—supports personal mobility through a unique personal number across wireless and wired-line networks

Spectrum Allocation

Frequency band (in MHz)		
From	To	Type of usage for the frequency allocation
0.535	1.705	AM Broadcast Standard North America AM
1.705	30.000	Shortwave, amateur, CB, ship-to-shore
30.010	37.000	Government, Army/Navy/Coast Guard, Commercial
37.020	37.420	Police mobile, police, public service
37.440	39.000	Government
39.020	39.980	Police, public service, police mobile
40.010	41.990	Government, Army/Navy/Air Force
42.010	42.940	Police/police mobile
42.960	44.600	Commercial
44.620	46.580	Fire, forestry conservation, highway maintenance, police, public service
46.610	47.000	Air Force/Army, government
47.020	47.400	Highway maintenance
47.420		Red Cross
47.440	49.580	Special industry, special emergency, power and water, forest products/ petroleum products
49.610	49.990	Air Force/Army, government
50.000	54.000	Amateur
54.000	72.000	VHF TV channels 2–4
72.000	73.000	Intersystem, paging, RC, astronomy
76.000	88.000	VHF TV channels 5–6
88.000	108.000	FM broadcast
108.020	132.000	Commercial aircraft
136.000	138.000	Satellite
150.775	160.200	Ambulance, fire, police, forest conservation, taxi base/mobile, commercial/ industrial, utilities, local government mobile, auto/special emergency, highway maintenance mobile, maritime
160.215	161.565	Railroad
174.000	216.00)	VHF TV channels 7–13
225.000	381.000	Military aircraft
406.000	408.000	Digital data transmission
420.000	470.000	Amateur, broadcast pickup, water and power, commercial/industrial, telephone maintenance, tax, motor carrier/railroad, auto club, press/newspapers, local government, public safety, commercial, medical, police
470.000	806.000	UHF TV channels 14–69
824.000	849.000	Cellular mobile telephone—mobile
850.000	870.000	Trunk radio systems

Frequency band (in MHz)

From	To	Type of usage for the frequency allocation
870.000	894.000	Cellular mobile telephone—cell site
902.000	928.000	Unlicensed operation of approved radio devices
928.013	932.000	Domestic public radio service
952.100	959.800	Private microwave service
959.863	959.988	Common carrier radio service wide area paging
1240.000	1300.000	Amateur
2400.000	2483.500	Unlicensed operation of approved radio devices
3300.000	3500.000	Amateur
5750.000	5825.000	Unlicensed operation of approved radio devices
5925.000	6875.000	Common carrier and fixed SAT
8800.000	8800.000	Airborne Doppler radar
10000.000	10500.000	Amateur
10525.000	10525.000	Police X-band radar gun
24000.000	24250.000	Amateur
24150.000	24150.000	Police k-band radar gun
34200.000	35200.000	Police Ka-band photo-radar gun
48000.000	50000.000	Amateur
71000.000	76000.000	Amateur
165000.000	170000.000	Amateur
240000.000	250000.000	Amateur
300000.000	and above	Amateur

SOURCE: *Wireless LANs* by Davis-McGuffin.

Mobile Computing Information Resources

Information resource	Type of information available
ARDIS 300 Knightsbridge Parkway Lincolnshire, IL 60069 700-913-1215, 700-913-1453	Developers guide, quarterly newsletter
Bishop Training Company 1125 East Milham Kalamazoo, MI 49002 800-520-7766, 616-381-9416	User training on wireless and mobile computing documentation
BRP Publications on Internet Search through LYCOS.COM	On-line newsletter
CDPD Forum Liz Durham, Marketing Manager 401 North Michigan Avenue Chicago, IL 60611-4267 800-335-CDPD	Specifications of the CDPD standard, CDPD scorecard, information on third-party software availability, etc., industry public relations and application references
Cellular Perspectives	Newsletter on cellular industry standards
CompuServe	Telecommunications Forum—Wireless Data section
Ericsson	*Mobile Data News,* a monthly newsletter (a good source of application references)
Ericsson—RAM Mobile Data	Mobitex software catalog
INTERNET—Academic Terri Watson at elf@cs.washington.edu (maintains a virtual library; many other libraries are growing, also)	Several universities are doing research in mobile computing. Noteworthy are Columbia, University of Lancaster (United Kingdom), Michigan, MIT, Purdue, Rutgers, Xerox PARC, Washington State, Waterloo (Canada)
Internet—Academic most@comp.lancs.ac.uk	Mobile computing bibliography from United Kingdom
Internet—Business www.mobileinfo.com	Information services for mobile-computing professional
INTERNET—www.volksware.com	Industry letter on pen-based computing
Internet Wireless Digest (www.wirelessinc.ca)	Weekly news on wireless

Information resource	Type of information available
Internet User Groups in Mobile Computing	User experiences and applications of mobile computing
Internet Newsletter on Wireless— www.access.digex.inc/~brpinc	From BRP Publications—industry newsletter on wireless
Magazines	*Communications Week*—regular articles and specials *Data Communications* magazine—McGraw-Hill *Mobile Office*—good source on products and applications *Pen Computing*—pen-based applications *Telecommunications* magazine—Norwood, Mass.
Malcolm, Miller & Associates, Inc. 1600 Golf Road, Suite 1200 Rolling Meadows, IL 60008 708-981-5128	GSM education
MMTF Task Force 505-271-9933 mmtf-request@www.epilogue.com	Mobile Management Task Force for SNMP-compatible MIB standards
Mobile Computing Information Services DOLNET Computer Communications Inc. 245 West Beaver Creek Road, Unit 9-B Richmond Hill, Ontario, Canada L4B 1L1 905-881-9070, 905-881-3589 (Fax) www.mobileinfo.com	Mobile computing technology information, mobile computing application profiles, multivendor mobile product catalog, mobile computing vendor product strategies, mobile computing RFP template, mobile computing project plan template, custom research into mobile computing technologies
"Mobile Data" Washington, DC	Report by Alan Reiter
Mobile Insights Cupertino, Calif.	Industry newsletter by Gerry Purdy
NPCPA National Portable Computer Professional Association 212-987-7119	Practical, solutions-oriented information to members
Seybold Outlook on Mobile Computing PO Box 917 Brookdale, CA 95007 408-338-7701, 408-338-7806 (Fax)	Current information on the industry, including technology trends
Vendor Home Pages on Internet	Current product information

Rule-of-Thumb Costs for Various Mobile Computing Components

Component	Rule-of-thumb cost
E-mail-related costs	■ See Table 2.1 in Chap. 2. ■ Also see ARDIS and RAM sections in Chap. 8 for costs of messaging services.
Sales force automation packaged software costs	■ $200 to $700 per seat for packaged software. ■ Ask an application development vendor to provide estimates for integration of sales force automation package to operational data base applications.
Custom application development costs	■ Ask an application development vendor or in-house development staff to provide estimates for custom application development.
Mobile end-user devices	■ (Source: Use *Mobile Office* magazine for current prices.) ■ Full-function notebooks: $3000 to $4000 for Pentium class, active matrix, 24 MB, 1.3 gigabyte hard disk and CD-ROM. ■ Ruggedized notebooks: add $1000 to the price of full-function notebooks. Note ruggedized notebooks do not have the same capacity as ordinary notebooks. ■ Handheld computers: $2500 to $4500 per unit (ask vendors such as Telxon and Norand). ■ PDAs: $400 to $800 per unit. ■ Cellular PC Card modems: $300 for 28,800 bps per unit. ■ ARDIS or Mobitex PCMCIA modems: $500 per unit. ■ CDPD PC Card modem: $800 to $1000 per unit (expected to fall rapidly to meet competition).
Wireless LAN–related costs	■ Typical Ethernet LAN wiring: $75 to $100 per node (100 ft average). ■ Wireless PC Card LAN adapter (Ethernet): $400 to $600. ■ Wireless LAN Access Point (Ethernet): $1500 to $1600. ■ Wireless PC Card LAN adapter (Token Ring): $500 to $700. ■ Wireless LAN Access Point (Token Ring): $1600 to $1800.
Wireless or switched network operational (monthly for usage) costs	■ Network operational costs may vary from $200 to $400 per month per user, depending on the application: e-mail to OLTP. Network usage costs may be as high as 20 to 50 percent of the total bundled operational costs (hardware lease, software, maintenance, support, etc.) per user.

Component	Rule-of-thumb cost
SMR network hardware cost	■ $60,000 to $150,000 (for each additional 5-channel upgrade, add $50,000). ■ See section 8.3 for more information.
Paging network costs	■ See Tables 8.4 and 8.5.
ARDIS network cost	■ See Table 8.12.
RAM Mobile Data network costs	■ See Sec. 8.6.2 and Table 8.15.
Satellite-based network costs	■ See Tables 8.20 and 8.21.
MCSS costs	■ Low-end (100 users): $25,000; medium-capacity (100 to 500 users): $50,000; high-capacity (1000 to 2000 users): $100,000 to $500,000.
RNA (Remote Network Access) costs	■ An 8-port configuration (for typical 100-mobile-user population) will cost $8000 to $10,000. ■ See Sec 10.7.2 for more details.
ISDN costs	■ See Table 10.2.
Annual maintenance costs	■ 7 to 12 percent of purchase costs for hardware; 15 to 25 percent of one-time license for software.
Installation-related costs	■ In-car installation at $200 to $500 per car for custom bracket and installation.
Application integration costs	■ Very difficult to estimate, since they depend on type and number of platforms, number of applications, communications interfaces supported. ■ Ask a seasoned consultant or in-house specialists to provide ball-park estimates based on experience, or a systems integration firm to quote on your specific requirements.
Network and technical support costs	■ Refer to studies by research firms such as Gartner, IDC, and Infonetics. ■ Our own research suggests that we should allow $1200 to $2400 per user on an annual basis.

Note: These costs are based on many assumptions and should be used only for a very preliminary cost estimate prepared during initial stages of a feasibility study. This estimate should be refined further after determining capacity requirements and contacting the vendors listed in App. B. Costs will generally decrease by 15 to 20 percent annually with the above numbers representing 1996 figures.

Glossary

This glossary draws on the following sources of information: *Implementing Wireless Networks* by Nemzov (McGraw-Hill, 1994); Ericsson's Guide to Mobitex; Research in Motion's (Waterloo, Ontario) *Mobitex Made Easy* booklet; LanSer's (a Canadian wireless services company) glossary on the Internet; ARDIS publications; and *Mobile Office,* 1995 issues.

100Base-T An IEEE 802.3 extension for providing Ethernet transmission at 100 Mbps on twisted-pair and powered signal-regenerative hubs.

100Base VG-AnyLAN An IEEE 802.12 committee extension for providing Ethernet transmission at 100 Mbps on twisted-pair with quartet (MLT-5) signaling, with a new MAC-layer protocol that does not support collision detection.

10Base-T Ethernet IEEE 802.3 standard for transmission at 10 Mbps specifically for twisted-pair wiring and connectors and signal-regenerative powered hubs.

10Base2 Ethernet standard for baseband networks with transmission rates of 10 Mbps using coaxial cable segment lengths of 2×100 meters (200 meters).

10Base5 Ethernet standard for baseband networks with transmission rates of 10 Mbps, using coaxial cable segment lengths of 5×100 meters (500 meters).

A band A non-wire-line radio frequency spectrum. *See also* B Band.

Access charge A flat monthly fee charged to a subscriber for the use of a cellular system (whether the subscriber makes or receives any calls or not).

Access number (*cellular term*) The phone number that must be dialed by someone calling you when you are roaming outside of the National Network, prior to dialing the number of your phone. The access number gives the caller access to the facilities of the system in which you are roaming.

Adapter A PC board, usually installed inside a computer, that provides network communication capabilities to and from that computer system. The term *adapter* is often used interchangeably with *NIC* (Network Interface Card).

Adjacent cell Two cells are adjacent if it is possible for an MES to maintain continuous service while switching from one cell to another.

ADPCM Adaptive Differential Pulse Code Modulation.

Agent   Generally, a software that performs tasks, processes queries, and returns replies on behalf of a client application. In the network management context, it gathers information about a network device, and executes commands in response to a management console's requests. In network management systems, agents reside in all managed devices and report the values of specified variables to management stations. In SNMP, the agent's capabilities are determined by the management information base.

Airlink interface   The network interface between a MES and the CDPD service provider network.

AirTrue (R)   A wireless transmission technology from Air Communication, designed to recognize errors on a cellular carrier network and to provide error correction enhancements to increase the probability of making a connection on the first attempt and maintaining that connection in spite of network events, noise, or bad data.

AMPS (Advanced Mobile Phone Service)   The official name for the first commercial cellular system, which used 666 channels (A Band 333 and B Band 333). The standard for the analog cellular telephone service in use in North America. Developed by AT&T, this is the current standard for all North American cellular systems.

Analog signal   A transmission in which information is represented as physical magnitudes of electrical signals.

ANI   Automatic Number Identification.

ANSI (American National Standards Institute)   The coordinating body for voluntary standards groups within the United States that is a member of the International Organization for Standards.

APCO 25 (Association of Public Safety Communications Officials)   A set of standards for private radio networks designed specifically for public safety (police, fire, emergency services) applications in North America. It includes standards for common air access, encryption, over-the-air reprogramming, data port interface, host data interface, trunking, interconnection to PSTN, and network management. APCO-25 was adopted in 1995 and is a digital standard, as compared to APCO-16, which was analog.

API   Application Programming Interface.

ARQ (Automatic Repeat Request)   Communication method where the receiver detects errors and request retransmissions.

ASCII   American Standard Code for Information Interchange.

Asynchronous   A process where overlapping communications operations can occur independently and do not have to wait for a previous operation to be finished.

ATMD (Asynchronous Time Division Multiplexing)   A method of sending information in which normal time division multiplexing (TDM) is used, except that slots are allocated as needed rather than to specific transmitters.

Attenuation   Loss of communication signal energy.

AuC   Authentication Center.

AUI (Attachment Unit Interface)   An IEEE 802.3 cable connecting the MAU (media access unit) to the network.

B band   The wire-line radio frequency spectrum (from the phone company). *See also* A band

Band   A portion of the radio frequency.

Bandwidth   (1) The range (band) of frequencies that is transmitted on a channel. The difference between the highest and lowest frequencies is expressed in hertz (Hz) or millions of hertz (MHz). (2) The range of frequencies on the electromagnetic spectrum allocated for wireless transmission. (3) The wire speed of the transmission channel.

Base station   The low-power transmitter/receiver and signal equipment located in each cell in a cellular service area.

Baseband   A transmission channel which carries a single communications channel on which only one signal can transmit at a given time.

Baud   (1) A unit of signaling speed represented by code elements (often bits) per second. (2) A French language term that represents the transfer of one bit.

BCHO (Base Controlled Handoff)   Cell transfer initiated by the network.

Block (of frequencies)   A group of radio frequencies within a band set aside for a particular purpose. Cellular telephony uses four blocks of frequencies within the 800-MHz portion of the UHF band. Non-wire-line and wire-line carriers are assigned separate blocks of frequencies. *See also* band

Block A   The block of 800-MHz cellular radio frequencies assigned to the non-wire-line or block A carrier.

Block B   The block of 800-MHz cellular radio frequencies assigned to the wire-line or block B carrier.

Blockage   This occurs when a subscriber tries to make a call but all channels at the nearest cell site are busy. Cell splitting becomes necessary when there is a high blockage percentage.

BOC (Bell Operating Companies)   The local telephone companies that existed prior to deregulation, under which AT&T was ordered by the courts to divest itself in each of the seven U.S. regions. *See also* RBOC

BPR (Business Process Reengineering)   Business Process Reengineering is the discipline of first analyzing and then redesigning current business processes and their components in terms of their effectiveness, efficiency, and added value contribution to the objectives of the business.

BPS (Bits per second)   Kbps is for kilobits per second, and Mbps stands for million bits per second.

BTS   Base Transceiver Station.

Call setup time   The time required to establish a switched call between DTE and devices.

Carrier   (1) A company that provides telephone (or another communications) service. Also, an unmodulated radio signal. (2) A signal suitable for modulation by another signal containing information to be transmitted.

CCITT (Consultative Committee for International Telephone and Telegraph) An international organization that makes recommendations for networking standards like X.25, X.400, and facsimile data compression standards. Now called the International Telecommunications Union Telecommunication Standardization Sector, this is abbreviated as ITU, ITU-T, or ITU-TSS.

CDMA   Code Division Multiple Access.

CDPD (Cellular Digital Packet Data)   Uses idle moments on voice channels to send pure data over the channel without affecting quality of voice transmissions.

Cell   The basic geographic unit of a cellular system and the basis for the generic industry term *cellular*. A geographical area is divided into small *cells,* each of which is equipped with a low-powered radio transceiver. The cells can vary in size depending on terrain and capacity demands. By controlling the transmission power and the radio frequencies assigned from one cell to another, a computer at the Mobile telephone switching office monitors the movement and transfers (or hands off) the phone call to another cell and another radio frequency as needed. The region in which RF transmission from one fixed transmission site can be received at acceptable levels of signal strength.

Cell splitting   Dividing one cell into two or more cells to provide additional capacity within the original cell's region of coverage.

Cellular   (1) Using cellular phone technology. (2) A reference to the wireless switched circuit network consisting of overlapping coverage cells that provides analog voice and CDPD.

Channel   An individual communication path that carries signals at a specific frequency. The term also is used to describe the specific path between large computers (e.g., IBM mainframes) and attached peripherals.

Channel bandwidth   The frequency range of an RF channel; for example, in a CDPD, it is 30 KHz.

Channel hop   The process of changing the RF channel supporting a channel stream to a different RF channel on the same cell.

Channel hopping   A radio frequency transmission method whereby transmissions *hop* from one channel to another. The channels are visited in a predefined order specified by a hopping sequence. Typically this uses the ISM band from 2.4000 to 2.4835 GHz with 85 one-megahertz channels or hops, but at least 50 different frequencies must be used by FCC regulation. Also, CDPD uses frequency hopping on analog cellular systems to take advantage of unoccupied voice channels.

CHAP   Challenge Handshake Authentication Protocol.

CID   Caller ID.

Circuit switching   An open-pipe technique that establishes a temporary dedicated connection between two points for the duration of the call. A switching system in which a dedicated physical circuit path must exist between sender and receiver for the duration of the call. Used heavily in the phone company network, circuit switching often is contrasted with *contention* and *token passing* as a channel-access method, and with *message switching* and *packet switching* as a switching technique.

Class of Service (COS)   An indication of how an upper-layer protocol wants a lower-layer protocol to treat messages. In SNA subarea routing, COS definitions are used by subarea nodes to determine the optimal route to establish a given session.

Code Division Multiple Access (CDMA)   (1) A division of the transmission spectrum into codes, effectively scrambling conversations. Several transmissions can occur simultaneously within the same bandwidth, with the mutual inference reduced by the degree of orthogonality of the unique codes used in each transmission. (2) Wireless transmission technology that employs a range of radio-frequency wavelengths to transport multiple channels of communication signals. *See also* spread-spectrum technology.

CO (Central Office)   The telephone-switching station nearest the customer's location. A local telephone company office to which all local loops in a given area connect and in which circuit switching of subscriber lines occurs.

CODEC   Coding/Decoding Device.

Crosstalk   A technical term indicating that stray signals from other wavelengths, channels, communication pathways, or twisted-pair wiring have polluted the signal. It is particularly prevalent in twisted-pair networks or when telephone and network communications share copper-base wiring bundles. A symptom of interference caused by two cell sites causing competing signals to be received by the mobile subscriber. This can also by generated by two mobiles causing competing signals that are received by the cellular base station. Crosstalk sounds like two conversations and often a distortion of one or the other or both.

CSDS (Circuit Switched Data Service, *a cellular term*)   Developed for delivery vehicles to track packages (used by UPS). Services that carry data over conventional cellular where circuits are switched from call to call.

CSMA/CA   Carrier Sense Multiple Access with Collision Avoidance.

CSMA/CD (Carrier Sense Multiple Access with Collision Detection)   A communications protocol in which nodes contend for a shared communications channel and all nodes have equal access to the network. Simultaneous transmissions from two or more nodes results in random restart of those transmissions. Used in Ethernet protocol.

CSU (Channel Service Unit)   A digital interface device that connects end-user equipment to the local digital telephone loop. The piece of equipment that terminates the long-distance circuit in the customer's location. It is often paired with a digital service unit.

CT-2   Digital cordless telephony (2d generation).

CT-3   Digital cordless telephony (3d generation).

CT2/Telepoint   A U.K. system that allows callers with CT2 phones who are within 200 yards of a base station (or site) to make outgoing calls.

CTI   Computer Telephone Integration.

CTIA (Cellular Telecommunications Industry Association)   The organization created in 1981 to promote the cellular industry, address the common concerns of cellular carriers and serve as a forum for the exchange of nonproprietary information.

DACS   Digital Access and Cross-Connect Systems.

Data compression   A reduction in the size of data by exploiting redundancy. Many modems incorporate MNP5 or V.42bis protocols to compress data before it is sent over the phone line.

dB (decibels)   A value expressed in decibels is determined as 10 times the logarithm of the value taken to base 10.

DCE   (1) Data Communications Equipment in telecommunications context. (2) In software architecture, it implies distributed computing environment.

Dead spot   A location in a radio/cellular system where, for one reason or another, signals do not penetrate.

Decompression   The restoration of redundant data that was removed through compression.

DECT (Digital European Cordless Telephone)   The specs for future European cellular, as yet not fully defined.

Dedicated channel   An RF channel that is allocated solely for the use of a particular user or service. For example, in CDPD, a channel may be dedicated to data.

De facto standard   A standard by usage rather than official decree; a default standard.

De jure standard   Literally, "from the law." A standard by official decree.

DES (Data Encryption Standard)   An encryption/decryption algorithm defined in FIPS Publication 46. The standard cryptographic algorithm developed by the National Institute of Standards and Technology.

Downlink   The process of receiving information from a source computer.

Driver   A software program that controls a physical computer device such as an NIC, printer, disk drive, or RAM disk.

DSU (Data Service Unit)   A device used in digital transmission for connecting data terminal equipment (DTE), such as a router, to a digital transmission circuit (DTC) or service.

DTE (Data Terminal Equipment)   A computer terminal that connects to a host computer. It may also be a software session on a workstation or personal computer attached to a host computer.

DTMF   Dual Tone Multifrequency.

Dual-mode New cellular phones that work with both digital and analog switching equipment. Digital cellular offers the benefits of more channels, clearer-sounding calls, and ensured privacy.

Dual-NAM (*cellular term*) Allows user to have two phone numbers with separate carriers.

Duplex (1) The method in which communication occurs, either two-way as in full-duplex, or unidirectional as in half-duplex. (2) Cellular phones, using separate frequencies for transmission and reception, allow for duplex communications by allowing both parties to talk and listen at once. Push-to-talk systems are not duplex.

ECC (Enhanced Control Cellular) Proprietary protocol from Motorola for cellular modems.

Electromagnetic radiation All forms of visible light, invisible light, radio signal, and other energy used for the wireless transmission of data communications.

EMF (Electromagnetic Field) The minute magnetic energy fields which surround *all things* in nature and are suspected of causing health problems in significant strengths.

EMI (Electromagnetic Interference) Interference by electromagnetic signals that can cause reduced data integrity and increased error rates on transmission channels. Signal noise pollution from radio, radar, fluorescent lights, or electronic instruments.

Encryption The processing of data under a secret key in such a way that the original data can only be determined by a recipient in possession of a secret key.

End-to-end delivery Delivery of data between a source and destination endpoint. In CDPD at least one endpoint must be an MES. The other may be an MES or an FES. In the application sense, it implies client and server applications in a client/server design.

Erlang 1 hour, 300 seconds, and 36 CCs. If a channel is occupied (used) constantly for 1 hour, that circuit has carried 1 Erlang of traffic. Also known as a carried load.

ESMR Extended Specialized Mobile Radio.

ETC (Enhanced Throughput Cellular) AT&T Paradyne protocol for data transmission over analog cellular connections consisting of enhancements to V.42 and V.32bis for compression, error detection, and error correction.

Fading The combination of out-of-phase multiple signals that results in a weaker or self-canceling data signal.

FCC Federal Communications Commission.

FDDI (Fiber Data Distributed Interchange) FDDI provides 125 Mbps signal rate with 4 bits encoded into 5-bit format for a 100-Mbit/s transmission rate. It functions on single- or dual-ring and star network with a maximum circumference of 250 km.

FDMA (Frequency Division Multiple Access)   The analog communications technique that uses a common channel for communication among multiple users allocating unique time slots to different users.

FES (Fixed-End System)   The nonmobile communication system (and software) that handles OSI transport and higher layers of CDPD transmission.

FHSS (Frequency Hopping Spread Spectrum)   IEEE 802.11 Wireless LAN Standards Committee approval for Lannair's concept for at least 2 Mbps transmission rate with dynamic data rate switching.

Firewall   A device, mechanism, bridge, router, or gateway which prevents unauthorized access by hackers, crackers, vandals, and employees from private network services and data.

Frequency hopping   A radio frequency transmission method. Typically this uses the ISM band from 2.4000 to 2.4835 GHz with 85 one-megahertz channels or *hops*. Also, CDPD uses frequency hopping on analog cellular systems to unoccupied voice channels. Transmissions hop from one channel to the other, staying only $\frac{1}{10}$ of a second on any given channel. The channels are visited in a predefined order specified by a hopping sequence.

Geosynchronous orbit   Orbit taken by satellites where the satellite's orbit velocity matches the rotation of the earth, causing the satellite to remain stationary relative to a position on the earth's surface. Geosynchronous orbit demands a position about 23,000 miles above the earth's surface over the equator.

GIS (Geographic Information System)   Generally refers to a database of geographical data. In some circles, it refers to Graphics database.

GLS (Global Locationing System)   A triangulation system used to locate a vehicle and convey that information to a central management facility.

GPS (Global Positioning System)   A satellite-based triangulation system used to ascertain current location.

GSM (Global System for Mobile Communications)   The pan-European digital cellular system standard.

Handoff   The transfer of responsibility for a call from one cell site to the next. The process by which the MTSO, sensing by signal strength that the cellular mobile is reaching the outer range of one cell, transfers or *hands off* the call to an adjacent cell with a stronger signal.

HLR (Home Location Register)   Database of information about each cellular subscriber.

Hopping sequence   The preset order in which frequency-hopping RF transmissions are distributed over the 82 channels of the assigned ISM band.

Host   Any computer, although typically a mainframe, midsized computer, minicomputer, or LAN server, servicing users and processing their requests at the central processor but distributing the results to terminal-based or client connections.

Hz (hertz)   Signal frequency use for voice, data, TV, and other forms of electronic communications represented by the number of cycles per second.

IEEE (Institute for Electrical and Electronic Engineers)   A membership-based organization based in New York City that creates and publishes technical specifications and scientific publications.

IEEE 802   An Institute of Electrical Engineering standard for interconnecting local area networking computer equipment. The IEEE 802 standard describes the physical and link layers of the OSI reference model.

IEEE 802.1   A specification for media-layer physical linkages and bridging.

IEEE 802.3   An Ethernet specification derived from the original Xerox Ethernet specifications. It describes the CSMA/CD protocol on a bus topology using baseband transmissions.

IEEE 802.4   Broadband and baseband bus using token passing as the access method and physical interface specifications.

IEEE 802.5   A Token-Ring specification derived from the original IBM Token-Ring LAN specifications. It describes the token protocol on a star/ring topology using baseband transmissions.

IEEE 802.6   A token bus specification for metropolitan area networks with star/ring topology using baseband transmissions.

IEEE 802.11   A physical- and MAC-layer specification for wireless network transmission based on direct and frequency hopping, SST, and infrared at transmission speeds from 1 to 4 Mbps. This specification includes the basic rate set for fixed bandwidths supported by all wireless stations (for compatibility) and an extended rate set for optimal speeds.

IEEE 802.12   A specification for wireless network transmission based on SST.

IMTS (Improved Mobile Telephone Service)   Cellular telephone predecessor that uses a single central transmitter and receiver to service a region. A two-way mobile phone system that generally uses a single high-power transmitter to serve a given area and is automatically interconnected with a land-line telephone system.

In-band signaling   Transmission within a frequency range normally used for information transmission. Contrasted with out-of-band signaling, which uses frequencies outside the normal range of information-transfer frequencies.

Infrared   Electromagnetic waves whose frequency range is above that of microwave but below the visible spectrum. LAN systems based on this technology represent an emerging technology.

Infrastructure   The physical and local components of a network. Typically, this includes wiring, wiring connections, attachment devices, network nodes and stations, interconnectivity devices (such as hubs, routers, gateways, and switches), operating environment software, and software applications.

IP   Internet Protocol.

IPX   Internet Packet Exchange (Novell Netware LAN protocol).

IrDA (Infrared Data Association) A group of wireless infrared product vendors that promotes serial infrared linkages and interoperability between vendor products.

IS 54 Interim Standard developed by CTIA for introduction of TDMA in conjunction with FDMA.

IS-41 TIA cellular standard for seamless roaming with intersystem handoff, call delivery, validation, and authentication.

IS-54 TIA cellular standard defining the air interface to TDMA and digital handsets to base station communications.

IS-95 TIA cellular standard defining the air interface to CDMA and digital handsets to base station communications.

ISDN Integrated Services Digital Network.

ISM Instrumentation, Scientific, and Medical band.

ISO International Standards Organization.

Ka-band A high-bandwidth satellite wireless communication frequency using the 30-GHz spectrum.

KHz (kilohertz) A measure of audio and radio frequency (a thousand cycles per second). The human ear can hear frequencies up to about 20 KHz. There are 1000 KHz in one MHz.

Laser Light amplification by stimulated emission of radiation.

Lease line A dedicated common carrier circuit providing point-to-point or multipoint network connection, reserved for the permanent and private use of a customer. Also called a *private line*.

LEO Low-Earth Orbit Satellite.

Line of sight The connection between communication devices characteristic of certain transmission systems (such as laser, microwave, and infrared systems) in which no obstructions on a direct path between transmitter and receiver may exist.

Local loop The line from a telephone subscriber's premises to the telephone company CO.

Location directory The repository of information specifying the current forwarding address of a collection of mobile hosts to be accessed by the redirectors.

MAN (Metropolitan Area Network) A network that spans buildings, or city blocks, or a college, or corporate campus. Optical fiber repeaters, bridges, routers, packet switches, and PBX services usually supply the network links.

MASC (Mobitex Asynchronous Communication) Asynchronous link-level protocol on Mobitex.

Mbps (megabytes per second) Speed of data transmission.

MCP/1 Mobitex compression protocol.

MCSS (Mobile Communications Server Switch) A hardware/software configuration that provides communications connection and message switching functionality. It sits between the wireless network and information servers.

MDBS (Mobile Data Base Station, a CDPD term) The hardware used by a cellular provider to convert the data streams into a valid signal and route cellular switched data calls through the wired phone network or to the cellular destination. This station manages and accesses the radio interface from the network side. It relays and retransmits packets sent from the mobile data intermediate system.

MES (Mobile End System) The portable wireless computing device that can roam from site to cell while communicating with the MDBS via CDPD. An end system that accesses the CDPD network through the airlink interface.

MHX (Mobitex Main Hierarchical Exchange) Part of Mobitex network hierarchy. Each MHX exchanges information with other MHXs.

MHz (megahertz) Measures frequency in million cycles per second.

MIB (Management Information Base) An SNMP term.

Microwave Electromagnetic waves in 1 to 30 GHz range.

MIN Mobile Identification Number.

MMTF (Mobile Management Task Force) A vendor organized body that has undertaken to create SNMP-based MIB for mobile network management.

MNP (Microcom Network Protocols) A set of modem-to-modem protocols that provide error correction and compression.

MNP5 Microcom Network Protocols with simple data compression. Dynamically arranges for commonly occurring characters to be transmitted with fewer bits than rare characters. It takes into account changing character frequencies as data flows. Also encodes long runs of the same character. Typical compresses text by 35 percent.

MNP10 Microcom Network Protocols for cellular or wireless transmission applying compression, error detection and error correction, data rate fallback, and readjustment.

MOX (Mobitex Area Exchange) A node in the internal Mobitex network.

MPAK Mobitex packet that is routed through the Mobitex network. A 512 octet of user data.

MPT/1 Mobitex Transport Protocol.

MSA (Metropolitan Statistical Area) The 30 U.S. urban areas (markets) as defined by the FCC, using SMSA (standard metropolitan statistical area) data. All are licensed for two cellular operators, and almost all have both operators on the area. MSAs comprise 76 percent of the U.S. population, but only 22 percent of its land surface area.

MTSO (Mobile Telephone Switching Office) The cellular system's switching computer, located between a cell site and a conventional telephone switching office.

NAMPS (Narrowband Advanced Mobile Phone System)   Using a radio frequency transmission on a single, preset frequency.

Narrowband   PCS frequency in the 900 to 931 MHz range for two-way paging.

NDIS (Network Device Independent Specification)   A Microsoft network interface specification for operating system and protocol-independent device drivers. An effort to create a standard for bridging different types of network adapter cards and multiple protocol stacks. This network-level protocol is supported by IBM LAN Manager and new Microsoft networking products, such as MS Windows for Workgroups and NTAS.

NIC (Network Interface Card)   The network access unit that contains the hardware, software, and specialized PROM information necessary for a station to communicate across the network. Usually referenced as *network interface controller*.

NOS (Network Operating System)   A platform for networking services that combines operating system software with network access.

ODI (Open Data-link Interface)   A protocol that supports media- and protocol-independent communication by providing a standard interface allowing network layer protocols to share hardware without conflict. Presently used in PC software, mostly.

OLTP   Online Transaction Processing.

PC Card standard   Latest PCMCIA specification PCMCIA 5.0. Adds support for low-voltage 3.3-volt operation, DMA, multifunction capability, and Card-BUS which provides higher performance, bus mastering, and 32-bit data path.

PCCA (Portable Computer and Communications Association)   A nonprofit association of vendors to develop and promote software and hardware for mobile computing applications.

PCMCIA (Personal Computer Memory Card International Association)   A standard for a computer plug-in, credit card-sized card that provides about 90 percent compatibility across various platforms, BIOS, and application software.

PCS/PCN (Personal Communications Services)/(Personal Communications Network)   A term used to describe emerging wireless/portable network technology where subscribers carry their own personal communication numbers with them, and the system locates them wherever they are.

POP   Point-of-Presence.

POPS   One unit of population. The POPS concept is used to measure relative market sizes.

POTS   Plain Old Telephone Service.

PSTN (Public Switched Telephone Network)   The telecommunications network traditionally encompassing local and long distance land-line carriers and now also including cellular carriers. Refers to the telephone network.

PTC (Portable Transaction Computer)   A term used in the manufacturing, distribution, and warehousing industries to denote portable computers equipped with scanners, bar-code readers, or a pen.

Registration (CDPD term, also used in PCS) The process whereby an MES signs on to a serving mobile data intermediate system.

RF Radio Frequency.

RJ-11 Standard four-wire connectors for phone lines.

RJ-22 Standard four-wire connectors for phone lines with secondary phone functions (such as call forward, voice mail, or dual lines).

RJ-45 Standard eight-wire connectors for networks. Also used as phone lines in some cases.

RNA (Remote Network Access) Terminology for the hardware/software used for connecting remote workers, offices, customers, and suppliers through nondedicated (dial-up/ISDN) connections. This terminology is based on current usage; it may include wireless networks in the future.

Roaming The ability to access a network anywhere and move freely while maintaining an active link through a wireless connection to a network. Roaming usually requires a handoff when a node (user) moves from one cell to another.

Router A device that interconnects networks that are either local area or wide area.

SMP (Symmetric Multiple Processing) Processor hardware architecture that allows multiple processors to share processing workload using common memory.

SMR Specialized Mobile Radio.

SNMP Simple Network Management Protocol.

SPX System Packet Exchange, a protocol used in Novell's Netware network operating system.

SS7 Signaling System 7, channel for network control.

SST Spread-Spectrum Technology, used in wireless LANs.

T1 (1) Bell technology referring to a 1.544-Mbps communications circuit provided by long-distance carriers for voice or data transmission through the telephone hierarchy. Since the required framing bits do not carry data, actual T1 throughput is 1.536 Mbps. T1 lines may be divided into twenty-four 64-Kbps channels. This circuit is common in North America. Elsewhere, the T1 is superseded by the ITU-TTS designation DS-1. (2) A 2.054-Mbps communications circuit provided by long-distance carriers in Europe for voice or data transmission.

T3 An AT&T standard for dial-up or leased-line circuits with a signaling speed of 44.736 Mbps per second. Superseded in Europe by the ITU (ITU-TTS) DS-3 designation.

TCP/IP Transaction Control Protocol/Internet Protocol.

TDMA Time Division Multiple Access.

TIA Telecommunications Industry Association.

V.32 An international standard for synchronous and asynchronous transfer of data of up to 9600 bps over dial-up telephone lines.

V.32bis An international standard for synchronous and asynchronous transfer of data of up to 14,400 bps over dial-up telephone lines.

V.34 An international standard for synchronous and asynchronous transfer of data of up to 28,800 bps over dial-up telephone lines.

V.42 An international error correction protocol that uses Link Access Procedure Modem (LAP-M) as the primary protocol, and MNP2-4 as back-up protocols.

V.42bis An international data compression protocol that can compress data by as much as 4 to 1.

VSAT Very-Small-Aperture Terminal.

WAN Wide Area Network.

Index

ABOUT THE AUTHOR

Chander Dhawan is president of DOLNET Mobile Info
Services in Toronto, Canada, and previously spent 13 years
with IBM Canada. An independent consultant and a
systems integration expert with over 25 years of experience
in information technology, he has designed and installed
some of Canada's largest computer networks, including a
sophisticated mobile workstation pilot for the Canadian
government.